MW01518838

MATRIX STRUCTURAL
ANALYSIS
and
DYNAMICS
Theory and Computation

Mario Paz
Speed Engineering School
University of Louisville
Louisville, KY

Computers and Structures, Inc., Berkeley, California

Computers and Structures, Inc., 1995 University Avenue, Berkeley, California 94704 USA web: www.csiberkeley.com

SAP2000® is a registered trademark of Computers and Structures, Inc. Considerable time, effort and expense have gone into the development and documentation of SAP2000®, including thorough testing and use. The user must accept and understand that no warranty is expressed or implied by the developers or the distributors on the accuracy or the reliability of the program. SAP2000® is a practical tool for the design/check of structures. The user must thoroughly read the manuals and must clearly recognize the aspects of design that the program algorithms do not address. The user must explicitly understand the assumptions of the programs and must independently verify the results.

Library of Congress Cataloging-in-Publication Data

Paz, Mario.
   Matrix structural analysis and dynamics : theory and computation / Mario Paz.
      p. cm.
   "Matrix structural analysis that integrates theoretical material with practical applications to engineering problems using advanced computer software. Presents solved analytical problems and illustrative examples, giving both hand calculations and computer solutions"--Provided by publisher.
   Includes bibliographical references and index.
   ISBN 978-0-923907-25-9 (hardcover : alk. paper)
   1. Structural analysis (Engineering)--Matrix methods--Textbooks. 2. Structural analysis (Engineering)--Matrix methods--Data processing--Textbooks. I. Title.

   TA642.P388 2008
   624.1'71--dc22

                                                                2008062319

Printed in China

10 9 8 7 6 5 4 3 2 1

כַּבֵּד אֶת־אָבִיךָ וְאֶת־אִמֶּךָ לְמַעַן יַאֲרִכוּן יָמֶיךָ עַל הָאֲדָמָה אֲשֶׁר־יְהֹוָה אֱלֹהֶיךָ נֹתֵן לָךְ׃

Honor your father and your mother, as the Lord your God
has commanded you, that you may long endure and
that you may fare well...

Exodus 20:12

TO THE MEMORY OF MY PARENTS

Benjamin Maman Paz
Sela Miari Paz

Acknowledgements

In the preparation of this book, I became indebted to many people from whom I received valuable assistance. First, I wish to express my appreciation for the discussions and comments offered by my colleague Dr. Michael A. Cassaro, Civil Engineering Department of the University of Louisville, who helped me in refining the presentation of material in this book. I also would like to thank Dr. Michael Day, Department of Mechanical Engineering of the University of Louisville, for his assistance in comparing various commercial computer applications for structural analysis. I am also particularly grateful to Dr. Robert Sennett, professor at California Polytechnic State University, who most generously exchanged information with me on the presentation of several topics in the book. I wish to thank my dear late friend Jack Bension for his professional help in editing selected sections of this book. I also wish to thank Professor Robert Matthew and his assistant David Haag of the Center for Professional and Continuing Education at the University of Louisville for their patient help in assisting me to become acquainted with the intricacies of various computer applications to properly incorporate figures with the text material of the book.

I am especially grateful to Dr. Ashraf Habibullah, president of Computers and Structures, Inc., who most kindly authorized me to include in this volume the student version of SAP2000®. In addition, Dr. Habibullah provided me with the full version of SAP2000® so I could solve problems beyond the capability of the student version. I am also most grateful to Syed Hasanain and G. Robert Morris from Computers and Structures, Inc., who most patiently tutored me and clarified for me many of the intricacies of using SAP2000®.

The preparation of this book would not have been possible without the competent and dedicated assistance of Annis Kay Young, who patiently transcribed the many drafts until it reached its final form. She also most diligently applied her expertise in AutoCAD to prepare and refine many of the figures in the book.

Finally, with great pride, I recognize the help of two of my granddaughters, Clara and Rachel, both medical students, who most graciously spent time during the December break to check the accuracy of the index of this book.

As the author has done in the past, he with great love, dedicates this book to the memory of his parents.

December 2007   Mario Paz

Preface

This new textbook in Matrix Structural Analysis integrates the theoretical material in the subject with practical applications to engineering problems using advanced computer software commercially available. It also extends the scope of Matrix Analysis to structures subjected to dynamic loading. This volume includes a CD-ROM with the student version of the program SAP2000®. This computer program developed and maintained by Computers and Structures, Inc., Berkeley, California, has a most powerful capability for the analysis and computer-aided design of structures under static or dynamic loads that may result in elastic or inelastic behavior of the structure. The CD-ROM included in this volume with the student version of SAP2000® has limitations in both the size and behavior of the structure. Even with these limitations, the student version of SAP2000® is extremely useful in introducing the student to a powerful structural program used by professional engineers worldwide. In particular, for applications to Matrix Structural Analysis, the only perceptible limitation of the student version is the size of the structure. The CD-ROM accompanying this book contains, in addition to the student version of SAP2000®, various tutorials, example problems and a complete set of the user manuals for SAP2000®.

In order to maintain the continuity of the text material, the development of long and complex mathematical formulas is presented at the end of each chapter in a section on Analytical (solved) Problems, which is followed by another section containing Practice Problems. Throughout the book numerous Illustrative Examples are given with detailed numerical hand-calculated solutions as well as solutions obtained using SAP2000®. In using SAP2000®, the student is guided (as he or she becomes acquainted with the program) with detailed documentation of the commands implemented as well as input/output tables and graphs.

The book is organized into nine chapters. The first three chapters deal with the analysis of beams, serving to introduce and acquaint the student with Matrix Analysis of Structures. The first and the second chapters present, respectively, the analysis of beams under static or dynamic loads, while the third chapter provides the software documentation and commands in SAP2000® for the solution of a large number of Illustrative Examples. The next five chapters, Chapters 4 through 8 present the analysis of frame type structures; Plane Frames, Grid Frames, Space Frames, Plane Trusses and Space Trusses. The final chapter of this book, Chapter 9, introduces the reader to the *Finite Element Method of Analysis* as a natural extension of Matrix Structural Analysis. Both the analysis of *Plane Elasticity Problems* for plates with forces applied on their plane as well as plate bending, the analysis of plates with forces normal to the plane of the plate, are presented.

The book contains four appendices. Appendix I provides the formulae of the Equivalent Nodal Forces for beams under some common loading. Appendix II contains the analytical expressions of the lateral displacements of fixed end beams with some common loading conditions. These analytical expressions are needed to evaluate the total lateral displacements in beams resulting from nodal displacements and from loads applied on the spans of a beam. Appendices III and IV contain brief presentations of the fundamentals of theoretical dynamics, respectively, for structures modeled as single degree of freedom systems and for structures modeled with multi degree of freedom systems. The book also includes a glossary that will serve the reader to quickly locate definitions, concepts, and methods used throughout the various chapters of the book.

The organization of this textbook also allows studying exclusively Structural Analysis by skipping Chapter 2 and the last sections in each chapter dealing with the analysis of structures under dynamic loading. This approach will not affect in any way the continuity and understanding of the subject. It will also allow returning to those sections on structural dynamics after studying the material in structural analysis, thus providing an alternative approach to study all the material in this textbook.

Finally, the author wishes to state that he believes that this textbook will prepare students who do not continue with graduate studies, to have attained the knowledge to equally be able to analyze structures subjected to static or dynamic loads. Also, the author expects that those students who continue with graduated studies will be able to master advanced subjects in structures with greater interest and ease, having been exposed in their undergraduate studies to an integrated approach in structural analysis.

December 2007   Mario Paz

Contents

MATRIX STRUCTURAL ANALYSIS
and
DYNAMICS
Theory and Computation

1 Beams: Structural Analysis

1.1 Introduction

Beams are defined as structures subjected to loads normal to their longitudinal direction, thereby producing lateral displacements and bending or flexural stresses. We restrict consideration to straight beams with symmetric cross-sectional areas; the loads are applied along the plane of symmetry of the beam. Figure 1.1 shows an example of a continuous beam, with two fixed supports at the ends (fixed for translation and for rotation) and two simple support rollers (fixed for lateral translation).

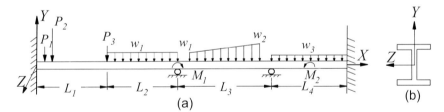

Fig. 1.1 (a) Example of a continuous beam; (b) Cross-sectional area

The beam in Figure 1.1 carries several distributed loads (w_i) that are expressed in units of force per unit of length (*kip/in*), for example, and several concentrated forces (P_i) and moments (M_i) applied at different locations along the beam. As shown in Figure 1.1(b), the beam has an "I" cross-sectional area with a vertical axis of symmetry forming a plane of symmetry along the length of the beam on which the external loads are applied. Coordinate axes *X, Y,* and *Z* are also shown in the figure. These coordinate axes are designated as the *global* or the *system coordinate axes* to distinguish this system of axes from the *member or element coordinate axes* as defined in Section 1.3.

1.2   Elements, Nodes and Nodal Coordinates

In preparation for *matrix analysis*, the structure is divided into *beam elements*, or simply *elements*, connected at *nodes* or *joints*. The beam of Figure 1.1 has been divided into four elements numbered consecutively as 1, 2, 3, and 4 connected at nodes, also numbered consecutively as ①, ②, ③, ④, and ⑤ as shown in Figure 1.2. Both the elements and the nodes may be numbered in any order. However, it is customary to number these items consecutively from the left to the right end of the beam.

The division of the beam into elements is generally arbitrary; however, this division aims to locate the nodes or connections between elements at points of support or at locations along the beam where there is a change of the cross-section properties.

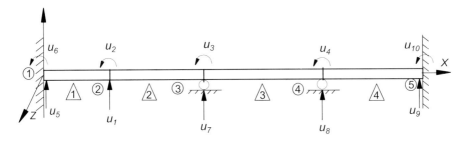

Fig. 1.2   Example of a beam showing the elements 1 through 4,
nodes ① through ⑤ and nodal coordinates u_1 through u_{10}

For structures modeled as beams, two possible displacements are considered at each node: a translatory displacement normal to the longitudinal direction of the beam and a rotational displacement around a perpendicular axis to the plane (X, Y) of the beam. These displacements are designated as *nodal coordinates,* which in Figure 1.2 are labeled u_1 through u_{10}.

Free nodal coordinates are those coordinates that are not restricted for displacement, while the fixed nodal coordinates are constrained to have no displacement. For convenience, the free nodal coordinates u_1 to u_4 were numbered first, followed by the fixed nodal coordinates u_5 through u_{10}. The nodal coordinates in Figure 1.2 are shown in the positive direction to agree with the direction of the Y-axis for nodal translations and with the Z-axis for nodal rotations according to the right-hand rule.

1.3   Shape Functions and Stiffness Coefficients

In Figure 1.3, we have isolated one element of the beam shown in Figure 1.2. The nodal coordinates at the two ends of this element are now designated δ_1, δ_2, δ_3, and δ_4, (linear or angular displacement). This figure also shows the end forces (or moments) P_1, P_2, P_3, and P_4 corresponding to the element nodal displacements δ_i. The local or element system of coordinate axes x, y, and z, which are fixed on the element with the origin at its left end, also are shown in Figure 1.3.

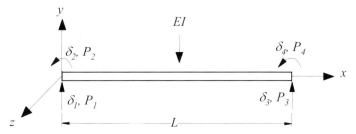

Fig. 1.3   Beam element showing the nodal displacement coordinates δ_1 through δ_4, nodal forces P_1 through P_4 and the element coordinate axes x, y, and z

The *stiffness coefficient* k_{ij} is defined as the force at the nodal coordinate i when a unit displacement is given at the nodal coordinate j (all other nodal coordinates are kept fixed). Figure 1.4(a) shows the stiffness coefficients for a beam element corresponding to a unit displacement given at the nodal coordinate 1, that is, for $\delta_1 = 1.0$. The shape of the curve due to this unit displacement is labeled in Figure 1.4 (a) as $N_1(x)$. Analogously, the other diagrams in Figure 1.4(b), (c), and (d) show the stiffness coefficients and the shape functions for a unit displacement $\delta_2 = 1$, $\delta_3 = 1$, and $\delta_4 = 1$, respectively.

The evaluation of the stiffness coefficients, k_{ij}, and of the *shape functions, $N_1(x)$, $N_2(x)$, $N_3(x)$,* and *$N_4(x)$* for the curves shown in Fig 1.4 corresponding to a unit displacement at the nodal coordinates δ_1, δ_2, δ_3, or δ_4, respectively, may be obtained by integrating the *differential equation of the beam,* which for a uniform beam is given by

$$\frac{d^4 y}{dx^4} = \frac{w(x)}{EI} \qquad (1.1)$$

in which
$$\begin{aligned} w(x) &= \text{external force per unit length} \\ E &= \text{modulus of elasticity} \\ I &= \text{cross-sectional moment of inertia} \end{aligned}$$

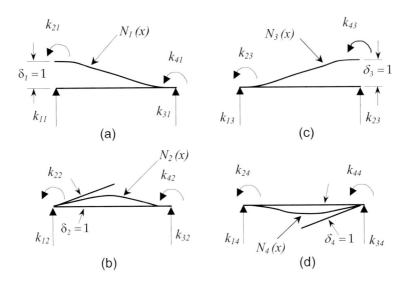

Fig. 1.4   Beam element showing in (a), (b), (c), and (d) the stiffness coefficients due to unit displacement at the nodal coordinates δ_1, δ_2, δ_3, and δ_4, respectively

For the beams shown in Figure 1.4 with no load on the span, that is, for $w(x) = 0$, eq. (1.1) reduces to

$$\frac{d^4y}{dx^4} = 0. \tag{1.2}$$

Integrating eq. (1.2) successively four times yields the equation for displacement of a beam in terms of the four constants of integration; C_1, C_2, C_3, C_4:

$$y = \frac{1}{6}C_1 x^3 + \frac{1}{2}C_2 x^2 + C_3 x + C_4. \tag{1.3}$$

The constants of integration C_1 through C_4 in eq. (1.3) are evaluated from the corresponding boundary conditions. For example, for the beam in Figure 1.4 (a), the boundary conditions are:

$$\text{at } x = 0 \quad y(0) = 1 \quad \text{and} \quad \frac{dy(0)}{dx} = 0$$

$$\tag{1.4}$$

$$\text{at } x = L \quad y(L) = 0 \quad \text{and} \quad \frac{dy(L)}{dx} = 0$$

in which L is the length of the beam element.

The use of these boundary conditions in eq. (1.3) results in an algebraic system of four equations to determine C_1, C_2, C_3, and C_4, which subsequently substituted into eq. (1.3) results in the equation of the shape function for the beam in Figure 1.4(a):

$$N_1(x) = 1 - 3\left(\frac{x}{L}\right)^2 + 2\left(\frac{x}{L}\right)^3 \tag{1.5a}$$

in which $N_1(x)$ is used instead of $y(x)$ to correspond to the condition $\delta_1 = 1$ imposed on this beam. Proceeding in analogous fashion, we obtain the equations of the other shape functions depicted in Figure 1.4 as:

$$N_2(x) = x\left(1 - \frac{x}{L}\right)^2 \tag{1.5b}$$

$$N_3(x) = 3\left(\frac{x}{L}\right)^2 - 2\left(\frac{x}{L}\right)^3 \tag{1.5c}$$

$$N_4(x) = \frac{x^2}{L}\left(\frac{x}{L} - 1\right). \tag{1.5d}$$

The displacement $y(x)$ at coordinate x due to arbitrary displacements δ_1, δ_2, δ_3, and δ_4 at the nodal coordinates of a beam element is then given by superposition as

$$y(x) = N_1(x)\,\delta_1 + N_2(x)\,\delta_2 + N_3(x)\,\delta_3 + N_4(x)\,\delta_4. \tag{1.6}$$

1.4   Element Stiffness Matrix

The general expression to calculate the *stiffness coefficients* for a beam element (see Analytical Problem 1.1, Section 1.16) is given by

$$k_{ij} = \int_0^L EI\, N_i''(x)\, N_j''(x)dx \tag{1.7}$$

in which $N_i''(x)$ and $N_j''(x)$ are the second derivatives of the shape functions [eqs. (1.5)] with respect to x; E is the modulus of elasticity; and I is the cross-sectional moment of inertia of the beam. For example, to calculate the coefficient k_{12}, we substitute into eq. (1.7) the second derivatives of the shape functions $N_1(x)$ and $N_2(x)$ given, respectively, by eqs. (1.5a) and (1.5b), to obtain

$$k_{12} = \int_0^L EI\left(\frac{-6}{L^2} + \frac{12x}{L^3}\right)\left(\frac{-4}{L} + \frac{6x}{L^2}\right)dx$$

which upon integration results in

$$k_{12} = \frac{6EI}{L^2}.$$

(1.8)

The stiffness coefficient k_{ij} has been defined as a force at the nodal coordinates i due to a unit displacement at the nodal coordinate j. Consequently, the forces at nodal coordinate 1 due to successive displacements δ_1, δ_2, δ_3, and δ_4 at the four nodal coordinates of the beam element are $k_{11}\,\delta_1$, $k_{12}\,\delta_2$, $k_{13}\,\delta_3$, and $k_{14}\,\delta_4$, respectively. Therefore, the total force P_1 at the nodal coordinate 1 resulting from these nodal displacements is obtained by the superposition of these four forces, that is,

$$P_1 = k_{11}\delta_1 + k_{12}\delta_2 + k_{13}\delta_3 + k_{14}\delta_4.$$

Analogously, the forces at the other nodal coordinates are:

$$
\begin{aligned}
P_2 &= k_{21}\delta_1 + k_{22}\delta_2 + k_{23}\delta_3 + k_{24}\delta_4 \\
P_3 &= k_{31}\delta_1 + k_{32}\delta_2 + k_{33}\delta_3 + k_{34}\delta_4 \\
P_4 &= k_{41}\delta_1 + k_{42}\delta_2 + k_{43}\delta_3 + k_{44}\delta_4
\end{aligned}
$$

(1.9)

The preceding equations may be conveniently written in matrix notation as

$$
\begin{Bmatrix} P_1 \\ P_2 \\ P_3 \\ P_4 \end{Bmatrix} =
\begin{bmatrix}
k_{11} & k_{12} & k_{13} & k_{14} \\
k_{21} & k_{22} & k_{23} & k_{24} \\
k_{31} & k_{32} & k_{33} & k_{34} \\
k_{41} & k_{42} & k_{43} & k_{44}
\end{bmatrix}
\begin{Bmatrix} \delta_1 \\ \delta_2 \\ \delta_3 \\ \delta_4 \end{Bmatrix}
$$

(1.10)

The use of eq. (1.7) in the manner shown to determine the coefficient k_{12} in eq. (1.8) will result in the evaluation of all the coefficients of the *element stiffness matrix* in eq. (1.10). The result for a uniform beam element is

$$
\begin{Bmatrix} P_1 \\ P_2 \\ P_3 \\ P_4 \end{Bmatrix} =
\frac{EI}{L^3}
\begin{bmatrix}
12 & 6L & -12 & 6L \\
6L & 4L^2 & -6L & 2L^2 \\
-12 & -6L & 12 & -6L \\
6L & 2L^2 & -6L & 4L^2
\end{bmatrix}
\begin{Bmatrix} \delta_1 \\ \delta_2 \\ \delta_3 \\ \delta_4 \end{Bmatrix}
$$

(1.11)

or in condensed notation as

$$\{P\} = [k]\{\delta\}$$

(1.12)

in which $[k]$ is the element stiffness matrix and $\{P\}$ and $\{\delta\}$ are, respectively, the force and the nodal displacement vectors defined in eq. (1.11).

1.5   Assemble System Stiffness Matrix

Thus far, we have established in eq. (1.11) or eq. (1.12) the stiffness equation for a uniform beam element; that is, we have obtained for a beam element the relationship between nodal displacements $\{\delta\}$ (linear and angular) and nodal forces $\{P\}$ (forces and moments). Our next objective is to obtain the same type of relationship for the entire structure between the nodal displacements $\{u\}$ and the nodal forces $\{F\}$. This relationship establishing the *system stiffness equation* for the entire structure may be expressed in condensed notation as

$$\{F\} = [K]\{u\} \tag{1.13}$$

in which $[K]$ is the system stiffness matrix. Furthermore, our aim is to obtain the system stiffness matrix from the stiffness matrices of the elements of the system. The procedure is perhaps better explained through a specific example, such as the beam shown with its load in Figure 1.1 and with the nodal coordinates numbered as shown in Figure 1.2. This beam has been divided into four beam elements, labeled 1 through 4. The nodal coordinates are indicated in this figure as u_1 through u_{10} with the four free coordinates conveniently numbered first, as already stated.

The procedure of assembling the system stiffness matrix consists of transferring and adding appropriately the coefficients in each element stiffness matrix to obtain the system stiffness matrix. This method of assembling the system stiffness matrix is called the *direct method*. In effect, any coefficient k_{ij} of the system may be obtained by adding together the corresponding element stiffness coefficients associated with those nodal coordinates. For example, to obtain the system stiffness coefficient k_{11}, it is necessary to add the stiffness coefficients of the beam elements 1 and 2 that correspond to the nodal coordinates u_1. These coefficients are designated, respectively, as $k_{33}^{(1)}$ and $k_{11}^{(2)}$ for these two beam elements. The upper indices (1) and (2) serve to identify the beam elements and the lower indices to locate the appropriate stiffness coefficients in the corresponding element stiffness matrix.

Illustrative Example 1.1
For the beam shown in Fig 1.5, which has been divided into four beam elements, obtain: (1) the stiffness matrix for each element and (2) the system stiffness matrix for the beam. The material modulus of elasticity is $E = 29,000$ *ksi* and the cross-sectional moment of inertia for the entire beam is 882 in^4.

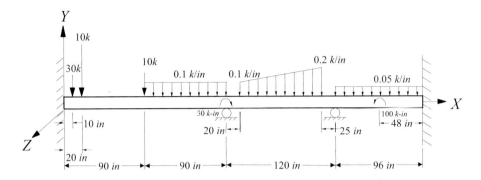

Fig. 1.5 Beam for Illustrative Examples 1.1 and 1.2

Solution:
1. Element Stiffness Matrices

The substitution of numerical values into the element stiffness matrix in eq. (1.11) for each of the four beam elements of this beam results in the following element stiffness matrices:

ELEMENT 1

$$[k]_1 = \frac{29 \times 10^3 \times 882}{90^3} \begin{bmatrix} 12 & 6 \times 90 & -12 & 6 \times 90 \\ 6 \times 90 & 4 \times 90^2 & -6 \times 90 & 2 \times 90^2 \\ -12 & -6 \times 90 & 12 & -6 \times 90 \\ 6 \times 90 & 2 \times 90^2 & -6 \times 90 & 4 \times 90^2 \end{bmatrix}$$

or

$$[k]_1 = \begin{matrix} & 5 & 6 & 1 & 2 & \\ & \begin{bmatrix} 420 & 18900 & -420 & 18900 \\ 18900 & 1134200 & -18900 & 567110 \\ -420 & -18900 & 420 & -18900 \\ 18900 & 567110 & -18900 & 1134200 \end{bmatrix} & \begin{matrix} 5 \\ 6 \\ 1 \\ 2 \end{matrix} \end{matrix}$$ (a)

ELEMENT   2

$$[k]_2 = \begin{array}{c} \\ \begin{bmatrix} 420 & 18900 & -420 & 18900 \\ 18900 & 1134200 & -18900 & 567110 \\ -420 & -18900 & 420 & -18900 \\ 18900 & 567100 & -18900 & 1134200 \end{bmatrix} \begin{array}{c} 1 \\ 2 \\ 7 \\ 3 \end{array} \end{array}$$

(b)

with column labels $\quad 1 \quad 2 \quad 7 \quad 3$

ELEMENT 3

$$[k]_3 = \begin{array}{c} \\ \begin{bmatrix} 177 & 10633 & -177 & 10633 \\ 10633 & 850667 & -10633 & 425333 \\ -177 & -10633 & 177 & -10633 \\ 10633 & 425333 & -10633 & 850667 \end{bmatrix} \begin{array}{c} 7 \\ 3 \\ 8 \\ 4 \end{array} \end{array}$$

(c)

with column labels $\quad 7 \quad 3 \quad 8 \quad 4$

ELEMENT 4

$$[k]_4 = \begin{array}{c} \\ \begin{bmatrix} 346 & 16615 & -346 & 16615 \\ 16615 & 1063333 & -16615 & 531667 \\ -346 & -16615 & 346 & -16615 \\ 16615 & 531667 & -16615 & 1063333 \end{bmatrix} \begin{array}{c} 8 \\ 4 \\ 9 \\ 10 \end{array} \end{array}$$

(d)

with column labels $\quad 8 \quad 4 \quad 9 \quad 10$

2.   System Stiffness Matrix

As previously stated, the system stiffness matrix is obtained by transferring and adding appropriately the coefficients of the element stiffness matrices. To perform this transfer, one should realize that the nodal coordinates for the beam elements should be labeled according to the system coordinates assigned to the two ends of a specific beam element. For example, for element 1, this assignment is u_5, u_6, u_1, and u_2, as shown in Figure 1.2, and not δ_1, δ_2, δ_3, and δ_4, as indicated in Figure 1.3 for an isolated beam element. A simple way to indicate this allocation of nodal coordinates when working by hand is to write at the top and on the right of the element stiffness matrix the coordinate numbers corresponding to the system nodal coordinates for the elements, as it is shown in the matrices (a), (b), (c), and (d) for the four beam elements of this structure.

Proceeding systematically to assemble the system stiffness matrix, we transfer each entry in matrices (a), (b), (c), and (d) to the row and column indicated respectively on the right and at the top of these matrices. Each transferred coefficient is added to the total accumulated at the transferred location of the system stiffness matrix. For example, the coefficient 420 in the third row and third column of matrix (a) is transferred to the location row 1 and column 1, of the system stiffness matrix. Also, the coefficient 420 in the first row and first column of matrix (b) is transferred to the same location—row 1 and column 1 of the system stiffness matrix—to give a total of 840 at this location. The final complete system stiffness matrix $[K]$, obtained after transferring all the coefficients of element stiffness matrices (a), (b), (c), and (d), is given by eq. (e).

$$[K] = \begin{bmatrix} 840 & 0 & 18900 & 0 & -420 & -18900 & -420 & 0 & 0 & 0 \\ 0 & 2268400 & 567110 & 0 & 18900 & 567110 & -18900 & 0 & 0 & 0 \\ 18900 & 567110 & 1984867 & 425333 & 0 & 0 & -8267 & -10633 & 0 & 0 \\ 0 & 0 & 425333 & 1914000 & 0 & 0 & 10633 & 5982 & -16615 & 531667 \\ -420 & 18900 & 0 & 0 & 420 & 18900 & 0 & 0 & 0 & 0 \\ -18900 & 567110 & 0 & 0 & 18900 & 1134200 & 0 & 0 & 0 & 0 \\ -420 & -18900 & -8267 & 10633 & 0 & 0 & 597 & -177 & 0 & 0 \\ 0 & 0 & -10633 & 5982 & 0 & 0 & -177 & 523 & -346 & 16615 \\ 0 & 0 & 0 & -16615 & 0 & 0 & 0 & -346 & 346 & -16615 \\ 0 & 0 & 0 & 531667 & 0 & 0 & 0 & 16615 & -16615 & 106333 \end{bmatrix} \quad (e)$$

The system stiffness matrix in eq. (e) has been partitioned to separate the first four free nodal coordinates from the last six fixed nodal coordinates. The separation of free and fixed coordinates will be needed to calculate the four unknown displacements at the free nodal coordinates.

Illustrative Example 1.2

Consider the cantilever beam shown in Figure 1.6, which has been modeled with three elements as shown in the figure. Determine:

      (1) the element's stiffness matrices and
      (2) the reduced system stiffness matrix in which only the free coordinates are considered.

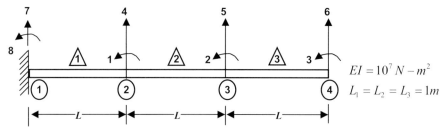

Fig. 1.6   Cantilever beam divided into three beam elements with numbered system nodal coordinates

Solution:

The element's stiffness matrices are obtained by substituting into eq. (1.11) the numerical values for the elements of the beam of this example. Since the three elements in which the beam has been divided are identical, the following element stiffness matrix would correspond to any of these three elements:

$$[k] = 10^7 \begin{array}{ccccc} & 5 & 2 & 6 & 3 \\ & 4 & 1 & 5 & 2 \\ & 7 & 8 & 4 & 1 \\ \begin{bmatrix} 12 & 6 & -12 & 6 \\ 6 & 4 & -6 & 2 \\ -12 & -6 & 12 & -6 \\ 6 & 2 & -6 & 4 \end{bmatrix} & \begin{matrix} 7\ 4\ 5 \\ 8\ 1\ 2 \\ 4\ 5\ 6 \\ 1\ 2\ 3 \end{matrix} \end{array} \qquad (a)$$

To transfer the coefficients in eq. (a) to the system stiffness matrix, we have conveniently indicated at the top and on the right of eq. (a) the nodal coordinates for each of the elements as assigned in Figure 1.6. Furthermore, to obtain the reduced system stiffness matrix, this transfer of coefficients is performed only for the free nodal coordinates, that is, only for the nodal coordinates 1 through 6 omitting the transfer corresponding to the fixed nodal coordinates 7 or 8.

The assemblage of the reduced system stiffness matrix $[K]_R$ in the manner described, results in the following 6×6 matrices:

$$[K]_R = 10^7 \begin{bmatrix} 8 & 2 & 0 & 0 & -6 & 0 \\ 2 & 8 & 2 & 6 & 0 & -6 \\ 0 & 2 & 4 & 0 & 6 & -6 \\ 0 & 6 & 0 & 24 & -12 & 0 \\ -6 & 0 & 6 & -12 & 24 & -12 \\ 0 & -6 & -6 & 0 & -12 & 12 \end{bmatrix} \qquad (b)$$

Equation (b) is thus the reduced system stiffness matrix for the cantilever beam shown in Figure 1.6, which has been segmented into three elements.

1.6   Static Condensation

A practical method of accomplishing the reduction of the stiffness matrix is to identify those degrees of freedom to be condensed as dependent or secondary degrees of freedom, and express them in the term of the remaining independent or primary degrees of freedom. The relationship between the secondary and primary degrees of freedom is found by establishing the static relation between them, hence the name *Static Condensation Method* (Guyan 1965).

In order to describe the Static Condensation Method, let us assume that, in modeling the structure, the secondary degrees of freedom to be reduced or condensed are arranged as the first's nodal coordinates, and the remaining (primary) degrees of freedom are the last p nodal coordinates. With this arrangement, the stiffness equation for the structure may be written using partition matrices as

$$\begin{bmatrix} [K_{ss}] \vdots [K_{sp}] \\ \cdots \quad \vdots \quad \cdots \\ [K_{ps}] \vdots [K_{pp}] \end{bmatrix} \begin{Bmatrix} \{u_s\} \\ \cdots \\ \{u_p\} \end{Bmatrix} = \begin{Bmatrix} \{0\} \\ \cdots \\ \{F_p\} \end{Bmatrix} \tag{1.14}$$

where $\{u_s\}$ is the displacement vector corresponding to the s degrees of freedom to be condensed and $\{u_p\}$ is the vector corresponding to the p primary degrees of freedom. In eq. (1.14), it has been assumed that the external forces were zero at the dependent (i.e., secondary) degrees of freedom; this assumption is not mandatory but serves to simplify explanations without affecting the final results. A simple multiplication of the matrices on the left side of eq. (1.14) expands this equation into two matrix equations, namely,

$$[K_{ss}]\{u_s\} + [K_{sp}]\{u_p\} = \{0\} \tag{1.15}$$

$$[K_{ps}]\{u_s\} + [K_{pp}]\{u_p\} = \{F_p\} \tag{1.16}$$

Equations (1.15) and (1.16) could be solved to determine the relationship between the secondary degrees of freedom $\{u_s\}$ and the primary degree of freedom $\{u_p\}$. Also, the solution of these equations would provide the relationship between the forces $\{F_p\}$ and the primary degrees of freedom $\{u_p\}$. Such solution would require the inconvenient determination of the inverse of the matrix $[K_{ss}]^{-1}$ of the matrix $[K_{ss}]$ defined in eq. (1.14). However, the practical application of the Static Condensation Method does not require a matrix inversion. Instead of solving eqs. (1.15) and (1.16), the standard Gauss-Jordan elimination process, as it is used in the solution of a system of linear equations, is applied systematically on the system's stiffness matrix $[K]$ up to the elimination of the secondary coordinates $\{u_s\}$. At this stage of the elimination process, the stiffness equation (1.14) has been reduced to

$$\begin{bmatrix} [I] & -[\overline{T}] \\ [0] & [\overline{K}] \end{bmatrix} \begin{Bmatrix} \{u_s\} \\ \{u_p\} \end{Bmatrix} = \begin{Bmatrix} \{0\} \\ \{F_p\} \end{Bmatrix} \tag{1.17a}$$

in which $-[\overline{T}]$ and $[\overline{K}]$ designate the sub-matrices resulting after the application of the elimination process and $[I]$ is the unit matrix.

The expansion of eq. (1.17a) yields

$$\{u_s\} = [\overline{T}]\{u_p\}$$
$$\{F_p\} = [\overline{K}]\{u_p\}$$

(1.17b)

It may be observed from of eqs. (1.17b) that the partition matrices $[\overline{T}]$ and $[\overline{K}]$ are, respectively, the transformation matrix and the reduced stiffness matrix. Thus, the Gauss-Jordan elimination process yields both the transformation matrix $[\overline{T}]$, relating secondary nodal coordinates $\{u_s\}$ and primary nodal coordinates $\{u_p\}$, and it also yields the condensed stiffness matrix $[\overline{K}]$, relating the primary coordinates $\{u_p\}$ and the forces $\{F_p\}$ at these coordinates. There is thus no need to calculate $[K_{ss}]^{-1}$ in order to condense the secondary coordinates of the system.

Illustrative Example 1.3

Use Static Condensation to eliminate the three rotational coordinates 1, 2, and 3 from the stiffness matrix given by eq. (b) for the cantilever beam in Illustrative Example 1.2.

Solution:

The stiffness matrix for this cantilever beam determined in Illustrative Example 1.2 is given by

$$[K]_R = 10^7 \begin{bmatrix} 8 & 2 & 0 & 0 & -6 & 0 \\ 2 & 8 & 2 & 6 & 0 & -6 \\ 0 & 2 & 4 & 0 & 6 & -6 \\ 0 & 6 & 0 & 24 & -12 & 0 \\ -6 & 0 & 6 & -12 & 24 & -12 \\ 0 & -6 & -6 & 0 & -12 & 12 \end{bmatrix}$$

(a)

in which the dotted lines partition the stiffness matrix to separate the secondary and primary nodal coordinates.

The following operations (omitting the factor 10^7) are performed to condense the first three nodal coordinates:

1. Divide the first row by the coefficient 8 (located in the first row and first column), then multiply successively the new first row by the coefficients in the first column (2, 0, 0, −6, 0) and subtract the product from the corresponding coefficients in the second, third, fourth, fifth, and sixth rows to

obtain zeros in the first column, except for the coefficients 1.0 in the first row. These operations result in the following matrix:

$$
\begin{bmatrix}
1 & 0.25 & 0 & \vline & 0 & -0.75 & 0 \\
0 & 7.5 & 2 & \vline & 6 & 1.5 & -6 \\
0 & 2 & 4 & \vline & 0 & 6 & -6 \\
\hline
0 & 6 & 0 & \vline & 24 & -12 & 0 \\
0 & 1.5 & 6 & \vline & -12 & 19.5 & -12 \\
0 & -6 & -6 & \vline & 0 & -12 & 12
\end{bmatrix}
\qquad (b)
$$

2. Repeat the above process to obtain 1.0 in the second row and second column, and zeros in the other locations of the second column. This process results in the following matrix:

$$
\begin{bmatrix}
1 & 0 & -0.067 & \vline & -0.2 & -0.8 & 0.2 \\
0 & 1 & 0.267 & \vline & 0.8 & 0.2 & -0.8 \\
0 & 0 & 3.466 & \vline & -1.6 & 5.6 & -4.4 \\
\hline
0 & 0 & -1.6 & \vline & 19.2 & -13.2 & 4.8 \\
0 & 0 & 5.6 & \vline & -13.2 & 19.2 & -10.8 \\
0 & 0 & -4.4 & \vline & 4.8 & -10.8 & 7.2
\end{bmatrix}
\qquad (c)
$$

3. Repeat the process to obtain the value 1 in the third row, third column, and zeros for the rest of the coefficients in the third column:

$$
\begin{bmatrix}
1 & 0 & 0 & \vline & -0.231 & -0.689 & 0.115 \\
0 & 1 & 0 & \vline & 0.923 & -0.231 & -0.461 \\
0 & 0 & 1 & \vline & -0.462 & 1.616 & -1.270 \\
\hline
0 & 0 & 0 & \vline & 18.461E7 & -10.613E7 & 2.768E7 \\
0 & 0 & 0 & \vline & -10.613E7 & 10.150E7 & -3.690E7 \\
0 & 0 & 0 & \vline & 2.768E7 & -3.690E7 & 1.612E7
\end{bmatrix}
\qquad (d)
$$

Finally, by comparing the partition matrices in eq. (d) with those in eq. (1.17a), we identify the transformation matrix $[\bar{T}]$ and the reduced stiffness system matrix $[\bar{K}]$ as

$$
[\bar{T}] =
\begin{bmatrix}
0.231 & 0.689 & -0.115 \\
-0.923 & 0.231 & 0.461 \\
0.462 & -1.616 & 1.270
\end{bmatrix}
\qquad (e)
$$

and

$$\left[\bar{K}\right]=10^7\begin{bmatrix} 18.461 & -10.613 & 2.768 \\ -10.613 & 10.150 & -3.690 \\ 2.768 & -3.690 & 1.612 \end{bmatrix} \qquad \text{(f)}$$

where the factor 10^7, which to simplify calculations was omitted in eq. (b), has been reintroduced in eqs. (d) and (f)

1.7   System Force Vector

The force vector $\{F\}$ in the system stiffness matrix [eq. (1.13)] consists of forces directly applied to nodal coordinates and forces denoted as *equivalent nodal forces*, for those forces applied along the span of the beam elements. These forces are equivalent in the sense that their application at the nodal coordinates of an element results in the same nodal displacements as those that would be produced by the actual forces applied on the span of the beam element.

The forces at the nodal coordinates of the structure are directly assigned to the corresponding coefficient in the force vector $\{F\}$ of eq. (1.13). On the other hand, for forces applied on the elements, it is necessary to determine the *equivalent forces* at the nodal coordinates of the element and then allocate those equivalent nodal forces to the corresponding entry of the system force vector $\{F\}$ in eq. (1.13).

The *equivalent nodal force Q_i*, at nodal coordinate "i", for a distributed load $w(x)$ applied on the span of a beam element is given, in general, by (see Analytical Problem 1.2, Section 1.16):

$$Q_i = \int_0^L N_i(x) \ w(x) \, dx \qquad \text{(1.18a)}$$

and for a distributed moment $m(x)$ by (see Analytical Problem 1.3, Section 1.16):

$$Q_i = \int_0^L N_i'(x) \, m(x) \, dx \qquad \text{(1.18b)}$$

in which

Q_i	=	equivalent force at the nodal coordinate i of the element
$w(x)$	=	distributed load on the beam element
		[$w(x)$ is positive for an upward load]
$m(x)$	=	distributed moment on the beam element
$N_i(x)$	=	shape function for nodal displacement $\delta_i = 1$, given by eqs. (1.5)
$N_i'(x)$	=	derivative of the shape function N_i, given by eqs. (1.5)

For example, for an element carrying a uniformly distributed load w as shown in Figure 1.7, the application of eq. (1.18a) to determine the equivalent nodal force Q_2 yields

$$Q_2 = \int_0^L w\, N_2(x)dx$$

and substituting $N_2(x)$ from eq. (1.5b)

$$Q_2 = \int_0^L w\, x\left(1-\frac{x}{L}\right)^2 dx$$

which, upon integration, results in

$$Q_2 = \frac{wL^2}{12}.$$

Analogously, substituting into eq. (1.18a) the expression for the shape functions $N_i(x)$ [given by eqs. (1.5)], we obtain all the equivalent nodal forces:

$$Q_1 = \frac{wL}{2} \qquad Q_2 = \frac{wL^2}{12} \qquad Q_3 = \frac{wL}{2} \qquad Q_4 = -\frac{wL^2}{12}$$

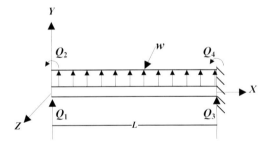

Fig. 1.7   Equivalent nodal forces Q_i for a beam element supporting a uniformly distributed load w

Alternatively to the use of eqs. (1.18), the equivalent nodal forces $\{Q\}$ may be evaluated by determining the *Fixed End Reactions* $\{FER\}$ for the loaded beam element by assuming that the beam element is completely fixed for translation or rotation at its two ends and then reversing the sense of these reactions. To demonstrate the validity of this method for determining the equivalent nodal forces, we consider in Figure 1.8(a) a simply supported beam divided into three elements carrying a general distributed load $w(x)$.

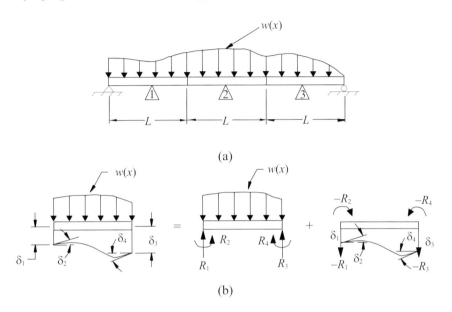

(a)

(b)

Fig. 1.8 (a) Simply supported beam divided into three elements carrying a load $w(x)$
(b) Element 2 equated to the superposition of the element supporting the load $w(x)$ and the FER plus the element loaded with these reactions acting in the opposite sense

Figure 1.8(b) shows element 2 of this beam equated to: (1) the element fixed at its two ends, loaded with the external force $w(x)$ and the Fixed End Reactions (FER), R_i; and (2) the same element supporting the equivalent end forces equal in magnitude to the FER but acting in the opposite sense. It may be observed from Figure 1.8(b) that since the fixed-end beam does not develop end displacements, the FER applied in the opposite sense produce in the beam element the actual nodal displacements. Therefore, these FER applied in opposite directions are precisely the equivalent nodal forces.

Appendix I provides the expressions of equivalent nodal forces for some common loads applied on the span of a beam element. The expressions for the nodal equivalent forces listed in Appendix I may be obtained by using eq. (1.18) or alternatively by calculating the element FER and reversing their directions. (See Analytical Problems 1.4 and 1.5, Section 1.16.)

Illustrative Example 1.4
For the loaded beam shown in Figure 1.5 of the Illustrative Example 1.1, determine:

(a) The equivalent nodal forces for each of the four elements of the beam.
(b) The assembled system force vector.

Solution:
(a) Equivalent Nodal Forces

ELEMENT 1

From Appendix I, Case (a), with the concentrated loads $W_1 = -30\ kip$ and $W_2 = -10\ kip$ applied from the left end of element 1 at 10 in and 20 in, respectively, we obtain

$$Q_1 = -\frac{30x80^2\left(3x10+80\right)}{90^3} - \frac{10x70^2\left(3x20+70\right)}{90^3} = -37.71\ kip$$

$$Q_2 = -\frac{30x10x80^2}{90^2} - \frac{10x20x70^2}{90^2} = -358.02\ kip\cdot in$$

Analogously, we obtain

$$Q_3 = -2.29\ kip$$
$$Q_4 = +\ 64.20\ kip\cdot in$$

ELEMENT 2

From Appendix I, Case (c), for a uniformly distributed load $w = -0.10\ kip/in$:

$$Q_1 = \frac{wL}{2} = -\frac{0.10x90}{2} = -4.50\ kip$$

$$Q_2 = \frac{wL^2}{12} = -\frac{0.10x90^2}{12} = -67.50\ kip\cdot in$$

$$Q_3 = \frac{wL}{2} = -4.50\ kip$$

$$Q_4 = -\frac{wL^2}{12} = +67.50\ kip\cdot in$$

ELEMENT 3

From Appendix I, Case (d), with $w_1 = -0.10\ kip/in$, $w_2 = -0.20\ kip/in$, $L_1 = 20\ in$, $L_2 = 25\ in$, and $L = 120\ in$:

$$Q_1 = -5.39\ kip$$
$$Q_2 = -142.82\ kip\cdot in$$
$$Q_3 = -5.86\ kip$$
$$Q_4 = 152.10\ kip\cdot in$$

ELEMENT 4

From Appendix I, Case (c), for a uniformly distributed load $w= -0.05$ kip/in and Case (b) for a concentrated moment $M = 100$ $kip{\cdot}in$ at a distance of 48 in from the left end of the element, we have

$$Q_1 = \frac{wL}{2} - \frac{6ML_1L_2}{L^3} = -\frac{0.05x96}{2} - \frac{6x100x48x48}{96^3} = -3.96 \ kip$$

$$Q_2 = \frac{wL^2}{12} + \frac{ML_2\left(L_2 - 2L_1\right)}{L^2} = -\frac{0.05x96^2}{12} + \frac{100x48\left(48 - 96\right)}{96^2} = -63.40 \ kip{\cdot}in$$

Analogously, we obtain

$$Q_3 = -0.84 \ kip$$
$$Q_4 = 13.40 \ kip{\cdot}in$$

(b) Assemble System Force Vector

The equivalent nodal forces for the four beam elements, as calculated previously, are arranged in the following vectors:

$$\{Q\}_1 = \begin{Bmatrix} -37.71 \\ -358.02 \\ -2.29 \\ +64.20 \end{Bmatrix} \begin{matrix} 5 \\ 6 \\ 1 \\ 2 \end{matrix} \quad \{Q\}_2 = \begin{Bmatrix} -4.50 \\ -67.50 \\ -4.50 \\ +67.50 \end{Bmatrix} \begin{matrix} 1 \\ 2 \\ 7 \\ 3 \end{matrix} \quad \{Q\}_3 = \begin{Bmatrix} -5.39 \\ -142.82 \\ -5.86 \\ +152.10 \end{Bmatrix} \begin{matrix} 7 \\ 3 \\ 8 \\ 4 \end{matrix} \quad \{Q\}_4 = \begin{Bmatrix} -3.96 \\ -63.40 \\ -0.84 \\ +13.40 \end{Bmatrix} \begin{matrix} 8 \\ 4 \\ 9 \\ 10 \end{matrix} \qquad (a)$$

The coefficients of these element force vectors are transferred to the system force vector according to the system nodal coordinates assigned to each beam element. For example, for element 1 the assignment of nodal coordinates is 5, 6, 1, and 2 as has been indicated in Figure 1.2. For convenience, the assignment of nodal coordinates for the first element force vector is indicated in eq. (a) on the right side of the equivalent nodal force vector $\{Q\}_1$ and corresponding assignments for the nodal vectors $\{Q\}_2$, $\{Q\}_3$, and $\{Q\}_4$. We proceed to transfer each coefficient in these vectors to the corresponding location in the system force vector $\{F\}$ as shown in the first force vector of eq. (b):

$$\{F\} = \begin{Bmatrix} -2.29 & -4.50 \\ +64.20 & -67.50 \\ +67.50 & -142.82 \\ +152.10 & -63.40 \\ \hline -37.71 \\ -358.02 \\ -4.50 & -5.39 \\ -5.86 & -3.96 \\ -0.84 \\ +13.40 \end{Bmatrix} + \begin{Bmatrix} -10 \\ 0 \\ -50 \\ 0 \\ \hline R_5 \\ R_6 \\ R_7 \\ R_8 \\ R_9 \\ R_{10} \end{Bmatrix} = \begin{Bmatrix} -16.79 \\ -3.30 \\ -125.32 \\ +88.70 \\ \hline R_5 - 37.71 \\ R_6 - 358.02 \\ R_7 - 9.89 \\ R_8 - 9.82 \\ R_9 - 0.84 \\ R_{10} + 13.40 \end{Bmatrix}$$ (b)

The second force vector in eq. (b) contains the forces applied directly to the free nodal coordinates and the reactions R_i at the fixed nodal coordinates labeled u_5 through u_{10} as shown for this beam in Figure 1.2.

1.8   System Nodal Displacements and Support Reactions

The unknown nodal displacements and support reactions are calculated by partitioning the system stiffness equation [eq. (1.13)], solving first for the unknown displacements and then calculating the unknown reactions. Therefore, eq. (1.13) is conveniently partitioned to separate the unknown free nodal displacements from the fixed displacements at the supports. Thus, by eq. (1.13)

$$\{F\} = [K]\{u\}$$ (1.13) repeated

or in partitioned form

$$\begin{Bmatrix} \{F\}_1 \\ \hline \{F\}_2 \end{Bmatrix} = \begin{bmatrix} [K]_{11} & | & [K]_{12} \\ \hline [K]_{21} & | & [K]_{22} \end{bmatrix} \begin{Bmatrix} \{u\}_1 \\ \hline \{u\}_2 \end{Bmatrix}$$ (1.19)

in which

$\{F\}_1$ = external force vector at the free nodal coordinates
$\{u\}_1$ = unknown displacements at the free nodal coordinates
$\{F\}_2$ = external force vector at the fixed nodal coordinates
$\{u\}_2$ = $\{0\}$ (no displacements at the fixed nodal coordinates)

After setting $\{u\}_2 = \{0\}$ and expanding the two matrix equations in eq. (1.19) we obtain

$$\{F\}_1 = [K]_{11} \{u\}_1 \tag{1.20}$$

$$\{F\} = [K]_{21} \{u\}_1 \tag{1.21}$$

Equation (1.20), which is referred to as the *reduced equation stiffness matrix*, is then solved for the unknown nodal displacements $\{u\}_1$, which are subsequently substituted into eq. (1.21) to yield the nodal forces at the fixed nodal coordinates, thus allowing the determination of the reactions at the supports. The following Illustrative Example 1.5 uses eqs. (1.20) and (1.21) to determine the displacements at the free nodal coordinates and the reactions at the supports.

Illustrative Example 1.5

Use the results in Illustrative Example 1.1 to establish eqs. (1.20) and (1.21). Then solve for the unknown displacements at the free nodal coordinates and calculate the reactions at the fixed nodal coordinates.

Solution:

Substituting into eq. (1.20) the first four forces in the partitioned force vector, eq. (b) of Illustrative Example 1.4, and the top-left partitioned stiffness matrix [eq. (e)] of Illustrative Example 1.1, we obtain

$$\begin{Bmatrix} -16.79 \\ -3.30 \\ -125.32 \\ +88.70 \end{Bmatrix} = \begin{bmatrix} 840 & 0 & 18900 & 0 \\ 0 & 2268400 & 567110 & 0 \\ 18900 & 567110 & 1984867 & 425333 \\ 0 & 0 & 425333 & 1914000 \end{bmatrix} \begin{Bmatrix} u_1 \\ u_2 \\ u_3 \\ u_4 \end{Bmatrix} \tag{a}$$

Solving eq. (a) for the unknown nodal displacements u_1 through u_4 results in

$$\begin{aligned}
u_1 &= -2.40E\text{--}2 \ in \\
u_2 &= -4.56E\text{--}5 \ rad \\
u_3 &= 1.77E\text{--}4 \ rad \\
u_4 &= 7.12E\text{--}6 \ rad
\end{aligned} \tag{b}$$

To solve for the reactions at the fixed supports, we substitute into eq. (1.21) the lower partition vector of eq. (b) in Illustrative Example 1.5 and the left lower partition of the stiffness matrix from eq. (e) of Illustrative Example 1.1:

$$
\begin{Bmatrix} R_5 & -37.71 \\ R_6 & -358.02 \\ R_7 & -9.89 \\ R_8 & -9.82 \\ R_9 & -0.84 \\ R_{10} & +13.40 \end{Bmatrix} = \begin{bmatrix} -420 & -18900 & 0 & 0 \\ -18900 & 567110 & 0 & 0 \\ -420 & 18900 & -8267 & 10633 \\ 0 & 0 & -10633 & 5982 \\ 0 & 0 & 0 & -16615 \\ 0 & 0 & 0 & 531667 \end{bmatrix} \begin{Bmatrix} -2.40E\!-\!2 \\ -4.56E\!-\!5 \\ 1.77E\!-\!4 \\ 7.12E\!-\!6 \end{Bmatrix} = \begin{Bmatrix} 9.20 \\ 427.09 \\ 9.54 \\ -1.83 \\ -0.12 \\ 3.78 \end{Bmatrix} \quad (c)
$$

Thus, the reactions $\{R\}$ at the supports of the beam are then calculated from eq. (c) as

$$
\begin{bmatrix} R_5 \\ R_6 \\ R_7 \\ R_8 \\ R_9 \\ R_{10} \end{bmatrix} = \begin{bmatrix} 9.20 \\ 427.09 \\ 9.54 \\ -1.83 \\ -0.12 \\ 3.78 \end{bmatrix} + \begin{bmatrix} 37.71 \\ 358.02 \\ 9.89 \\ 9.82 \\ 0.84 \\ -13.40 \end{bmatrix} = \begin{bmatrix} 46.91\,kip \\ 785.11\,kip\!\cdot\!in \\ 19.43\ kip \\ 7.99\ kip \\ 0.72\ kip \\ -9.62\ kip\!\cdot\!in \end{bmatrix} \quad (d)
$$

Alternatively, the reactions at the fixed nodal coordinates also could be determined by calculating the element end forces, as presented in the following Section 1.9.

1.9   Element End Forces

The *element end forces* $\{P\}$ at the nodal coordinates of a beam element are given by the superposition of the forces $[k]\{\delta\}$ due to the element nodal displacements and the Fixed-End Reactions $\{FER\}$ at the nodal coordinates of the element due to the applied loads on the span of the element; that is,

$$\{P\} = [k]\{\delta\} + \{FER\} \tag{1.22}$$

or

$$\{P\} = [k]\{\delta\} - \{Q\} \tag{1.23}$$

because, as shown in Section 1.7, the fixed-end reactions $\{FER\}$ are equal to the equivalent nodal forces $\{Q\}$, but with opposite sign.

The following symbols are used in eqs. (1.22) and (1.23):

$[k]$ = element stiffness matrix
$\{\delta\}$ = displacement vector at the element nodal coordinates
$\{FER\}$ = Fixed End Reactions at the element nodal coordinates due to applied loads
$\{Q\}$ = equivalent forces at the nodal coordinates of a loaded beam element

Illustrative Example 1.6

For the beam in Illustrative Example 1.1, determine:

(a)   displacements at the nodal coordinates for each element
(b)   end forces at the nodal coordinates for each element of the beam
(c)   support reactions

Solution:

(a)   Element Nodal Displacements

The nodal displacement vectors for each element of the beam are identified from Figure 1.5 as follows:

$$\{\delta\}_1 = \begin{Bmatrix} 0 \\ 0 \\ u_1 \\ u_2 \end{Bmatrix} \quad \{\delta\}_2 = \begin{Bmatrix} u_1 \\ u_2 \\ 0 \\ u_3 \end{Bmatrix} \quad \{\delta\}_3 = \begin{Bmatrix} 0 \\ u_3 \\ 0 \\ u_4 \end{Bmatrix} \quad \{\delta\}_4 = \begin{Bmatrix} 0 \\ u_4 \\ 0 \\ 0 \end{Bmatrix} \tag{a}$$

or substituting numerical values for u_1, u_2, u_3, and u_4 calculated in Illustrative Example 1.5:

$$\{\delta\}_1 = \begin{Bmatrix} 0 \\ 0 \\ -2.4E\text{-}2 \\ -4.56E\text{-}5 \end{Bmatrix} \quad \{\delta\}_2 = \begin{Bmatrix} -2.40E\text{-}2 \\ -4.56E\text{-}5 \\ 0 \\ 1.77E\text{-}4 \end{Bmatrix} \quad \{\delta\}_3 = \begin{Bmatrix} 0 \\ 1.77E\text{-}4 \\ 0 \\ 7.12E\text{-}6 \end{Bmatrix} \quad \{\delta\}_4 = \begin{Bmatrix} 0 \\ 7.12E\text{-}6 \\ 0 \\ 0 \end{Bmatrix} \tag{b}$$

(b) Element End Forces

Substituting into eq. (1.23) the following: (1) the numerical values for the element stiffness matrix $[k]_1$ of element 1, determined in Illustrative Example 1.1, (2) the displacement vector $\{\delta\}_1$, given in eq. (b) of Illustrative Example 1.6 and (3) the equivalent nodal force vector $\{Q\}_1$ calculated in Illustrative Example 1.4, we obtain

$$\begin{Bmatrix} P_1 \\ P_2 \\ P_3 \\ P_4 \end{Bmatrix}_1 = \begin{bmatrix} 420 & 18900 & -420 & 18900 \\ 18900 & 1134200 & -18900 & 567100 \\ -420 & -18900 & 420 & -18900 \\ 18900 & 567110 & -18900 & 1134200 \end{bmatrix} \begin{Bmatrix} 0 \\ 0 \\ -2.40E\text{-}2 \\ -4.56E\text{-}5 \end{Bmatrix} - \begin{Bmatrix} -37.71 \\ -358.02 \\ -2.29 \\ +64.20 \end{Bmatrix} = \begin{Bmatrix} 46.91 \\ 785.04 \\ -6.91 \\ 336.97 \end{Bmatrix}$$

$$\tag{c}$$

or

$$P_{11} = 46.91 \ kip$$
$$P_{21} = 785.04 \, kip \cdot in$$
$$P_{31} = -6.91 \ kip$$ (d)
$$P_{41} = 336.9 \ kip \cdot in$$

Analogously, for elements 2, 3, and 4, we obtain:

$$P_{12} = -3.08 \ kip \qquad P_{13} = 7.34 \ kip \qquad P_{14} = 4.08 \ kip$$

$$P_{22} = -336.97 \ kip \cdot in \qquad P_{23} = 296.01 \ kip \cdot in \qquad P_{24} = 70.97 \ kip \cdot in \qquad (e)$$

$$P_{32} = 12.09 \ kip \qquad P_{33} = 3.91 \ kip \qquad P_{34} = 0.72 \ kip$$

$$P_{42} = -346.01 \ kip \cdot in \qquad P_{43} = -70.97 \ kip \cdot in \qquad P_{44} = -9.62 \ kip \cdot in$$

In eqs. (d) and (e) the second sub-index of P_{ij} serves to identify the beam element number. The end forces for each of the four elements of this beam are shown in their proper direction in Figure 1.9; thus allowing to check the equilibrium conditions in each of these four beam elements.

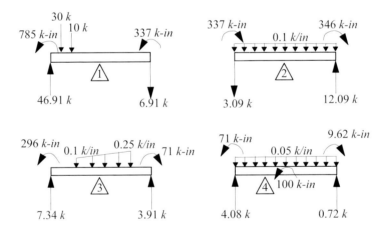

Fig. 1.9   Beam elements of the beam in Fig. 1.5 showing the applied loads and the calculated end forces

(c) Reactions at the Supports

As stated in Section 1.8, the reactions at the fixed nodal coordinates could also be determined from the calculated element end forces. For the beam in Illustrative Example 1.1, the reactions R_5 and R_6 at the fixed support of joint 1 are identified as the forces at the nodal coordinates 1 and 2 of element 1. Likewise, the reactions R_9 and R_{10} at the fixed support of joint 5 are identified as the forces at the nodal coordinate 3 and 4 of element 4. That is, from eq. (d) we have:

$$R_s = P_{11} = 46.91 \ kip \qquad R_6 = P_{21} = 785.04 \ kip \cdot in$$

and from eq. (e)

$$R_9 = P_{34} = 0.72 \ kip \qquad R_{10} = P_{44} = -9.62 \ kip \cdot in$$

At the interior support at joint 3, the reaction R_7 is equal to the addition of the end forces at this joint for elements 2 and 3. Analogously, the reaction R_8 at the interior support at joint 4 is equal to the sum of the end forces at this joint for elements 3 and 4. That is,

$$R_7 = P_{32} + P_{13} = 12.09 + 7.34 = 19.43 \ kip$$

and

$$R_8 = P_{33} + P_{14} = 3.91 + 4.08 = 7.99 \ kip$$

The values for these reactions obtained from element end forces are certainly equal to those calculated in eq. (d) of Illustrative Example 1.5.

1.10   End Releases in Beam Elements

The stiffness matrix method of analysis for beams presented in the preceding sections is based on the condition that each beam element is continuously connected at the nodes or joints at both ends. This method of analysis has to be modified for cases in which elements of the beam are connected at the nodes through hinges. When an element of a beam is connected to the adjacent node by a hinge, the moment at the hinge end must be zero. Such connections through hinges are often referred to as *member* or *element releases*.

The effect of end releases in a beam element can be conveniently considered in the stiffness method of analysis by modifying both the element stiffness matrix and the equivalent force vector. Only released moments are considered in this section; these are the most common releases in structural systems such as beams and frames. Other types of releases could also be considered in the development of the element stiffness matrix.

The substitution of the stiffness matrix for a beam element from eq. (1.11) into eq. (1.23) followed by the matrix multiplication $[k]\{\delta\}$ yields

$$P_1 = \frac{EI}{L^3}\left[12\delta_1 + 6L\delta_2 - 12\delta_3 + 6L\delta_4\right] - Q_1 \tag{1.24a}$$

$$P_2 = \frac{EI}{L^3}\left[6L\delta_1 + 4L^2\delta_2 - 6L\delta_3 + 2L^2\delta_4\right] - Q_2 \tag{1.24b}$$

$$P_3 = \frac{EI}{L^3}\left[-12\delta_1 - 6L\delta_2 + 12\delta_3 - 6L\delta_4\right] - Q_3 \tag{1.24c}$$

$$P_4 = \frac{EI}{L^3}\left[6L\delta_1 + 2L^2\delta_2 - 6L\delta_3 + 4L^2\delta_4\right] - Q_4 \tag{1.24d}$$

We distinguish three cases of element end releases as shown in Figure 1.10

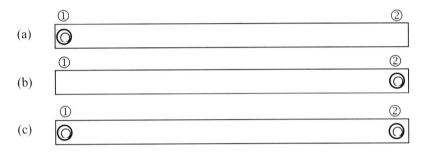

Fig.1.10   Element end releases: (a) Case 1 (hinge at node 1); (b) Case 2 (hinge at node 2) ; and (c) Case 3 (hinges at the two nodes of the beam element)

Case 1: Beam element with a hinge at the first node 1

When a beam element has a hinge at node 1 as shown in Figure 1.10(a), the bending moment at this end as expressed by eq. (1.24b) is equal to zero. Thus, after equating eq. (1.24b) to zero and solving for the rotational displacement δ_2, we obtain

$$\delta_2 = -\frac{3}{2L}\delta_1 + \frac{3}{2L}\delta_3 - \frac{1}{2}\delta_4 + \frac{L}{4EI}Q_2 \tag{1.25}$$

Equation (1.25) shows, that in this case (a hinge at the first node of a beam element), the angular displacement δ_2 at this node is not an independent nodal coordinate, but a function of the nodal displacements δ_1, δ_3, and δ_4. Therefore, the number of independent nodal displacements for the hinged beam element is reduced to three. We can then eliminate the rotational displacement δ_2 from the element stiffness equation by substituting δ_2 from eq. (1.25) into eqs. (1.24) to obtain

$$P_1 = \frac{EI}{L^3}(3\delta_1 - 3\delta_3 + 3L\delta_4) - \left(Q_1 - \frac{3}{2L}Q_2\right)$$

$$P_2 = 0$$

$$P_3 = \frac{EI}{L^3}(-3\delta_1 + 3\delta_3 - 3L\delta_4) - \left(Q_3 + \frac{3}{2L}Q_2\right) \qquad (1.26)$$

$$P_4 = \frac{EI}{L}(3L\delta_1 - 3L\delta_3 + 3L^2\delta_4) - \left(Q_4 - \frac{1}{2}Q_2\right)$$

Equations (1.26), which establish the modified relationship between nodal forces and nodal displacement for a beam element released at node 1, can be expressed in matrix form as

$$
\begin{Bmatrix} P_1 \\ P_2 \\ P_3 \\ P_4 \end{Bmatrix} = \frac{EI}{L^3}
\begin{bmatrix}
3 & 0 & -3 & 3L \\
0 & 0 & 0 & 0 \\
-3 & 0 & 3 & -3L \\
3L & 0 & -3L & 3L^2
\end{bmatrix}
\begin{Bmatrix} \delta_1 \\ \delta_2 \\ \delta_3 \\ \delta_4 \end{Bmatrix}
- \begin{Bmatrix} Q_1 - \dfrac{3}{2L}Q_2 \\ 0 \\ Q_3 + \dfrac{3}{2L}Q_2 \\ Q_4 - \dfrac{1}{2}Q_2 \end{Bmatrix} \qquad (1.27)
$$

or in condensed notation

$$\{P\} = [k]_m \{\delta\} - \{Q\}_m$$

in which the modified element stiffness matrix $[k]_m$ is given by

$$
[k]_m = \frac{EI}{L^3}
\begin{bmatrix}
3 & 0 & -3 & 3L \\
0 & 0 & 0 & 0 \\
-3 & 0 & 3 & -3L \\
3L & 0 & -3L & 3L^2
\end{bmatrix} \qquad (1.28)
$$

and the modified equivalent force vector $\{Q\}_m$ by

$$\{Q\}_m = \begin{Bmatrix} Q_1 - \dfrac{3}{2L}Q_2 \\[2ex] 0 \\[2ex] Q_3 + \dfrac{3}{2L}Q_2 \\[2ex] Q_4 - \dfrac{1}{2}Q_2 \end{Bmatrix}. \tag{1.29}$$

Case 2: Beam element with a hinge at node 2

Proceeding as in Case 1, we set the moment P_4 in eq. (1.24d) equal to zero followed by the calculation of the angular displacement δ_4 and its subsequent substitution into eqs. (1.24) to obtain for this case the modified element stiffness matrix $[k]_m$ and the modified equivalent force vector $\{Q\}_m$ as:

$$[k]_m = \dfrac{EI}{L^3} \begin{bmatrix} 3 & 3L & -3 & 0 \\ 3L & 3L^2 & -3L & 0 \\ -3 & -3L & 3 & 0 \\ 0 & 0 & 0 & 0 \end{bmatrix} \tag{1.30}$$

and

$$\{Q\}_m = \begin{Bmatrix} Q_1 - \dfrac{3}{2L}Q_4 \\[2ex] Q_2 - \dfrac{1}{2}Q_4 \\[2ex] Q_3 + \dfrac{3}{2L}Q_4 \\[2ex] 0 \end{Bmatrix} \tag{1.31}$$

Case 3: Beam element with hinges at both ends

Finally, for the case in which both ends of the beam element are released (hinges at both ends), the modified stiffness matrix and the modified equivalent force vector calculated as in Cases 1 and 2 are

$$[k]_m = \begin{Bmatrix} 0 & 0 & 0 & 0 \\ 0 & 0 & 0 & 0 \\ 0 & 0 & 0 & 0 \\ 0 & 0 & 0 & 0 \end{Bmatrix} \tag{1.32}$$

and

$$\{Q\}_m = \begin{Bmatrix} Q_1 - \dfrac{1}{L}(Q_2 + Q_4) \\ 0 \\ Q_3 + \dfrac{1}{L}(Q_2 + Q_4) \\ 0 \end{Bmatrix}. \tag{1.33}$$

Illustrative Example 1.7
Determine the system stiffness matrix and the system force vector for the beam of Illustrative Example 1.1 after releasing the left end of element 4 by introducing a hinge at this location, as shown in Fig 1.11

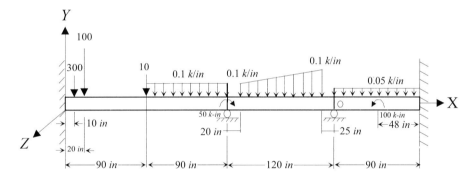

Fig. 1.11   Beam for Illustrative Example 1.7 with a hinge at the left end of element 4

Solution:
The stiffness matrix for elements 1, 2, and 3 has been calculated in Illustrative Example 1.1. Only the modified matrix and the modified equivalent force vector for element 4 must be determined. In this case (hinge at the first node), the modified element stiffness matrix is obtained from eq. (1.28) as

$$[K]_m = \frac{29 \times 10^3 \times 882}{96^3} \begin{bmatrix} 3 & 0 & -3 & 3x96 \\ 0 & 0 & 0 & 0 \\ -3 & 0 & 3 & -3x96 \\ 3x96 & 0 & -3x96 & 3x96^2 \end{bmatrix}$$

or

$$
[k]_m =
\begin{array}{cccc}
8 & 4 & 9 & 10 \\
\end{array}
\begin{bmatrix}
86.53 & 0 & -86.53 & 8307 \\
0 & 0 & 0 & 0 \\
-86.53 & 0 & 86.53 & -8307 \\
8307 & 0 & -8307 & 797500 \\
\end{bmatrix}
\begin{array}{c}
8 \\ 4 \\ 9 \\ 10
\end{array}
\tag{a}
$$

The modified equivalent force vector for element 4 is then obtained by substituting into eq. (1.29) the values calculated in Illustrative Example 1.5 for Q_1, Q_2, Q_3, and Q_4:

$$
\{Q\}_m =
\begin{Bmatrix}
-3.96 + \dfrac{3}{2\times 96}(63.40) \\[2mm]
0 \\[2mm]
-0.84 - \dfrac{3}{2\times 96}(63.40) \\[2mm]
+13.40 + \dfrac{1}{2}(63.40)
\end{Bmatrix}
=
\begin{Bmatrix}
-2.97 \\ 0 \\ -1.83 \\ +45.10
\end{Bmatrix}
\begin{array}{c}
8 \\ 4 \\ 9 \\ 10
\end{array}.
\tag{b}
$$

The system stiffness matrix is then assembled from the stiffness matrices developed in Illustrative Example 1.1 for elements 1, 2, and 3 and the modified stiffness matrix determined in eq. (a) for element 4. It should be realized that only the sub-matrix 4 x 4 corresponding to the first four free nodal coordinates is needed for further calculation. Thus the reduced system stiffness matrix $[K]_R$ assembled from stiffness matrices for elements 1, 2, 3 (calculated in Illustrative Example 1.1) and the modified stiffness matrix [given by eq. (a) for element 4] results in

$$
[K]_R =
\begin{Bmatrix}
840 & 0 & 18900 & 0 \\
0 & 2268400 & 567110 & 0 \\
18900 & 567110 & 1984889 & 425333 \\
0 & 0 & 425333 & 850667
\end{Bmatrix}
\tag{c}
$$

The system force vector is obtained by assembling from eq. (a) of Illustrative Example 1.5 the force vectors for elements 1, 2, 3, and the modified equivalent force vector for the released element 4 of this beam calculated in eq. (b). The reduced system force vector is then obtained as

$$\{F\}_R = \begin{Bmatrix} -2.29 - 4.50 \\ +64.20 - 67.50 \\ +67.50 - 142.82 \\ +152.10 - 0 \end{Bmatrix} + \begin{Bmatrix} -10 \\ 0 \\ -50 \\ 0 \end{Bmatrix} = \begin{Bmatrix} -16.79 \\ -3.30 \\ -125.32 \\ +152.10 \end{Bmatrix}. \qquad (d)$$

1.11 Support Displacements

When displacements at supports do occur due to yielding at the foundation or to imposed displacements at the nodal coordinates, the values of these displacements may be introduced in the corresponding entries of the vector $\{u\}$ of the system nodal displacements. Alternatively, the effects of small imposed support or nodal displacements are introduced in the analysis as equivalent forces at the nodal coordinates.

Consider a beam element having both displacements δ_1, δ_2, δ_3, and δ_4 due to the applied loads and displacements Δ_1, Δ_2, Δ_3, and Δ_4 due to imposed displacements, as shown in Figure 1.12(a) and (b), respectively. In this case, the element nodal forces $\{P\}$ resulting from the total nodal displacement of $[\{\delta\}+\{\Delta)\}]$ are calculated using eq. (1.23) as

$$\{P\} = [k]\left[\{\delta\} + \{\Delta\}\right] - \{Q\} \qquad (1.34)$$

in which

$$
\begin{aligned}
[k] \ &= \ \text{element stiffness matrix} \\
\{\delta\} \ &= \ \text{element nodal displacements resulting from external loads} \\
\{\Delta\} \ &= \ \text{imposed element nodal displacements} \\
\{Q\} \ &= \ \text{vector of equivalent nodal forces for load applied on the span of} \\
& \quad\ \ \text{of the element}
\end{aligned}
$$

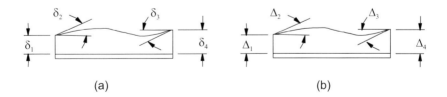

(a)   (b)

Fig. 1.12 Beam element showing: (a) nodal displacement δ_1, δ_2, δ_3 and δ_4, resulting from applied loads; and (b) imposed nodal displacements Δ_1, Δ_2, Δ_3, and Δ_4.

From eq. (1.34) we have

$$\{P\} = [k]\{\delta\} + [k]\{\Delta\} - \{Q\}$$

or

$$\{P\} = [k]\{\delta\} - \left[\{Q\} + \{Q\}_\Delta\right] \tag{1.35}$$

in which the equivalent nodal force vector $\{Q\}_\Delta$ resulting from the imposed displacements $\{\Delta\}$ is given by

$$\{Q\}_\Delta = -[k]\{\Delta\}. \tag{1.36}$$

It is evident from eq. (1.35) that the effect of the displacements $\{\Delta\}$ imposed at the nodal coordinates may be considered in the analysis by introducing for each element of the beam the equivalent nodal forces $\{Q\}_\Delta$ given in eq. (1.36).

As presented in Section 1.5, the system stiffness matrix is assembled by transferring the coefficients of the element stiffness matrices $[k]$ to the proper locations in the reduced system stiffness matrix $[K]_R$. The elements of the equivalent nodal forces $\{Q\}_\Delta$ due to imposed support or nodal displacements $\{\Delta\}$ also are transferred to the proper locations in the reduced system force vector $\{F\}_R$. The following example illustrates the necessary calculations.

Illustrative Example 1.8

Assume that the beam analyzed in Illustrative Example 1.1 is subjected to a displacement of 1.0 *in* downward at joint 3 and of 2.0 *in* downward at joint 4. Determine:

(a) Displacements at the free nodal coordinates.
(b) Element end forces.
(c) Reactions at the fixed nodal coordinates.

Solution:

(a) Displacements at the Free Nodal Coordinates

The reduced system stiffness matrix and the reduced system force vector resulting from the applied loads are given by the reduced stiffness matrix $[K]_R$ and the reduced force vector $\{F\}_R$ in eq. (a) of Illustrative Example 1.5 as

$$[K]_R = \begin{bmatrix} 840 & 0 & 18900 & 0 \\ 0 & 2668400 & 567110 & 0 \\ 18900 & 567110 & 1984867 & 425333 \\ 0 & 0 & 425333 & 1914000 \end{bmatrix} \quad \text{(a)}$$

and

$$\{F\}_R = \begin{Bmatrix} -16.79 \\ -3.30 \\ -125.32 \\ +88.70 \end{Bmatrix} \begin{matrix} 1 \\ 2 \\ 3 \\ 4 \end{matrix} \quad \text{(b)}$$

To include the effect of the imposed nodal displacements, it is necessary to recalculate the equivalent force vectors for the elements 2, 3, and 4, since these elements are affected by the displacements and supports 3 and 4. The application of eq. (1.36) to the elements 2, 3, and 4 for which the stiffness matrices are given, respectively, by eqs. (b), (c), and (d) of Illustrative Example 1.1 results in

ELEMENT 2

$$\begin{Bmatrix} Q_1 \\ Q_2 \\ Q_3 \\ Q_4 \end{Bmatrix}_{(2)} = - \begin{bmatrix} 420 & 18900 & -420 & 18900 \\ 18900 & 1134200 & -18900 & 567110 \\ -420 & -18900 & 420 & -18900 \\ 18900 & 567110 & -18900 & 1134200 \end{bmatrix} \begin{Bmatrix} 0 \\ 0 \\ -1 \\ 0 \end{Bmatrix} = \begin{Bmatrix} -420 \\ -18900 \\ 420 \\ -18900 \end{Bmatrix}_{(2)} \begin{matrix} 1 \\ 2 \\ 7 \\ 3 \end{matrix} \quad \text{(c)}$$

ELEMENT 3

$$\begin{Bmatrix} Q_1 \\ Q_2 \\ Q_3 \\ Q_4 \end{Bmatrix}_{(3)} = - \begin{bmatrix} 177 & 10633 & -177 & 10633 \\ 10633 & 850667 & -10633 & 425333 \\ -177 & -10633 & 177 & -10633 \\ 10633 & 425333 & -10633 & 850667 \end{bmatrix} \begin{Bmatrix} -1 \\ 0 \\ -2 \\ 0 \end{Bmatrix} = \begin{Bmatrix} -177 \\ -10633 \\ 177 \\ -10633 \end{Bmatrix}_{(3)} \begin{matrix} 7 \\ 3 \\ 8 \\ 4 \end{matrix} \quad \text{(d)}$$

ELEMENT 4

$$\begin{Bmatrix} Q_1 \\ Q_2 \\ Q_3 \\ Q_4 \end{Bmatrix}_{(4)} = - \begin{bmatrix} 346 & 16615 & -346 & 16615 \\ 16615 & 1063333 & -16615 & 531667 \\ -346 & -16615 & 346 & -16615 \\ 16615 & 531667 & -16615 & 1063333 \end{bmatrix} \begin{Bmatrix} -2 \\ 0 \\ 0 \\ 0 \end{Bmatrix} = \begin{Bmatrix} 692 \\ 33230 \\ -692 \\ 33230 \end{Bmatrix}_{(4)} \begin{matrix} 8 \\ 4 \\ 9 \\ 10 \end{matrix} \quad \text{(e)}$$

The equivalent nodal forces calculated for elements 2, 3, and 4 are then transferred and added to the reduced system force vector [in eq. (a) of Illustrative Example 1.5] at locations corresponding to the system nodal coordinates as indicated on the right of eqs. (c), (d), and (e):

$$
\begin{Bmatrix} F_1 \\ F_2 \\ F_3 \\ F_4 \end{Bmatrix}
\begin{bmatrix}
-16.79 & -420 & & \\
 & -3.30 & -18900 & \\
-125.32 & -18900 & -10633 & \\
+88.70 & -10633 & +33230 &
\end{bmatrix}
=
\begin{Bmatrix}
-436.79 \\
-18903.30 \\
-29658.32 \\
+22685.70
\end{Bmatrix}
\tag{f}
$$

Substituting the system force vector calculated in eq. (f) for the force vector in the system stiffness equation [eq. (a) of Illustrative Example 1.5] results in

$$
\begin{Bmatrix}
-436.79 \\
-18903.30 \\
-29658.32 \\
+22685.70
\end{Bmatrix}
=
\begin{bmatrix}
840 & 0 & 18900 & 0 \\
0 & 2268445 & 567110 & 0 \\
18900 & 567110 & 1984867 & 425333 \\
0 & 0 & 425333 & 1914000
\end{bmatrix}
\begin{bmatrix} u_1 \\ u_2 \\ u_3 \\ u_4 \end{bmatrix}
\tag{g}
$$

The solution of eq. (g) yields

$$u_1 = -0.177 \ in. \qquad\qquad u_2 = -4.53E{-}3 \ rad$$
$$u_3 = -1.52E{-}2 \ rad \qquad u_4 = 1.52E{-}2 \ rad$$

(b)   Element End Forces:

The element's end forces are calculated using eq. (1.35), which for convenience is repeated here:

or

$$\{P\} = [k]\{\delta\} - \big[\{Q\} + \{Q\}_\Delta\big] \tag{1.35 repeated}$$

in which

$$\{Q\}_\Delta = -[k]\,\{\Delta\} \tag{1.36 repeated}$$

The element nodal displacements, excluding imposed displacements (the effects of these imposed displacements are considered as equivalent nodal forces) are identified from the solution of eq. (g) as

$$\{\delta\}_1 = \begin{Bmatrix} 0 \\ 0 \\ -0.177 \\ -4.53x10^{-3} \end{Bmatrix} \qquad \{\delta\}_2 = \begin{Bmatrix} -0.177 \\ -4,53x10^{-3} \\ 0 \\ -1.52x10^{-2} \end{Bmatrix}$$

$$\{\delta\}_3 = \begin{Bmatrix} 0 \\ -1.52x10^{-2} \\ 0 \\ 1.52x10^{-2} \end{Bmatrix} \qquad \{\delta\}_4 = \begin{Bmatrix} 0 \\ 1.52x10^{-2} \\ 0 \\ 0 \end{Bmatrix}$$

(h)

The total element equivalent nodal forces are then obtained by adding the equivalent forces due to imposed nodal displacements as given by eqs. (c) through (e) to the nodal forces as calculated in eq. (a) of Illustrative Example 1.4, namely,

$$\{Q\}_1 + \{Q\}_{\Delta 1} = \begin{Bmatrix} -37.71 \\ -358.02 \\ -2.29 \\ +64.20 \end{Bmatrix} \qquad \{Q\}_3 + \{Q\}_{\Delta 2} = \begin{Bmatrix} -4.50-420 \\ -67.50-18900 \\ -4.5+420 \\ +67.50-18900 \end{Bmatrix}$$

(i)

$$\{Q\}_3 + \{Q\}_{\Delta 3} = \begin{Bmatrix} -5.39-177 \\ -142.82-106.33 \\ -5.86+177 \\ 152.10-10633 \end{Bmatrix} \qquad \{Q\}_4 + \{Q\}_{\Delta 4} = \begin{Bmatrix} -3.96+692 \\ -63.40+33230 \\ -0.84-692 \\ +13.40+33230 \end{Bmatrix}$$

Finally, introducing into eq. (1.35) the nodal displacement vector $\{\delta\}_1$ from eqs. (h), the equivalent nodal force vector $\{Q\}_1 + \{Q\}_{\Delta 1}$ from eqs. (i) and the element stiffness matrix $[k]_1$ calculated in Illustrative Example 1.1 (for element 1) results in

$$\{P\}_1 = \begin{bmatrix} 420 & 18900 & -420 & 18900 \\ 18900 & 1134200 & -18900 & 567110 \\ -420 & -18900 & 420 & -18900 \\ 18900 & 567110 & -18900 & 1134200 \end{bmatrix} \begin{Bmatrix} 0 \\ 0 \\ -0.177 \\ -4.53(10^{-3}) \end{Bmatrix} - \begin{Bmatrix} -37.71 \\ -358.02 \\ -2.29 \\ 64.20 \end{Bmatrix} = \begin{Bmatrix} 26.63 \\ 1143.71 \\ 13.37 \\ -1846.66 \end{Bmatrix}_1 \quad (j)$$

Analogously, we obtain from elements 2, 3, and 4 the end forces:

$$\{P\}_2 = \begin{Bmatrix} -23.36 \\ 1846.66 \\ 32.36 \\ -4354.61 \end{Bmatrix}_2 \qquad \{P\}_3 = \begin{Bmatrix} 182.72 \\ 4304.61 \\ -171.47 \\ 16965.57 \end{Bmatrix}_3 \qquad \{P\}_4 = \begin{Bmatrix} -435.18 \\ -16965.57 \\ 439.98 \\ -25142.47 \end{Bmatrix}_4$$

(k)

(c) Reactions at the Fixed Nodal Coordinates

The reactions R_5 and R_6 at the fixed support of joint 1 are identified as the forces P_{11} and P_{21} at the nodal coordinates of joint 1 of element 1. Analogously, the reactions R_9 and R_{10} at the fixed support of joint 5 are identified as the forces P_{34} and P_{44} of element 4. Thus, from eqs. (j) and (k), we have:

$$R_5 = P_{11} = 26.63 \ kip \qquad R_6 = P_{21} = 1143.71 \ kip \cdot in$$

and

$$R_9 = P_{34} = 439.98 \ kip \qquad R_{10} = P_{44} = -25142.47 \ kip \cdot in$$

At the interior support joint 3, the reaction R_7 is equal to the sum of the end forces at this joint for elements 2 and 3. Analogously, the reactions R_8 at the interior support at joint 4 is equal to the sum of the end forces at this joint for elements 3 and 4. Thus, from eqs. (j) and (k), we have:

$$R_7 = P_{32} + P_{13} = \quad 32.36 \ + \ 182.72 \ = \ 215.08 \ kip$$
$$R_8 = P_{33} + P_{14} = -171.47 \ + \ -435.18 = -606.65 \ kip$$

1.12 Element Displacement Functions

The displacement function $y(x)$ due to the nodal displacements δ_1, δ_2, δ_3, and δ_4 for an element of the beam is given by eq. (1.6) in which the coefficients $N_1(x)$, $N_2(x)$, $N_3(x)$, and $N_4(x)$ are given by eqs. (1.5). The total displacement $y_T(x)$ at any section x of a beam element is then calculated as the superposition of $y(x)$ due to the element nodal displacements and the displacement $y_L(x)$ resulting from the loads applied to the beam element assumed to be fixed at its two nodes; that is,

$$y_T(x) = y(x) + y_L(x). \tag{1.37}$$

The determination of the displacement function $y_L(x)$ due to an applied load on a beam element may be obtained by integrating twice the *differential equation for a beam*:

$$\frac{d^2y(x)}{dx^2} = \frac{M(x)}{E \, I} \tag{1.38}$$

in which

$M(x) = $ bending moment
$E \quad = $ modulus of elasticity

$$I \quad = \quad \text{cross-sectional moment of inertia}$$

The determination of the displacement function $y_L(x)$ due to an applied load on a fixed-end beam element supporting a uniformly distributed load is illustrated in the following example.

Illustrative Example 1.9

Determine the displacement function for a fixed-end beam element with an applied uniformly distributed load w as shown in Figure 1.13(a).

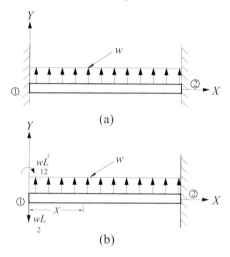

(a)

(b)

Fig. 1.13   (a) Fixed-end beam element supporting a uniformly distributed load
(b) Beam element showing the fixed-end reactions at node 1.

Solution:

Consider in Figure 1.13(a) a fixed-end beam element supporting a uniformly distributed load w applied in the positive direction of axis y and in Figure 1.13(b) the support reactions at node 1 for this beam element. These support reactions are numerically equal to the equivalent nodal forces but with opposite signs as given in Appendix I, Case c. The bending moment at any section x, using the strength of materials convention for positive moment, is then given from Figure 1.13(b) by

$$M(x) = \frac{wL^2}{12} - \frac{wL}{2}x + \frac{w}{2}x^2. \tag{1.39}$$

The substitution of eq. (1.39) into eq. (1.38) followed by two successive integrations results in

$$EI\frac{d^2 y_L(x)}{dx^2} = \frac{wL^2}{12} - \frac{wL}{2}x + \frac{w}{2}x^2$$

$$EI\frac{dy_L(x)}{dx} = \frac{wL^2}{12}x - \frac{wL}{4}x^2 + \frac{w}{6}x^3 + C_1 \qquad (1.40)$$

$$EI y_L(x) = \frac{wL^2}{24}x^2 - \frac{wL}{12}x^3 + \frac{w}{24}x^4 + C_1 x + C_2$$

The constants of integration C_1 and C_2 are evaluated from the boundary conditions of zero displacement and zero slope at $x = 0$, giving $C_1 = C_2 = 0$. Thus from eq. (1.40), the displacement function y_L for a fixed-end beam element with a uniformly distributed force is:

$$y_L(x) = \frac{1}{EI}\left[\frac{wL^2}{24}x^2 - \frac{wL}{12}x^3 + \frac{w}{24}x^4 \right] \qquad (1.41)$$

Analogously, the displacement functions for other common loads applied to a fixed-end beam element may be determined by integrating eq. (1.38) for each specific load, as it has been presented for a uniformly distributed force in Illustrative Example 1.9. Appendix II provides displacement functions of fixed-end beams for some common loads.

1.13 Shear Force and Bending Moment Functions

The shear force and bending moment at a section of beam element may readily be found from static equations of equilibrium. Consider in Figure 1.14(a) a loaded beam element showing the end element forces and in Figure 1.14(b) the free body diagram of a segment of this beam element.

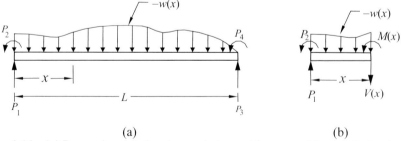

(a)   (b)

Fig. 1.14   (a) Beam element showing end element forces and the applied external forces, and (b) Free body diagram up to section x of the beam

By equating to zero the sum of the forces and also by equating to zero the sum of the moments of forces in the free body diagram shown in Fig 1.14(b), we obtain:

$$-V(x) + P_1 - \int_0^x w(x_1)dx_1 = 0$$

and

$$M(x) + P_2 - P_1 x + \int_0^x (x - x_1) w(x_1)dx_1 = 0$$

or

$$V(x) = P_1 - \int_0^x w(x_1)dx_1 \qquad (1.42)$$

and

$$M(x) = -P_2 + P_1 x - \int_0^x (x - x_1)w(x_1)dx_1 \qquad (1.43)$$

Equations (1.42) and (1.43) provide, respectively, the functions to calculate the shear force $V(x)$ and the bending moment $M(x)$ for a beam element in terms of the element end forces P_1 and P_2 and of the load $w(x)$ applied on the element.

1.14   Temperature Effect

Temperature changes, like support displacements, can cause large stresses in statically indeterminate structures, but not in statically determined structures in which members are free to expand or contract due to changes in temperature. In the case of beams, a uniform change in temperature in the cross-section does not induce stresses, since the analysis of beams does not consider axial deformations along the longitudinal axis of the beam. However, stresses are developed in beams that are subjected to a temperature variation across the depth of the beam. As in the case of loads applied on the beam elements, the effect of a temperature change in the cross-section of the beam is considered in the analysis by introducing equivalent forces at the nodes of the beam elements subjected to temperature changes. Consider in Figure 1.15(a) a beam element that is subjected to a linear temperature change over the depth of the cross-sectional area with a temperature $T_1{}^1$ at the bottom and $T_2{}^*$ at the top of the beam element. Assuming that $T_2 > T_1$, an isolated beam element will curve as shown in Figure 1.15(a). For a symmetric cross-sectional area with respect to the centroidal axis, the expression for the temperature function $T(y)$ in the cross-section of the beam is obtained from in Figure 1.15(b).

---

[1] T_1 and T_2 are temperatures above the ambient temperature at which the beam is under no strain. However, because the final expression for the equivalent nodal bending moment includes the factor $(T_2 - T_1)$, actual temperature values could as well be used for T_1 and T_2.

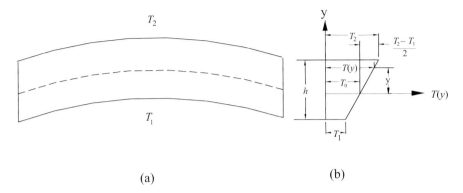

(a)   (b)

Fig. 1.15 (a) Beam element subjected to a linear temperature change in the cross-section
  (b) Temperature function *T(y)* along the depth of the beam

The temperature T_0 at the center of the cross-sectional area of the beam is given by the average of the temperatures at the two faces of the beam.

$$T_0 = \frac{T_1 + T_2}{2} \qquad (1.44)$$

Then, from Figure 1.15(b)

$$T(y) = T_0 + (T_2 - T_1)\frac{y}{h}$$

or substituting T_0 from eq. (1.44)

$$T(y) = \frac{T_1 + T_2}{2} + (T_2 - T_1)\frac{y}{h}. \qquad (1.45)$$

The strain, $\varepsilon(y)$, in the cross-section of any point at a distance y from the centroid is $\varepsilon(y) = \alpha T(y)$ where α is the coefficient of thermal expansion. Then the corresponding stress is $\sigma(y) = E\alpha T(y)$. The thermal axial force P_T in the axially constrained beam element will then be

$$P_T = \int_{(A)} \sigma(y)\, dA = \int_{(A)} E\alpha\, T(y)\, dA \qquad (1.46)$$

and the thermal bending moment

$$M_T = \int_{(A)} y\,\sigma(y)\,dA = \int_{(A)} E\,\alpha\; yT(y)\,dA \tag{1.47}$$

in which A is the cross-sectional area of the beam.

For a beam having a rectangular cross-sectional area of width b and height h, eqs. (1.46) and (1.47) become

$$P_T = \int_{-h/2}^{h/2} E\alpha\; T(y)\,b\,dy = E\alpha A T_0$$

and

$$M_T = \int_{-h/2}^{h/2} \alpha\,E\,T(y)\,b\,y\,dy = \alpha\frac{E\,I}{h}\,(T_2 - T_1) \tag{1.48}$$

The integral in eq. (1.48) was executed after substituting $T(y)$ for the expression given in eq. (1.45).

The equivalent nodal forces for a beam element subjected to a linear temperature change in its cross-section are then given by

$$Q_1 = 0 \quad Q_2 = \frac{\alpha E I}{h}(T_2 - T_1) \quad Q_3 = 0 \quad Q_4 = -\frac{\alpha E I}{h}(T_2 - T_1) \tag{1.49}$$

or in the vector notation

$$\{Q\}_T = \left\{ \begin{array}{c} 0 \\[4pt] +\dfrac{\alpha E I}{h}(T_2 - T_1) \\[4pt] 0 \\[4pt] -\dfrac{\alpha E I}{h}(T_2 - T_1) \end{array} \right\} \tag{1.50}$$

These equivalent nodal forces for a beam element subjected to a linear temperature variation along the height of the cross-section are shown in Figure 1.16.

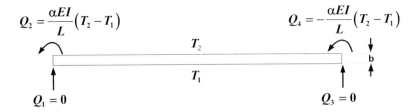

$$Q_2 = \frac{\alpha EI}{L}(T_2 - T_1) \qquad\qquad Q_4 = -\frac{\alpha EI}{L}(T_2 - T_1)$$

Fig. 1.16 Beam element showing the equivalent nodal forces due to a linear
temperature variation in the cross-section

Illustrative Example 1.10

Element 3 of the beam shown in Figure 1.5 is now subjected only to a linear
temperature change, with $T_1 = 70°F$ at the bottom face and $T_2 = 120°F$ at the top
face. The coefficient of thermal expansion is $\alpha = 6.5 \times 10^{-6}/°F$ and the height of the
cross-sectional area of the beam is $h = 10\ in$. Determine the nodal displacements
resulting from this temperature variation.

Solution:

The element stiffness matrices [eq. (a) through eq. (d)] and the system stiffness
matrix [eq. (e)] for this beam have been determined in Illustrative Example 1.1.
The equivalent nodal force vector for element 3 subjected to a linear variation of
temperature is given by eq. (1.50), which after substituting numerical values, results
in

$$\{Q\}_T = \left\{ \begin{matrix} 0 \\ 831 \\ 0 \\ -831 \end{matrix} \right| \begin{matrix} 7 \\ 3 \\ 8 \\ 4 \end{matrix} \tag{a}$$

The reduced system force vector, that is, considering only the coefficient of the free
nodal coordinates as shown in Figure 1.5, is obtained by transferring the coefficients
in eq. (a) to locations indicated on the right side of this equation. Thus resulting in

$$\{F\}_R = \left\{ \begin{matrix} 0 \\ 0 \\ 831 \\ -831 \end{matrix} \right| \begin{matrix} 1 \\ 2 \\ 3 \\ 4 \end{matrix} \tag{b}$$

Then the substitution of the force vector in eq. (a) of Illustrative Example 1.5 for the
force vector $\{F\}_R$ determined in eq. (a) yields

$$
\{F\}_R = \begin{Bmatrix} 0 \\ 0 \\ 831 \\ -831 \end{Bmatrix} = \begin{bmatrix} 840 & 0 & 18900 & 0 \\ 0 & 2268400 & 567110 & 0 \\ 18900 & 567110 & 1984867 & 425333 \\ 0 & 0 & 425333 & 1914000 \end{bmatrix} \begin{Bmatrix} u_1 \\ u_2 \\ u_3 \\ u_4 \end{Bmatrix}
\tag{c}
$$

The solution of eq. (b) for the unknown displacements yields

$$
\begin{aligned}
u_1 &= -0.00946 \ in & u_3 &= -4.658E{-}4 \ rad \\
u_2 &= 1.150E{-}4 \ rad & u_4 &= 5.130E{-}4 \ rad
\end{aligned}
$$

1.15 Elastic Supports

Linear elastic supports on a beam, whether resulting from axial or torsional springs,
may be considered in the analysis by simply adding the value of the spring constant
(stiffness coefficient) to the corresponding coefficient in the diagonal of the system
stiffness matrix. In the following Illustrative Example 1.11, the stiffness method of
analysis is applied to a beam having an elastic support.

Illustrative Example 1.11
Consider in Figure 1.17 a fixed-end beam supported by a spring at its center
having an axial stiffness $k_a = 100 \ kip/in$ and a torsional stiffness $k_t = 1000$
$kip{\cdot}in/rad$. For this beam determine:

(a) Vertical and rotational displacements at the central section
(b) End forces on the two elements of the beam
(c) Reactions at the supports
(d) Forces in the axial and torsional springs

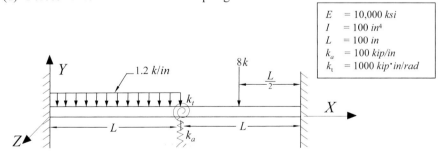

$$
\begin{aligned}
E &= 10{,}000 \ ksi \\
I &= 100 \ in^4 \\
L &= 100 \ in \\
k_a &= 100 \ kip/in \\
k_t &= 1000 \ kip{\cdot}in/rad
\end{aligned}
$$

Fig. 1.17 Beam of Illustrative Example 1.11

1. Analytical model

The analytical model for this beam is shown in Figure 1.18. It has two beam elements, three nodes, and two free nodal coordinates labeled u_1 and u_2.

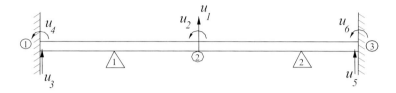

Fig. 1.18   Analytical model for the beam of Illustrative Example 1.11

2. Element stiffness matrices:

Substituting numerical values into the element stiffness matrix eq. (1.11) results in

$$
[k]_1 = [k]_2 = \frac{10^4 \times 10^2}{(100)^3}
\begin{array}{cc}
\begin{array}{cccc}
1 & 2 & 5 & 6 \\
3 & 4 & 1 & 2
\end{array} & \\
\begin{bmatrix}
12 & 600 & -12 & 600 \\
600 & 40000 & -600 & 20000 \\
-12 & -600 & 12 & -600 \\
600 & 20000 & -600 & 40000
\end{bmatrix} &
\begin{array}{cc}
3 & 1 \\
4 & 2 \\
1 & 5 \\
2 & 6
\end{array}
\end{array}
\tag{a}
$$

3. Reduced system stiffness matrix:

The reduced system stiffness matrix is assembled by transferring to this matrix the coefficients of the matrix in eq. (a) to locations indicated for the two elements of this beam at the top and on the right side of this matrix. The axial stiffness $k_a = 100$ *kip/in* and the torsional stiffness $k_t = 1000$ *kip·in/rad* are then added, respectively, to the diagonal coefficients on rows 1 and 2 of the reduced system stiffness matrix, namely,

$$
[K]_R =
\begin{bmatrix}
12 + 12 + 100 & -600 + 600 \\
-600 + 600 & 40000 + 40000 + 1000
\end{bmatrix}
$$

or

$$
[K]_R =
\begin{bmatrix}
124 & 0 \\
0 & 81000
\end{bmatrix}
\tag{b}
$$

4.   Element equivalent force vectors:

ELEMENT 1:

From Case (C) in Appendix I (with $w = -1.2$ *kip/in*):

$$Q_1 = \frac{wL}{2} = -\frac{1.2 \times 100}{2} = -60 \ kip$$

$$Q_2 = \frac{wL^2}{12} = -\frac{1.2 \times 100^2}{12} = -1000 \ kip \cdot in/rad$$

$$Q_3 = \frac{wL}{2} = -\frac{1.2 \times 100}{2} = -60 \ kip$$

$$Q_4 = -\frac{wL^2}{12} = \frac{1.2 \times 100^2}{12} = 1000 \ kip \cdot in/rad$$

or in vector form

$$\{Q\}_1 = \begin{Bmatrix} -60 \\ -1000 \\ -60 \\ 1000 \end{Bmatrix} \begin{matrix} 3 \\ 4 \\ 1 \\ 2 \end{matrix} \qquad (c)$$

ELEMENT 2:

From Case (a) in Appendix I (with $W = -8$ *kip*):

$$Q_1 = \frac{W}{2} = -\frac{8}{2} = -4 \ kip$$

$$Q_2 = \frac{WL}{8} = -\frac{8 \times 100}{8} = -100 \ kip \cdot in/rad$$

$$Q_3 = \frac{W}{2} = -\frac{8}{2} = -4 \ kip$$

$$Q_4 = -\frac{WL}{8} = \frac{8 \times 100}{8} = 100 \ kip \cdot in/rad$$

or in vector notation:

$$\{Q\}_2 = \begin{Bmatrix} -4 \\ -100 \\ -4 \\ 100 \end{Bmatrix} \begin{matrix} 1 \\ 2 \\ 5 \\ 6 \end{matrix} \qquad (d)$$

5. Reduced system force vector:

    The coefficients of equivalent force vectors in eqs. (c) and (d) are transferred to reduced system force vector in the location indicated (for the free nodal coordinates 1 and 2) on the right of these vectors, namely,

    $$\{F\}_R = \begin{Bmatrix} -60 - 4 \\ 1000 - 100 \end{Bmatrix} = \begin{Bmatrix} -64 \\ 900 \end{Bmatrix} \qquad (e)$$

6. Reduced system stiffness equation:

    The reduced system stiffness matrix is given by

    $$\{F\}_R = [K]_R \{u\}$$

    Substitution of $\{F\}_R$ from eq. (e) and $[K]_R$ from eq. (b) results in

    $$\begin{Bmatrix} -64 \\ 900 \end{Bmatrix} = \begin{bmatrix} 124 & 0 \\ 0 & 81000 \end{bmatrix} \begin{Bmatrix} u_1 \\ u_2 \end{Bmatrix} \qquad (f)$$

7. Solution of eq. (f):

    $$u_1 = -0.516 \; in \qquad and \qquad u_2 = 0.0111 \; rd$$

8. Element end forces:

    Element end forces are given by eq. (1.23) as

    $$\{P\} = [k]\{\delta\} - \{Q\} \qquad (1.23) \text{ repeated}$$

    The nodal displacements for elements 1 and 2 are identified from Figure 1.18 as

    $$\{\delta\}_1 = \begin{Bmatrix} u_3 \\ u_4 \\ u_1 \\ u_2 \end{Bmatrix} = \begin{Bmatrix} 0 \\ 0 \\ -0.516 \\ 0.0111 \end{Bmatrix} \qquad \{\delta\}_2 = \begin{Bmatrix} u_1 \\ u_2 \\ u_5 \\ u_6 \end{Bmatrix} = \begin{Bmatrix} -0.516 \\ 0.0111 \\ 0 \\ 0 \end{Bmatrix} \qquad (g)$$

ELEMENT 1:

The substitution of eqs. (a), (c), and (g) into eq. (1.23) results in

$$\begin{Bmatrix} P_1 \\ P_2 \\ P_3 \\ P_4 \end{Bmatrix}_1 = \begin{bmatrix} 12 & 600 & -12 & 600 \\ 600 & 40000 & -600 & 20000 \\ -12 & -600 & 12 & -600 \\ 600 & 20000 & -600 & 40000 \end{bmatrix} \begin{Bmatrix} 0 \\ 0 \\ -0.516 \\ 0.0111 \end{Bmatrix} - \begin{Bmatrix} -60 \\ -1000 \\ -60 \\ 1000 \end{Bmatrix} \tag{h}$$

ELEMENT 2:

Analogously, for element 2, we obtain

$$\begin{Bmatrix} P_1 \\ P_2 \\ P_3 \\ P_4 \end{Bmatrix}_2 = \begin{bmatrix} 12 & 600 & -12 & 600 \\ 600 & 40000 & -600 & 20000 \\ -12 & -600 & 12 & -600 \\ 600 & 20000 & -600 & 40000 \end{bmatrix} \begin{Bmatrix} -0.516 \\ 0.0111 \\ 0 \\ 0 \end{Bmatrix} - \begin{Bmatrix} -4 \\ -100 \\ -4 \\ 100 \end{Bmatrix} \tag{i}$$

Final result from eqs. (h) and (i)

$$\begin{Bmatrix} P_1 \\ P_2 \\ P_3 \\ P_4 \end{Bmatrix}_1 = \begin{Bmatrix} 72.85 \\ 1531.6 \\ 47.15 \\ -245.4 \end{Bmatrix}_1 \quad \text{and} \quad \begin{Bmatrix} P_1 \\ P_2 \\ P_3 \\ P_4 \end{Bmatrix}_2 = \begin{Bmatrix} 4.47 \\ 234.40 \\ 3.53 \\ -187.6 \end{Bmatrix}_2$$

9. Reactions:

$$R_3 = P_{11} = \quad 72.85 \ kip$$
$$R_4 = P_{21} = 1531.6 \ kip{\cdot}in$$
$$R_5 = P_{32} = \quad 3.53 \ kip$$
$$R_6 = P_{42} = -187.6 \ kip{\cdot}in$$

Notation: P_{ik} = Force at nodal coordinate i of beam element k

10. Forces at the elastic support:

Spring axial force:

$$S_a = P_{31} + P_{12}$$
$$= 47.15 + 4.47$$
$$= 51.62 \ kip \ \text{(Compression)}$$

Spring torsional moment S_t:

$$S_t = P_{41} + P_{22}$$
$$= -245.4 + 234.4$$
$$= 11.0 \ kip{\cdot}in$$

11. Alternatively, the forces S_a or S_t may be calculated as the product of the spring constant and the corresponding displacement. Namely,

$$S_a = (1000 \ kip/in)(-0.516 \ in) = -51.6 \ kip \quad \text{(Compression)}$$

and

$$S_t = (100 \ kip{\cdot}in/rad)(0.0111 \ rad) = 11.1 \ kip{\cdot}in.$$

1.16 Analytical Problems

Analytical Problem 1.1
Derive the general expression [eq. (1.7)] to calculate the stiffness coefficient for a beam element.

$$k_{ij} = \int_0^L EI \ N_i^{''}(x) \ N_j^{''}(x) \, dx \qquad \text{(1.7) repeated}$$

in which $N_i^{''}(x)$ and $N_j^{''}(x)$ are the second derivatives of the shape functions given by eqs. (1.5); I is the cross-sectional moment of inertia; and E the modulus of elasticity.

Solution:
Consider in Fig .P1.1(a), a beam element and in (b) and (c) the plots, respectively, of the shape function corresponding to unit displacements $\delta_1 = 1.0$ and $\delta_2 = 1.0$.

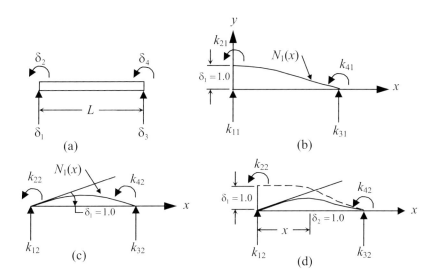

Fig. P1.1 Beam element: (a) nodal coordinates; (b) stiffness coefficients for a unit
displacement $\delta_1 = 1.0$; (c) stiffness coefficients for $\delta_2 = 1.0$; and (d)
virtual displacement $\delta_1 = 1$ given to beam at (c)

The displacement functions $N_1(x)$ and $N_2(x)$ corresponding to these unit displacements are given by eqs. (1.5a) and (1.5b). These functions are repeated here for
convenience:

$$N_1(x) = 1 - 3\left(\frac{x}{L}\right)^2 + 2\left(\frac{x}{L}\right)^3 \qquad \text{(1.5a) repeated}$$

$$N_2(x) = x\left(1 - \frac{x}{L}\right)^2 \qquad \text{(1.5b) repeated}$$

We assume that the beam element in Figure P1.1(c) undergoes an additional
displacement equal to the deflected curve shown in Figure P1.1(b), as it is shown in
Figure P1.1(d). We then apply the *Principle of Virtual Work,* which states that for
an elastic system in equilibrium, the work done by the external forces is equal to the
work of the internal forces during the virtual displacement. In order to apply this
principle, we note that the external work W_E is equal to the product of the force k_{12}
displaced by $\delta_1 = 1.0$, that is

$$W_E = k_{12}\delta_1 = k_{12}. \qquad \text{(a)}$$

This work, as stated, is equal to the internal work W_i performed by the elastic forces during the virtual displacement. Considering the work performed by the bending moment, we obtain for the internal work

$$W_i = \int_0^L M(x)\ d\theta \tag{b}$$

in which $M(x)$ is the bending moment at section x of the beam and $d\theta$ the relative angular displacement at this section. For the virtual displacement under consideration, the transverse deflection of the beam is given by eq. (1.5b), which is related to the bending moment through the differential equation [eq. (1.38)]

$$\frac{d^2 y(x)}{dx^2} = \frac{M(x)}{EI}. \tag{1.38 repeated}$$

Substitution of the second derivative $N_2''(x)$ from eq. (1.5b) into eq. (1.38) results in

$$EI\ N_2''(x) = M(x) \tag{c}$$

with the notation $y = N_2(x)$ for $\delta_2 = 1.0$.

The angular deflection $d\theta$ produced during this virtual displacement is related to the transverse deflection $N_1(x)$ by

$$\frac{d\theta}{dx} = \frac{d^2 N_1(x)}{dx^2} = N_1''(x)$$

or

$$d\theta = N_1''(x)\ dx \tag{d}$$

Substituting into eq. (b), $M(x)$ and $d\theta$ from eqs. (c) and (d), respectively, and then equating the external work and the internal work results in the stiffness coefficient k_{12}:

$$k_{12} = \int_0^L EI\ N_1''(x)\ N_2''(x)\ dx.$$

Therefore, in general, any stiffness coefficient associated with beam flexure may be expressed as

$$k_{ij} = \int_0^L EI\ N_i''(x)\ N_j''(x)\ dx. \tag{Q.E.D.}$$

Analytical Problem 1.2

Demonstrate that the equivalent nodal force Q_i for a distributed force $w(x)$ applied to a beam element is given by

$$Q_i = \int_0^L N_i(x)\, w(x)\ dx \qquad\qquad \text{(1.18a) repeated}$$

where $N_i(x)$ is the shape function given by eqs. (1.5).

Solution:

Consider in Figure P1.2 (a) a beam element showing a general applied force $w(x)$ and in Figure P1.2(b) the beam element showing the nodal equivalent forces Q_1 through Q_4. We give to both beam elements a unit virtual displacement at nodal coordinate 1 resulting in the shape function $N_1(x)$ given by eq. (1.5a), which is repeated here for convenience:

$$N_1(x) = 1 - 3\left(\frac{x}{L}\right)^2 + 2\left(\frac{x}{L}\right)^3. \qquad\qquad \text{(1.5a) repeated}$$

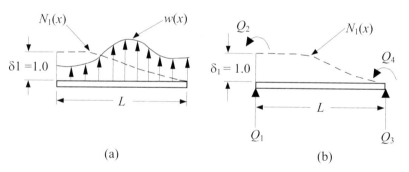

(a)   (b)

Fig. P1.2   (a) Beam element acted on by a distributed force $w(x)$ undergoing a virtual displacement $\delta_1 = 1.0$, resulting in the shape function $N_1(x)$, and (b) the beam element acted upon by the equivalent nodal forces Q_I undergoing the same virtual displacement, $\delta_1 = 1.0$

Now we require that the work performed as a result of this virtual displacement by the equivalent forces Q_i be equal to the work performed by the externally applied forces $w(x)$. The work W_E' of the equivalent forces Q_i is simply

$$W_E' = Q_1 \delta_1 = Q_1 \qquad\qquad \text{(a)}$$

since Q_1 is the only equivalent force undergoing displacement and $\delta_1 = 1.0$.

The work performed by the applied force $w(x)\,dx$ during the virtual displacement on the differential beam segment of length dx is $N_1(x)\,w(x)\,dx$. The total work is then

$$W_E'' = \int_0^L N_1(x)\ w(x)\ dx. \tag{b}$$

Equating the work of the equivalent force in eq. (a) with the work done by the applied load, eq. (b) results in the following expression for the equivalent nodal force Q_1:

$$Q_1 = \int_0^L N_1(x)\ w(x)\ dx.$$

Then, in general, the expression for the equivalent virtual nodal force Q_i is given by

$$Q_i = \int_0^L N_i(x)\ w(x)\ dx. \hspace{2cm} Q.E.D.$$

Analytical Problem 1.3
Demonstrate that the equivalent nodal forces Q_i at the nodal coordinates of a beam element acted upon by a distributed moment $m(x)$ [Figure P1.3] is given by

$$Q_i = \int_0^L N_1'(x)\ m(x)\ dx \hspace{2cm} \text{(1.18b) repeated}$$

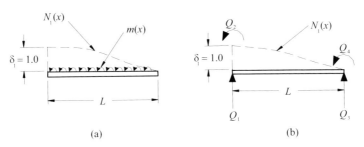

(a)   (b)

Fig. P1.3   Beam element undergoing a virtual displacement $\delta_1 = 1.0$
(a) Supporting a distributed moment $m(x)$, and
(b) Acted upon by nodal equivalent force Q_i

Solution:
Analogous to the demonstration in Analytical Problem 1.2, to determine the equivalent nodal force Q_1, we give a unit virtual displacement $\delta_1 = 1.0$ to both beam elements shown in Figure P1.3, resulting in displacements along these beams $N_1(x)$, given by the shape function in eq. (1.5a). For this virtual displacement, the external work W_E' performed by the equivalent nodal forces in Figure P1.4(b) is

$$W_E' = Q_1\delta_1 = Q_1 \qquad\qquad (a)$$

and the work done by the distributed moment $m(x)$ on a differential element beam segment of length dx is $N_I'(x)m(x)\ dx$ where $N_I'(x)$ is the derivative of the shape function $N_I(x)$. Thus $N_I'(x)$ is the slope of this function. Therefore, the total work W_E'' is given by

$$W_E'' = \int_0^L N_1'(x)\ m(x)\ dx. \qquad\qquad (b)$$

Finally, equating the expressions in (a) and (b) results in

$$Q_1 = \int_0^L N_1'(x)\ m(x)\ dx. \qquad\qquad (c)$$

Then, the generalization of eq. (c) gives the expression to calculate the equivalent nodal force Q_i as

$$Q_i = \int_0^L N_1'(x)\ m(x)\ dx. \qquad\qquad Q.E.D.$$

Analytical Problem 1.4
Determine the equivalent nodal forces for a fixed-end beam element supporting a concentrated force W as shown in Figure P1.4.

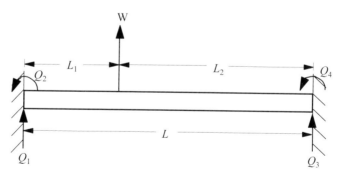

Fig. P1.4 Beam element showing a concentrated load W and the equivalent nodal forces Q_1, Q_2, Q_3, and Q_4

Solution:
The use of eq. (1.18a) provides the general expression to determine the equivalent nodal force Q_i as

$$Q_i = \int_0^L N_i(x) \; w(x) \; dx \qquad\qquad \text{(1.18a) repeated}$$

where $w(x)$ is the force applied to the beam and $N_i(x)$ is the shape function given by eq. (1.5). To evaluate Q_1, we substitute into eq. (1.18a) $N_1(x)$ given by eq. (1.5a) and $w(x)$ for W applied at location $x = L_1$. We then obtain:

$$Q_1 = W\left[1 - 3\left(\frac{L_1}{L}\right)^2 + 2\left(\frac{L_1}{L}\right)^3\right]$$

$$Q_1 = \frac{W \, L_2^2}{L^3}\left(3L_1 + L_2\right).$$

Analogously, the successive substitution of eqs. (1.5b), (1.5c), and (1.5d) into eq. (1.18a) gives, respectively,

$$Q_2 = \frac{W \, L_1 \, L_2^2}{L^2}$$

$$Q_3 = \frac{W \, L_1^2}{L^3}\left(L_1 + 3L_2\right)$$

$$\text{and} \qquad Q_4 = -\frac{W \, L_1^2 \, L_2}{L^2}$$

Analytical Problem 1.5

Consider a beam element in Figure P1.5 supporting a flexural moment M at the distance L_1 from the left end of the beam. Determine the expression for the equivalent nodal force Q_i.

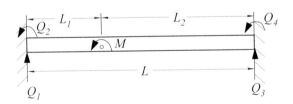

Fig. P1.5 Fixed-end beam supporting moment M, showing equivalent nodal forces Q_i

Solution:

The use of eq. (1.18b), repeated here for convenience, provides the expression to calculate the equivalent nodal forces:

$$Q_i = \int_0^L N_i'(x)\ m(x)\ dx \qquad\qquad \text{(1.18b) repeated}$$

where N_i' is the derivative of the shape function, eq. (1.5) and $m(x)$ is the applied moment along the beam. In this case, to calculate Q_1 we substitute into eq. (1.18b) the derivative of the shape function $N_1(x)$ given by eq. (1.5a) to obtain

$$Q_1 = \int_0^L \left(\frac{-6}{L^2} x + \frac{6}{L^3} x^2 \right) m(x)\ dx.$$

Since the external moment M exists only at $x = L_1$, we obtain

$$Q_1 = \{ -\frac{6}{L^2} L_1 + \frac{6}{L^3} L_1^2 \} M$$

or

$$Q_1 = -\frac{6M\ L_1\ L_2}{L^3}.$$

1.17 Practice Problems

Problem 1.6
For the beam shown in Figure P1.6 determine:

(a) Displacements at nodes 1 and 2
(b) End forces on the beam elements
(c) Reactions at the supports

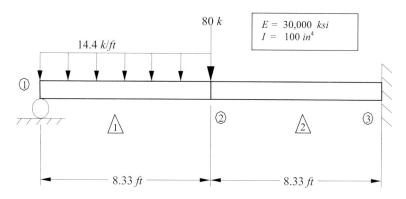

Fig. P1.6

Problem 1.7
For the beam shown in Fig P1.7 determine

(a) Displacements at node 2
(b) End forces on the beam elements
(c) Reactions at the supports

$E = 30,000\ ksi$
$I = 200\ in^4$

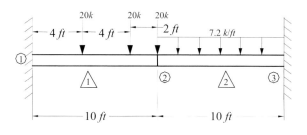

Fig P1.7

Problem 1.8
For the beam shown in Figure P1.8 determine:

(a) Displacement at the nodes 2 and 3
(b) End forces on the beam elements
(c) Reactions at the supports

$E = 200\ GPa$
$I = 40,000\ cm^4$

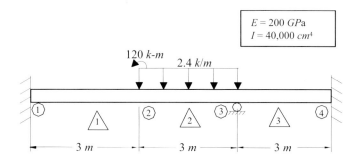

Fig. P1.8

Problem 1.9
For the beam shown in Figure P1.9 determine:

(a) Displacement at node 2
(b) End forces on the beam elements
(c) Reactions at the supports

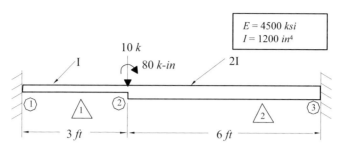

Fig. P1.9

Problem 1.10

For the beam shown in Figure P1.10 determine:

(a) Displacements at nodes 2 and 3,
(b) End forces on the beam elements, and
(c) Reactions at the supports.

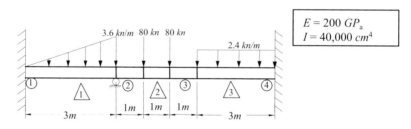

Fig. P1.10

Problem 1.11

For the beam shown in Figure P1.11 determine:

(a) Displacements at nodes 1, 2 and 3,
(b) End forces on the beam elements, and
(c) Reactions at the supports.

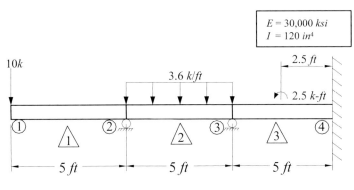

Fig. P1.11

Problem 1.12
Solve Problem 1.6 assuming that a hinge exists at the right end of beam element 1.

Problem 1.13
Solve Problem 1.7 assuming a hinge at the left end of beam element 2.

Problem 1.14
Solve Problem 1.8 assuming a hinge at the right end of beam element 2.

Problem 1.15
Solve Problem 1.9 assuming a hinge at node 2 (right end of beam element 1 and left end of element 2).

Problem 1.16
Solve Problem 1.10 assuming a hinge at the left end of beam element 2.

Problem 1.17
Solve Problem 1.11 assuming hinges at the two ends of beam element 3.

Problem 1.18
Solve Problem 1.6 knowing that the support at node 1 underwent a downward displacement of 1.0 *in* and rotational displacement of 5 degrees in the clockwise direction.

Problem 1.19
Solve Problem 1.8 knowing that the support at node 3 underwent a downward displacement of 2.0 *in.*

Problem 1.20
Solve Problem 1.8 knowing that the support at node 1 underwent a downward displacement of 0.5 *in* and clockwise rotation of 10 degrees.

Problem 1.21
Solve Problem 1.7 knowing that the temperature at the top side of beam element 2 is $30°F$ and at the bottom side is $80°F$ ($h = 2$ *in*).

Problem 1.22
Solve Problem 1.8 knowing that the temperature along the top of the beam is $40°F$ and at its bottom side is $100°F$.

Problem 1.23
For the beam shown in Figure P1.23, determine:

(a) Displacements at nodes 1 and 2.
(b) The force and the moment of the springs at node 2

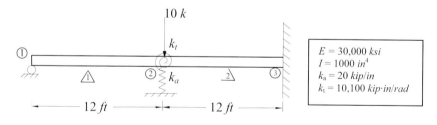

Fig. P1.23

Problem 1.24
For the beam shown in Figure P1.24 determine:

(a) Displacements at nodes 2 and 3.
(b) Force in the spring at node 2.
(c) Element and forces
(d) Reaction at the supports

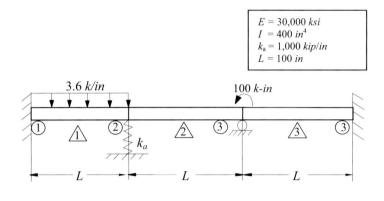

Fig. P1.24

Problem 1.25
Solve Problem 1.23 assuming that the support at node 3 underwent a 1.0 *in* displacement downwards and 0.05 *rad* counterclockwise rotation.

Problem 1.26
Solve Problem 1.23 assuming that a hinge exists at node 2 between elements 1 and 2.

2 Beams: Structural Dynamics

2.1   Introduction

In Chapter 1, the theory for the matrix structural analysis of beams was presented. As it was explained, such analysis usually requires the following steps:

1.   Modeling the beam into elements to define the nodes between elements and the nodal coordinates at those nodes. A distinction is made between (1) free nodal coordinates with unknown displacements to be determined and (2) fixed nodal coordinates that do not experience displacements or at which known displacements are imposed.

2.   The determination of the element's stiffness matrices $[k]$ followed by the assemblage of the system stiffness matrix $[K]$.

3.   The determination of the equivalent forces $\{Q\}$, at the nodal coordinates of the elements for those forces applied on the element (concentrated and distributed forces), followed by the assembly of those forces, including the forces directly applied at the nodes, to form the system force vector $\{F\}$.

4.   The solution for the free nodal coordinates $\{u\}$ of the linear system of equations relating, through the system stiffness matrix $[K]$, the external forces $\{F\}$ and the nodal displacements $\{u\}$, namely,

$$\{F\} = [K]\,\{u\}$$

(1.13) repeated

5. The calculation of the element's nodal forces $\{P\}$ due to the element nodal displacements $\{\delta\}$ and to the vector for the equivalent forces $\{Q\}$ for loads applied on the element, namely,

$$\{P\} = [k] \{\delta\} - \{Q\}$$ (1.23) repeated

where $\{\delta\}$ is the vector containing the element nodal displacements identified for an element from the values of the system nodal displacements $\{u\}$ calculated in eq. (1.13).

6. Determination of the total displacements $y_T(x)$ at any section of a beam element by adding the effects due to element nodal displacements $y(x)$ and the displacements resulting from the loads applied to the element $y_L(x)$, namely,

$$y_T(x) = y(x) + y_L(x)$$ (1.37) repeated

in which

$$y(x) = N_1(x)\, \delta_1 + N_2(x)\, \delta_2 + N_3(x)\, \delta_3 + N_4(x)\, \delta_4$$ (1.6) repeated

where the shape functions $N_i(x)$ are given by eqs. (1.5).

7. The calculation of the function for the shear force, $V(x)$, and the function for the bending moment, $M(x)$, are given, respectively, by

$$V(x) = P_1 - \int_0^x w(x_1)\,dx_1$$ (1.42) repeated

and

$$M(x) = -P_2 + P_1 x - \int_0^x (x - x_1)\, w(x_1)\,dx_1$$ (1.43) repeated

Equations (1.42) and (1.43) provide, respectively, the functions to calculate the shear force $V(x)$ and the bending moment $M(x)$ for a beam element in terms of the element end forces P_1 and P_2 and of the load $w(x)$ applied on the element.

2.2   Dynamic Analysis of Beams

The analysis of beams or other structures subjected to dynamic forces, that is, forces that are time dependent, requires the consideration of inertial forces in addition to the elastic forces and external forces to establish the equations of motion and to calculate the unknown nodal displacements. The inertial forces are determined as in the case of the elastic forces by calculating first the element's mass matrices $[m]$ followed by the assemblage of the system mass matrix $[M]$.

For a linear elastic structural system, the relationship between the inertial nodal forces $\{P\}$ and the element nodal acceleration $\{\ddot{\delta}\}$ is given by

$$\{P\} = [m]\{\ddot{\delta}\} \tag{2.1}$$

In the notation used in eq. (2.1), the double dots on the nodal displacement vector $\{\ddot{\delta}\}$ denotes the second derivative with respect to time, that is, the acceleration at the element nodal coordinates.

2.2.1   Inertial Properties: Lumped Mass

The simplest method for considering the inertial properties for dynamic analysis is to assume that the mass of the elements of the structure is lumped or concentrated at the nodal coordinates where translation displacements are defined, hence the name *lumped mass method*. The usual procedure is to allocate the mass of each element to the nodes of the element. This allocation is determined by static. Figure 2.1 shows, for beam segments of length L and distributed mass $\overline{m}$ (x) per unit of length, the nodal mass allocation for uniform, triangular, and general mass distribution along the beam element. The assemblage of the system mass matrix for the entire structure will be a simple matter of adding the contributions of lumped masses at the nodal coordinates defined as translations.

In the lumped mass method, the inertial effect associated with the rotational degree of freedom is usually assumed to be zero, although a finite value may be associated for the rotational degrees of freedom by calculating the mass moment of inertia about the nodal points of a fraction of the beam segment. For example, for a uniform beam element of length L and mass per unit length, $\overline{m}$, this calculation would result in determining the mass moment of inertia of half of the beam segment about each node, that is

$$I_A = I_B = \frac{1}{3}(\frac{\overline{m}L}{2})(\frac{L}{2})^2 \tag{2.2}$$

For the uniform cantilever beam shown in Figure 2.2 in which only translation mass effects are considered, the mass matrix of the system would be the diagonal matrix given by eq. (2.3).

Mass distribution	Lumped mass
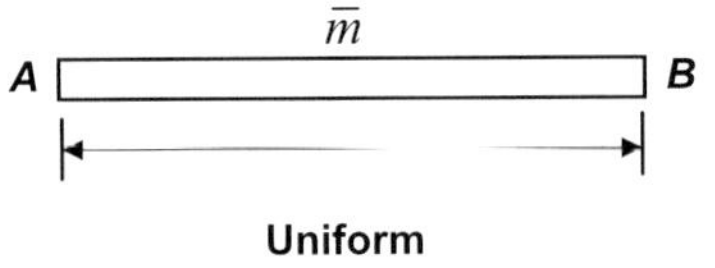 $\overline{m}$ A ▭ B Uniform	$m_A = \dfrac{\overline{m}L}{2}$ $m_B = \dfrac{\overline{m}L}{2}$
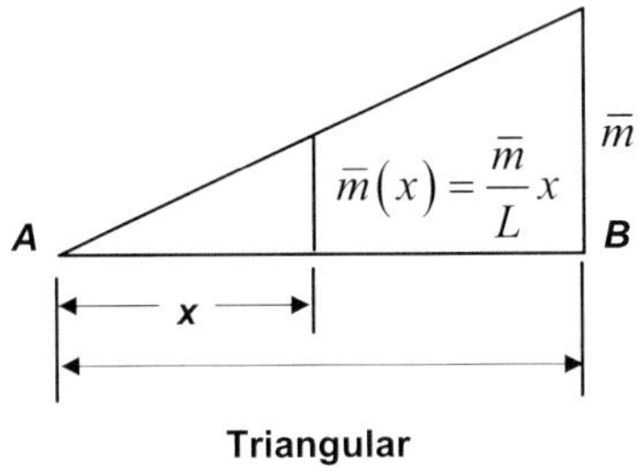 $\overline{m}(x) = \dfrac{\overline{m}}{L}x$ $\overline{m}$ A ◺ B x Triangular	$m_A = \dfrac{\overline{m}L}{6}$ $m_B = \dfrac{\overline{m}L}{3}$
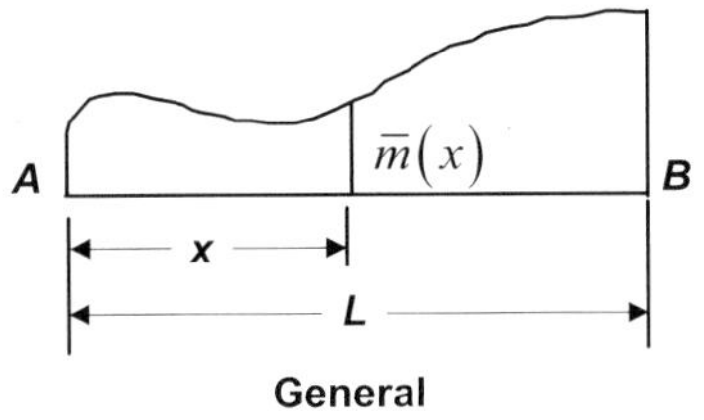 A ▭ B $\overline{m}(x)$ x L General	$m_A = \dfrac{\displaystyle\int_0^L (L-x)\overline{m}(x)\,dx}{L}$ $m_B = \dfrac{\displaystyle\int_0^L x\,\overline{m}(x)\,dx}{L}$

Fig. 2.1   Lumped mass for beam elements with distributed mass

$$[M] = \begin{matrix} & 1 & 2 & 3 & 4 & 5 & 6 & \\ \left[\begin{matrix} 0 & & & & & \\ & 0 & & & & \\ & & 0 & & & \\ & & & m_4 & & \\ & & & & m_5 & \\ & & & & & m_6 \end{matrix}\right. & & & & & & \left.\begin{matrix} \\ \\ \\ \\ \\ \\ \end{matrix}\right] & \begin{matrix} 1 \\ 2 \\ 3 \\ 4 \\ 5 \\ 6 \end{matrix} \end{matrix} \qquad (2.3)$$

in which

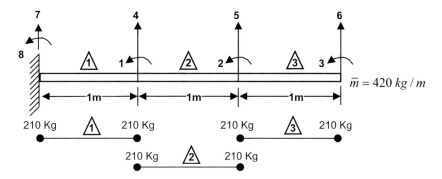

Figure 2.2   Cantilever beam divided into three beam segments showing thee element's lumped nodal masses

$$m_4 = \frac{\overline{m}L_1}{2} + \frac{\overline{m}L_2}{2}$$

$$m_5 = \frac{\overline{m}L_2}{2} + \frac{\overline{m}L_3}{2}$$

$$m_6 = \frac{\overline{m}L_3}{2}$$

and L_1, L_2, L_3 the length of the beam elements.

Using a special symbol $\lceil \ \rfloor$ for diagonal matrices, eq. (2.3) may be written as

$$[M] = \lceil 0 \quad 0 \quad 0 \quad m_4 \quad m_5 \quad m_6 \rfloor \tag{2.4}$$

2.2.2   Inertial Properties: Consistent Mass

It is possible to evaluate the mass coefficients corresponding to the nodal coordinates of a beam element by a procedure similar to the determination of element stiffness coefficients. First, we define the mass coefficient m_{ij} as the force at nodal coordinate i due to a unit acceleration at nodal coordinate j while all other nodal coordinates are maintained at zero acceleration. Similar to eq. (1.7) of Chapter 1, for calculating the stiffness coefficients, the mass coefficients m_{ij} using the consistent mass method are determined by the following formula (see Analytical Problem 2.1, Section 2.5):

$$m_{ij} = \int_0^L \overline{m}(x)N_i(x)N_j(x)dx \tag{2.5}$$

in which $\overline{m}(x)$ is the mass per unit of length and $N_i(x)$, $N_j(x)$ are the shape functions given by eqs. (1.5) of Chapter 1.

In the consistent mass method, it is assumed that the displacements resulting from a unit dynamic displacements at the nodal coordinates of the beam element are given by the same functions $N_1(x)$, $N_2(x)$, $N_3(x)$, and $N_4(x)$ of eqs. (1.5), which were obtained from static considerations. In practice, the cubic equations (1.5) are used in calculating the mass coefficients of any straight beam element. For the special case of a uniform beam with uniformly distributed mass $\overline{m}$ per unit length, the application of eq. (2.5) results in the element mass matrix relating in the following equation the element nodal forces $\{P\}$ and the nodal accelerations $\{\ddot{\delta}\}$:

$$
\begin{Bmatrix} P_1 \\ P_2 \\ P_3 \\ P_4 \end{Bmatrix} = \frac{\overline{m}L}{420}
\begin{bmatrix}
156 & 22L & 54 & -13L \\
22L & 4L^2 & 13L & -3L^2 \\
54 & 13L & 156 & -22L \\
-13L & -3L^2 & -22L & 4L^2
\end{bmatrix}
\begin{Bmatrix} \ddot{\delta}_1 \\ \ddot{\delta}_2 \\ \ddot{\delta}_3 \\ \ddot{\delta}_4 \end{Bmatrix}
\tag{2.6}
$$

or in condensed notation

$$\{P\} = [m]\,\{\ddot{\delta}\}$$

where $\ddot{\delta}_i$ denotes the acceleration, which is given by the second derivative with respect to time of the displacement function δ_i.

After the mass matrix in eq. (2.6) has been evaluated for each element of the beam, the mass matrix for the entire system is assembled using exactly the same procedure (direct method) as described in Chapter 1 to assemble the system stiffness matrix. The resulting mass matrix will in general have the same arrangement of nonzero terms as the stiffness matrix.

The dynamic analysis using the lumped mass matrix requires considerably less computational effort than the analysis using the consistent mass method because the lumped mass method results in a diagonal matrix, whereas the consistent mass matrix has many off diagonal terms that are called mass coupling. Also, the lumped mass matrix contains zeros in its main diagonal due to assumed zero rotational inertial forces. This fact requires the elimination by static condensation of the rotational degrees of freedom, thus reducing the dimension of the dynamic problem. Static condensation has been presented in Section 1.6 of Chapter 1.

Illustrative Example 2.1

For the cantilever beam shown in Fig. 2.3 determine: (a) the lumped mass and (b) the consistent mass matrices. Assume that the beam has uniformly distributed mass, $\bar{m} = 420 \ kg/m$.

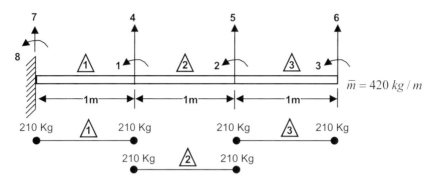

Figure 2.3   Element's lumped masses for Illustrative Example 2.1

Solution

(a) Lumped Mass Matrix:

The lumped mass at each node of any of the three beam elements, into which the cantilever beam has been divided, is simply half of the mass of the element. In the present case, the lumped mass at each node of an element is 210 *kg*, as shown in Figure 2.3. The lumped mass matrix $[M]_L$ for this structure is a diagonal matrix of dimension 6×6, which, using the special symbol for a diagonal matrix, may be expressed as follows:

$$\lceil M \rfloor_L = \lceil 0 \ 0 \ 0 \ 420 \ 420 \ 210 \rfloor_L.$$ (a)

The matrix in eq. (a) is readily condensed by eliminating the rows and columns of the first three rows and first three columns, to obtain

$$[\overline{M}]_L = \begin{bmatrix} 420 & 0 & 0 \\ 0 & 420 & 0 \\ 0 & 0 & 210 \end{bmatrix}$$ (b)

(b) Consistent Mass Matrix.

The consistent mass matrix for a uniform beam segment is given by eq. (2.6). The substitution of numerical values for this example $L = 1 \ m$ and $= 420 \ kg/m$ into eq. (2.6) results in the consistent mass matrix $[m]_C$ for any of the three beam elements:

$$
\begin{array}{cccc}
5 & 2 & 6 & 3 \\
4 & 1 & 5 & 2 \\
7 & 8 & 4 & 1
\end{array}
\qquad
\begin{array}{ccc}
7 & 4 & 5 \\
8 & 1 & 2 \\
4 & 5 & 6 \\
1 & 2 & 3
\end{array}
$$

$$
[m]_C = \begin{bmatrix}
156 & 22 & 54 & -13 \\
22 & 4 & 13 & -3 \\
54 & 13 & 156 & -22 \\
-13 & -3 & -22 & 4
\end{bmatrix}
\qquad (c)
$$

The system mass matrix is assembled from the element's mass matrices in exactly the same procedure as the assemblage of the system stiffness matrix from the element's stiffness matrices; that is, the coefficients in the element's mass matrices are allocated to appropriate entries in the system mass matrix. For the first beam element in Figure 2.3, this allocation corresponds to the system nodal coordinates 7, 8, 4, 1; for the second segment to 4, 1, 5, 2 and for the third segment to 5, 2, 6, 3. To facilitate this transfer, the appropriate nodal coordinates are indicated at the top and at the right side of the matrix in eq. (c). The reduced assembled (only the free nodal coordinates are considered) consistent mass matrix $[M]_C$ for this example results in

$$
\begin{array}{cccccc}
1 & 2 & 3 & 4 & 5 & 6
\end{array}
$$

$$
[M]_C = \begin{bmatrix}
8 & -3 & 0 & 0 & 13 & 0 \\
-3 & 8 & -3 & -13 & 0 & 13 \\
0 & -3 & 4 & 0 & -13 & -22 \\
0 & -13 & 0 & 312 & 54 & 0 \\
13 & 0 & -13 & 54 & 312 & 54 \\
0 & 13 & -22 & 0 & 54 & 156
\end{bmatrix}
\begin{array}{c}
1 \\ 2 \\ 3 \\ 4 \\ 5 \\ 6
\end{array}
\qquad (d)
$$

We note that the consistent mass matrix $[M]_C$ is symmetric and also banded, as in the case of the stiffness matrix for this system. These facts are of great importance in developing computer programs for structural analysis, since it is possible to perform the necessary calculations by storing in the memory of the computer only the non-zero elements to one side of the main diagonal.

2.3   Free Vibration: Natural Frequencies and Mode Shapes

The equations of motion for a structure undergoing free vibration, that is, motion started by initial conditions (given displacements and velocities at the nodal coordinates), is obtained by adding the inertia force $\{F\}_I = [M]\{\ddot{u}\}$ and the elastic forces $\{F\}_E = [K]\{u\}$, namely.

$$[M]\{\ddot{u}\} + [K]\{u\} = \{0\}. \tag{2.7}$$

The following trial solution is introduced into eq. (2.7)

$$\{u\} = \{a\}\sin\omega t \tag{2.8}$$

in which $\{a\}$ is the vector containing the amplitude components of the motion and ω the frequency of the vibration.

The sine function has a period 2π, that is,

$$\omega T = 2\pi$$

then, solving for the period of vibration T results in

$$T = \frac{2\pi}{\omega}\left(\frac{\text{sec}}{\text{cycle}}\right). \tag{2.9}$$

The reciprocal value of the period T is the *frequency* f (referred to as the natural frequency). Then from eq. (2.9)

$$f = \frac{1}{T} = \frac{\omega}{2\pi}. \tag{2.10}$$

The *natural frequency* f is usually expressed in hertz or cycles per second (*cps*). Because the quantity ω differs from the natural frequency f only by the constant factor 2π, ω also is referred to as the natural frequency. To distinguish between these two expressions for natural frequency, ω may be called the *circular or angular frequency*. Most often, the distinction is understood from the context or from the units. The natural frequency f is measured in *cps* as indicated, while the circular frequency ω should be given in radians per second (*rad/sec*).

Introducing the trial solution, eq. (2.8) into eq. (2.7), results in the following system of linear equations:

$$\left[[K] - \omega^2[M]\right]\{a\} = 0 \tag{2.11}$$

To obtain a non-trivial solution of eq. (2.11) (not all of the values for a_i equal 0), it is required that the determinant of the coefficient of $\{a\}$ be equal to zero, that is,

$$\left|[K] - \omega^2[M]\right| = \{0\} \tag{2.12}$$

The expansion of the determinant in eq. (2.12) leads to a polynomial equation of degree N (the number of unknown displacements) in the ω^2. The solution of this polynomial equation provides N roots ($\omega_1^2, \omega_2^2...\omega_N^2$), which are the squared values of the natural frequencies of the structure.

Because of the condition imposed by eq. (2.12), the number of independent equations has been reduced to $(N-1)$. Consequently, for each value of ω^2 introduced into eq. (2.11), this equation can be solved only for relative values of the unknowns a_i to obtain for each value calculated for, ω_i^2 the vectors

$$\{a\}_1, \ \{a\}_2, \ \{a\}_3, \ ... \{a\}_N \tag{2.13}$$

The vectors $\{a\}_i$ in eq. (2.13) are known as the *modal shapes* of the structure in free vibration, which conveniently are normalized and arranged in the columns of the modal matrix designated by $[\Phi]$ [see eqs. (IV-8) and (IV-9) in Appendix IV]:

$$[\Phi] = \begin{bmatrix} \phi_{11} & \phi_{12} & .. & .. & \phi_{1N} \\ \phi_{21} & \phi_{22} & .. & .. & \phi_{2N} \\ .. & .. & .. & .. & .. \\ .. & .. & .. & .. & .. \\ \phi_{N1} & \phi_{N2} & .. & .. & \phi_{NN} \end{bmatrix} \tag{2.14}$$

In eq. (2.14), ϕ_{ij} is the "I" component of the normalized mode shape "j".

Illustrative Example 2.2

Determine the natural frequencies and corresponding mode shapes for the cantilever beam shown in Figure 2.3. Use static condensation to reduce the number of free coordinates to the three translatory coordinates labeled in this figure as nodal coordinates 4, 5, and 6 (use the lumped mass method).

Solution:

The condensed stiffness matrix, $[\overline{K}]$ for this structure considering only the three translatory nodal coordinates labeled as 4, 5, and 6 in Figure 2.3 and the corresponding transformation matrix $[\overline{T}]$ have been calculated in the Illustrative Example 1.3 of Chapter 1. These matrices are:

$$[\overline{K}] = 10^7 \begin{bmatrix} 18.461 & -10.613 & 2.768 \\ -10.613 & 10.150 & -3.690 \\ 2.768 & -3.690 & 1.612 \end{bmatrix} \tag{a}$$

$$\left[\overline{T}\right] = \begin{bmatrix} 0.231 & 0.689 & -0.115 \\ -0.923 & 0.231 & 0.461 \\ 0.462 & -1.616 & 1.270 \end{bmatrix} \qquad \text{(b)}$$

The condensed lumped mass matrix for this beam has been obtained in Illustrative Example 2.1, eq. (b) as

$$[\overline{M}] = \begin{bmatrix} 420 & 0 & 0 \\ 0 & 420 & 0 \\ 0 & 0 & 210 \end{bmatrix} \qquad \text{(c)}$$

The natural frequencies calculated by substituting into eq. (2.12) the matrices $[\overline{K}]$ and $[\overline{M}]$ that are given, respectively, by eqs. (a) and (c) followed by the solution of the resulting cubic polynomial in $(\omega)^2$ results in:

$$(\omega_1)^2 = 3.290\text{E}3 \qquad (\omega_2)^2 = 1.048\text{E}5 \quad \text{and} \quad (\omega_3)^2 = 6.501\text{E}5 \qquad \text{(d)}$$

The natural frequencies in (rd/sec) are then:

$$\omega_1 = 57.36 \; rad/sec \qquad \omega_2 = 323.80 \; rad/sec \quad \text{and} \quad \omega_3 = 806.23 \; rad/sec \qquad \text{(e)}$$

and in cps $(f_i = \omega_i / 2\pi)$:

$$f_1 = 9.13 \; cps \qquad f_2 = 51.83 \; cps \quad \text{and} \quad f_3 = 128.30 \; cps \qquad \text{(f) Ans.}$$

The mode shapes of vibration are then calculated after successively substituting $(\omega)^2$ in eq. (2.11) for the values of $(\omega_1)^2$, $(\omega_2)^2$ and $(\omega_3)^2$ calculated in eq. (d), setting arbitrarily the first component of a mode shape equal to 1.000, and then solving two of the equations in (2.11) for the second and third components of the mode shapes, to obtain

$$a_{41} = 1.000 \qquad\qquad a_{42} = \;\; 1.000 \qquad\qquad a_{43} = \;\; 1.000$$

$$a_{51} = 3.338 \qquad\qquad a_{52} = \;\; 0.967 \qquad\qquad a_{53} = -0.716 \qquad \text{(g)}$$

$$a_{61} = 6.181 \qquad\qquad a_{62} = -1.369 \qquad\qquad a_{63} = \;\; 0.450$$

The mode shapes in eq. (g) are conveniently normalized by dividing these mode shapes using the following formula [see eq. (IV-5b) in Appendix IV]:

$$\sqrt{\sum_{i=1}^{3} m_i a_{ij}^2} \;. \qquad \text{(h)}$$

For the first mode shape, this factor is then calculated as

$$\sqrt{420x1.0000^2 + 420x3.338^2 + 210x6.181^2} = 114.56.$$

Analogously, the normalized factors are then calculated for the second and third modes, which together with the first factor are:

$$114.56, \qquad 34.7315 \qquad \text{and} \qquad 26.0335. \qquad (i)$$

The division of each of the mode shapes by the corresponding factor calculated in eq. (i) results in the normalized mode shapes:

$$
\begin{array}{lll}
\Phi_{41} = 0.008729 & \Phi_{42} = 0.0287923 & \Phi_{43} = 0.038412 \\
\Phi_{51} = 0.029141 & \Phi_{52} = 0.027851 & \Phi_{53} = -0.027496 \qquad (j) \\
\Phi_{61} = 0.053954 & \Phi_{62} = -0.039402 & \Phi_{63} = 0.017272
\end{array}
$$

The normalized modal shapes corresponding to the condensed coordinates 1, 2, and 3 are then calculated by multiplying successively the transformation matrix, eq. (b), by the primary modal shapes calculated in eq. (j), to obtain:

$$
\begin{array}{lll}
\Phi_{11} = 0.015964 & \Phi_{22} = 0.030472 & \Phi_{13} = -0.012163 \\
\Phi_{21} = 0.023570 & \Phi_{22} = -0.038336 & \Phi_{23} = -0.033834 \qquad (k) \\
\Phi_{31} = 0.025434 & \Phi_{32} = -0.081712 & \Phi_{33} = 0.084069
\end{array}
$$

The normalized modal shapes in eqs. (i) and (k) are then conveniently arranged in the columns of the modal matrix, $[\Phi]$:

$$
[\Phi] =
\begin{bmatrix}
0,015964 & 0.030472 & -0.012163 \\
0.023570 & -0.038336 & -0.033834 \\
0.025434 & -0.081712 & 0.084069 \\
0.008729 & 0.028792 & 0.038412 \\
0.029141 & 0.027851 & -0.027496 \\
0.053954 & -0.039402 & 0.017272
\end{bmatrix}
\qquad (l)
$$

2.4 Forced Motion: Modal Superposition Method

The differential equation of motion of a structure (neglecting damping) subjected to forces $\{F(t)\}$ applied at the nodal coordinates is

$$[M]\,\{\ddot{u}\} + [K]\,\{u\} = \{F(t)\}. \qquad (2.15)$$

For an elastic linear system, the solution of eq. (2.15) may be obtained by the *modal superposition method*. In applying this method, the following linear transformation is introduced in eq. (2.15):

$$\{u\} = [\Phi]\{z\} \tag{2.16}$$

in which $[\Phi]$ is the modal matrix given by eq. (2.14) and $\{z\}$ is a vector containing the new variables introduced by the transformation of eq. (2.16).

The substitution into eq. (2.15) of $\{u\}$ and $\{\ddot{u}\}$ defined by the transformation in eq. (2.16) followed by the pre-multiplication of the resulting equation by the transpose of the modal matrix, $[\Phi]^T$, results in

$$[\Phi]^T[M][\Phi]\{\ddot{z}\} + [\Phi]^T[K][\Phi]\{z\} = [\Phi]^T\{F(t)\}. \tag{2.17}$$

Because of the orthogonal condition between normal modes [see eqs. (IV-8) and (IV-9) of Appendix IV], the system of differential equations in eq. (2.17) is reduced to a set of *uncoupled equations* in which each equation is a function of a single variable, namely,

$$\ddot{z}_1 + \omega_1^2 z_1 = P_1(t)$$
$$\ddot{z}_2 + \omega_2^2 z_2 = P_2(t)$$
$$\dots\dots\dots\dots\dots\dots\dots\dots \tag{2.18}$$
$$\ddot{z}_j + \omega_j^2 z_j = P_j(t)$$
$$\dots\dots\dots\dots\dots\dots\dots\dots$$
$$\ddot{z}_N + \omega_N^2 z_N = P_N(t)$$

in which the modal force $P_j(t)$ is given by

$$P_j(t) = \sum_{i=1}^{N} \phi_{ij} F_i(t) \tag{2.19}$$

where ϕ_{ij} is the normalized i^{th} component of the mode shape j and $F_i(t)$ is the external force applied at the nodal coordinate i.

It follows that each of the equations (2.18) can then be solved independently by any of the methods available to solve a second order differential equation, such as *Duhamel's Integral* or the *direct method* described in Appendix III.

Damping that accounts for the mechanical energy losses may conveniently be introduced in the modal eqs. (2.18) by simply inserting in these equations the damping term, to obtain

$$\ddot{z}_j + 2\omega_j\xi_j\dot{z}_j + \omega_j^2 z_j = P_j(t) \qquad (j = 1, 2, \dots N) \qquad (2.20)$$

in which ξ is the modal damping ratio coefficient defined in Appendix III as $\xi = c/c_{cr}$. For most structural systems, the damping ratio may be estimated to have a value of about 5% ($\xi = 0.05$).

Illustrative Example 2.3

The cantilever beam of Illustrative Example 2.2, which for convenience is shown again in Figure 2.4, is subjected to force $F = 10,000\ N$ suddenly applied at its free end.

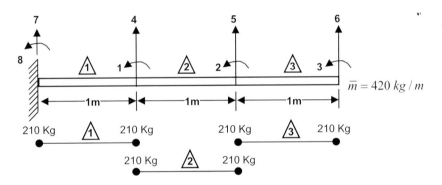

Fig. 2.4   Cantilever beam for Illustrative Example 2.3

Use the lumped mass method and determine:

(a) The modal equations of motion
(b) Solution of the modal equations
(c) Displacements at the nodal coordinates
(d) Maximum displacement at the free end of the beam
(e) Maximum shear force and bending moment at the support (reactions)

Modulus of elasticity: $E = 2\text{x}10^{10}\ N/m^2$
Cross-section moment of inertia $I = 5\text{x}10^{-4}\ m^4$ (Neglect damping)

Solution:
(a) The modal equations of motion:

The modal equations, neglecting damping, are given by eqs.(2.18). Introducing in these equations the values of ω_i^2 and ϕ_{ij} calculated, respectively, in eqs. (d) and (j) of Illustrative Example 2.2 results in:

$$
\begin{aligned}
\ddot{z}_1 + 3.29x10^3 z_1 &= 0.053954x10^5 \\
\ddot{z}_2 + 1.05x10^5 z_2 &= -0.039402x10^5 \\
\ddot{z}_3 + 6.501x10^5 z_3 &= 0.017272x10^5
\end{aligned}
\tag{a}
$$

(b) Solution of the modal equations:

The general solution of modal eq. (a) may be accomplished using Duhamel's Integral (see Appendix III), as follows:

$$
z_j = \frac{P_j}{\omega_j^2} \int_0^t \sin \omega_j (t - \tau) d\tau
$$

then

$$
z_j = \frac{P_j}{\omega_j^2} (1 - \cos \omega_j t) \qquad (j = 1, 2, 3)
\tag{b}
$$

in which P_j is given by eq. (2.19) as

$$
P_j(t) = \sum_{i=1}^{N} \phi_{ij} F_i(t) .
\tag{2.19 repeated}
$$

In this example, all of the applied forces at the nodal coordinates are equal to 0, except $F_6 = 10,000$ N. Substituting into eq. (b) the numerical values for ω_j^2 and for $P_j(t)$ ($i = 1, 2, 3$) results in

$$
\begin{aligned}
z_1 &= \frac{5395}{3.29x10^3} (1 - \cos 57.36t) \\
z_2 &= \frac{-3940}{1.05x10^5} (1 - \cos 323.8t) \\
z_3 &= \frac{1727}{6.50x10^5} (1 - \cos 806.29t)
\end{aligned}
\tag{c}
$$

(c) Displacements at the nodal coordinates

The displacements at the nodal coordinates are then determined using eq. (2.16). For this example, the modal matrix $[\Phi]$ is given by eq. (l) of Illustrative Example 2.2 and the vector of the modal solution $\{z\}$ by eq. (c). Then, the use of eq. (2.16) yields:

$u_1 = 0.0245 - 0.0257 \cos 1.095t + 0.00115 \cos 3.358t + 0.000032 \cos 1.982t$
$u_2 = 0.0394 - 0.0380 \cos 1.095t - 0.00144 \cos 3.358t + 0.000090 \cos 1.982t$
$u_3 = 0.0415 - 0.0410 \cos 1.095t - 0.00310 \cos 3.358t - 0.000720 \cos 1.982t$
$u_4 = 0.0131 - 0.0141 \cos 1.095t + 0.00108 \cos 3.358t - 0.000102 \cos 1.982t$ (d)
$u_5 = 0.0459 - 0.0470 \cos 1.095t + 0.00105 \cos 3.358t + 0.000073 \cos 1.982t$
$u_6 = 0.0885 - 0.0870 \cos 1.095t - 0.00148 \cos 3.358t - 0.000046 \cos 1.982t$

Note: The arguments of the cosine functions have been reduced by multiples of the factor 2π.

(d) Maximum displacement at the free end of the beam

The maximum displacement at the free end, which corresponds to the maximum value of the nodal coordinate u_6, may be estimated by introducing into the last equation of eqs. (d) the maximum possible value for the cosine functions (± 1). Thus resulting in

$$(u_6)_{max} = 0.08885 + 0.0870 + 0.00148 + 0.000046$$
$$= 0.177 \; m$$

(e) Maximum shear force and bending moment at the support (reactions)

The end forces for an element are calculated using eq. (1.23) of Chapter 1:

$$\{P\} = [k] \{\delta\} - \{Q\} \qquad\qquad\qquad (1.23) \text{ repeated}$$

in which

$\{P\}$ = element end forces
$[k]$ = element stiffness matrix
$\{\delta\}$ = element nodal displacements, and
$\{Q\}$ = element equivalent nodal forces for the loads applied on the beam element (zero for the elements of this beam)

The application of eq. (1.23) to element 1 results in

$$\begin{Bmatrix} P_1 \\ P_2 \\ P_3 \\ P_4 \end{Bmatrix} = 10^7 \begin{bmatrix} 12 & 6 & -12 & 6 \\ 6 & 4 & -6 & 2 \\ -12 & -6 & 12 & -6 \\ 6 & 2 & -6 & 4 \end{bmatrix} \begin{Bmatrix} 0 \\ 0 \\ u_4 \\ u_1 \end{Bmatrix} - \begin{Bmatrix} 0 \\ 0 \\ 0 \\ 0 \end{Bmatrix} \qquad (e)$$

The maximum values for the end forces are then estimated by introducing into eq. (e) the maximum values for u_4 and u_1 obtained in eqs. (d) after letting the cosine functions attain their maximum possible values of (± 1):

$$(u_4)_{max} = 0.0284 \ m \qquad \text{and} \qquad (u_1)_{max} = 0.053 \ m \qquad (f)$$

The substitution into eq. (e) of these maximum values $(u_4)_{max}$ and $(u_1)_{max}$ yields:

$(P_1)_{max} = 0.02256 \text{x} 10^7 \ N$ $(P_3)_{max} = 0.02256 \text{x} 10^7 \ N$
$(P_2)_{max} = 0.06428 \text{x} 10^7 \ N\text{-}m$ $(P_4)_{max} = 0.04172 \text{x} 10^7 \ N\text{-}m$

The maximum shear force V_{max} and maximum bending moment M_{max} at the fixed support of the beam (reactions) are given, respectively, by $(P_1)_{max}$ and $(P_2)_{max}$. Thus:

$$V_{max} \text{ (at support)} = (P_1)_{max} = 225,600 \ N \text{ and}$$
$$M_{max} \text{ (at support)} = (P_2)_{max} = 642,800 \ N\text{-}m$$

2.5   Analytical Problems

Analytical Problem 2.1

Derive the general expression given by eq. (2.5) to calculate the consistent mass coefficients m_{ij} for a beam element.

$$m_{ij} = \int_0^L \overline{m}(x) N_i(x) N_j(x) dx \qquad (2.5) \text{ repeated}$$

in which

$\overline{m}(x)$ = mass per unit of length

$N_i(x), \ N_j(x)$ = mode shape resulting from a unit displacement given at a nodal coordinate of a beam element [eqs. (1.5)].

Solution:

Consider the beam element shown in Figure P2.1(a), which has distributed mass $\overline{m}(x)$ per unit of length. In the consistent mass method, it is assumed that the deflections resulting from unit dynamic displacements at a nodal coordinate of the beam element are given by the same functions $N_1(x)$, $N_2(x)$, $N_3(x)$, and $N_4(x)$ in eqs. (1.5), which were obtained from static considerations.

If the beam element is subjected to a unit nodal acceleration at one of the nodal coordinates, say, $\ddot{\delta}_2 = 1.0$, the transverse acceleration developed along the length of the beam is given by the second derivative with respect to time of the displacement function, which is given by eq. (1.6).

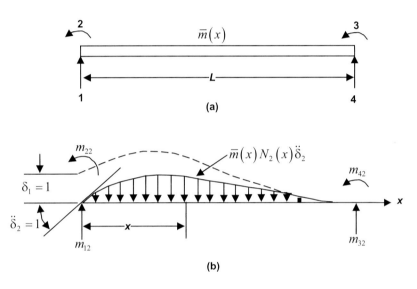

Fig. P2.1 (a) Beam element with distributed mass showing the nodal coordinates

(b) Beam element supporting inertial load due to acceleration $\ddot{\delta}_2 = 1$,

undergoing virtual displacement $\delta_1 = 1$

In this case, with $\ddot{\delta}_1 = \ddot{\delta}_3 = \ddot{\delta}_4 = 0$, we obtain

$$\ddot{u}_2(x) = N_2(x)\,\ddot{\delta}_2. \tag{2.21}$$

The inertial force $f_1(x)$ per unit of length along the beam due to this acceleration is then given by

$$f_1(x) = \overline{m}(x)\ddot{u}_2(x)$$

or using eq. (2.21) by

$$f_1(x) = \bar{m}(x)N_2\ddot{\delta}_2$$

or, since $\ddot{\delta}_2 = 1.0$

$$f_1(x) = \bar{m}(x)N_2(x). \qquad (2.22)$$

Now, to determine the mass coefficient m_{12}, we give to the beam in Figure P2.1(b), which is in dynamic equilibrium with the forces resulting from the unit acceleration $\ddot{\delta}_2 = 1$, a virtual displacement corresponding to a unit displacement at coordinate 1, $\delta_1 = 1$, and proceed to apply the Principle of Virtual Work for an elastic system (external virtual work equal to internal virtual work). The virtual work, δW_E of the external force is simply

$$\delta W_E = m_{12}\delta_1 = m_{12} \qquad (2.23)$$

since the only external force undergoing virtual displacement is the inertial force reaction m_{12} with a virtual displacement, $\delta_1 = 1$.

By eq. (1.6), the displacement at the coordinate x along the beam resulting from this virtual displacement $\delta_1 = 1$ with $\delta_2 = \delta_3 = \delta_4 = 0$ is equal to $N_1(x)$. Thus, the virtual work, δW_1 of the internal forces per unit length along the beam segment is

$$\delta W_1 = f_1(x)N_1(x)$$

or by eq. (2.22)

$$\delta W_1 = \bar{m}(x)N_2(x)N_1(x)$$

and for the entire beam

$$W_1 = \int_0^L \bar{m}(x)N_2(x)N_1(x)dx. \qquad (2.24)$$

Equating the external and internal virtual work given, respectively, by eqs. (2.23) and (2.24) results in

$$m_{12} = \int_0^L \bar{m}(x)N_2(x)N_1(x)dx \qquad (2.25)$$

which is the expression for the consistent mass coefficient m_{12}.

Then, in general, a consistent mass coefficient may be calculated from

$$m_{ij} = \int_0^L \bar{m}(x)N_i(x)N_j(x)dx \qquad (2.26)$$

Q.E.D.

2.6   Practice Problems

Problem 2.2
A uniform beam of flexural stiffness $EI = 10^9$ *lb in*^{2} and length 300 *in* has one end fixed and the other simply supported. Determine the system stiffness matrix considering three beam elements with the free nodal coordinates 4 and 5 after condensing the nodal coordinates 1, 2, and 3 as shown in Figure P2.2.

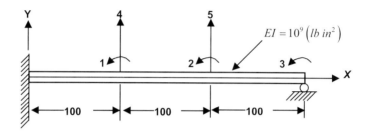

Fig. P2.2.

Problem 2.3
Assuming that the beam shown in Figure P2.2 carries a uniform weight per unit length $q = 3.86$ *lb/in*, determine the system mass matrix corresponding to the lumped mass formulation.

Problem 2.4
Determine the system mass matrix for Problem 2.3 using the consistent mass method.

Problem 2.5
For the beam in Problems 2.2 and 2.3, use static condensation to eliminate the massless degrees of freedom and to determine the transformation matrix and the reduced system stiffness and mass matrices.

Problem 2.6
For the beam in Problems 2.2 and 2.4, use static condensation to eliminate the rotational degrees of freedom. Find the transformation matrix and the reduced system stiffness and mass matrices.

Problem 2.7
Determine the natural frequencies and corresponding normal modes using the reduced stiffness and mass matrices obtained in Problem 2.5.

Problem 2.8
Determine the stiffness matrix for a beam element in which the flexural stiffness has a linear variation as shown in Figure P2.8.

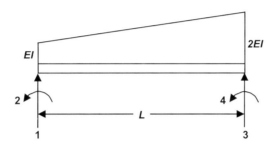

Fig. P2.8

Problem 2.9
Determine the lumped mass matrix for a beam element in which the mass has a linear distribution as shown in Figure P2.9.

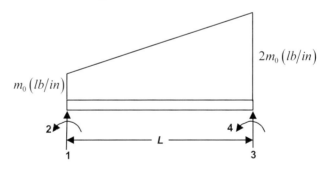

Fig. P2.9

Problem 2.10
Determine the consistent mass matrix for the beam element shown in Figure P2.9.

Problem 2.11

The uniform beam shown in Figure P2.11 is subjected to a constant force of 5,000 *lb* suddenly applied along the nodal coordinate 4. Use the results obtained in Problem 2.5 to determine the response by the modal superposition method. (Use only the two translatory modes left by the static condensation.)

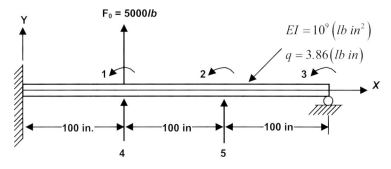

Fig. P2.11

Problem 2.12

Solve Problem 2.11 using the results obtained in Problem 2.6, which are based on the consistent mass formulation.

Problem 2.13

Determine the steady-state response for the beam shown in Figure P2.13, which is acted upon by a harmonic force $F(t) = 5,000 \sin 30t$ (*lb*) as shown in the figure. Use the lumped mass method. Neglect damping in the system.

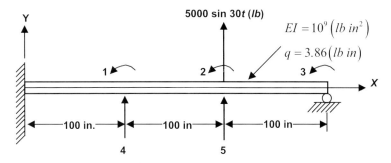

Fig. P2.13

Problem 2.14
Determine the natural frequencies and corresponding normal modes for the beam shown in Figure P2.13 after condensing the three rotational nodal coordinates. (Use the consistent mass method.)

Problem 2.15
Determine the response for the beam shown in Figure P2.13. Use the consistent mass method and condense the rotational degrees of freedom. Neglect damping.

Problem 2.16
Repeat Problem 2.15 assuming 10% damping in all the modes.

Problem 2.17
Determine the steady-state response for the beam shown in Figure P2.13 when subjected to a harmonic force $F(t) = \sin 30t$ as shown in the figure. Use the consistent mass matrix method. Neglect damping in the system.

3 Beams: Computer Applications

3.1   Computer Program

The computer program used throughout this book is the educational version of the powerful structural analysis program SAP2000®. In its advanced version, the program includes, in addition to static and dynamic analysis of structures (frames, plates, shells, and so on), analysis of structures with non-linear behavior.

The CD-ROM accompanying this volume includes user manuals and a tutorial in addition to the educational version of the program SAP2000®. The reader may use or install the educational version of SAP2000® on his or her computer. The following examples describe in detail the use of SAP2000® to analyze the beams of the Illustrative Examples presented in Chapters 1 and 2.

3.2   Applications in Structural Analysis

Illustrative Example 3.1

Use SAP2000 to analyze the steel beam W14X82 of Illustrative Example 1.1 of Chapter 1. For convenience, the beam of this example is reproduced in Fig 3.1 with the axis Z in the upward direction to match the coordinate system used by SAP2000. The edited input data for this example is given in Table 3.1 (material modulus of elasticity $E = 29000$ ksi).

Table 3.1   Edited Input Data for Illustrative Example 3.1 (Units: kips, inches)

JOINT DATA

JOINT	GLOBAL-X	GLOBAL-Y	GLOBAL-Z	RESTRAINTS *
1	0.00000	0.00000	0.00000	1 1 1 1 1 1
2	90.00000	0.00000	0.00000	0 0 0 0 0 0
3	180.00000	0.00000	0.00000	0 0 1 0 0 0
4	300.00000	0.00000	0.00000	0 0 1 0 0 0
5	396.00000	0.00000	0.00000	1 1 1 1 1 1

FRAME ELEMENT DATA

FRAME	JNT-1	JNT-2	SECTION	ANGLE	RELEASES	SEGMENTS	LENGTH
1	1	2	W14X82	0.000	000000	4	90.000
2	2	3	W14X82	0.000	000000	4	90.000
3	3	4	W14X82	0.000	000000.	4	120.000
4	4	5	W14X82	0.000	000000	4	96.000

JOINT FORCES Load Case LOAD1

JOINT	GLOBAL-X	GLOBAL-Y	GLOBAL-Z	GLOBAL-XX	GLOBAL-YY	GLOBAL-ZZ
2	0.000	0.000	−10.000	0.000	0.000	0.000
3	0.000	0.000	0.000	0.000	50.000	0.000

FRAME SPAN DISTRIBUTED LOADS Load Case LOAD1

FRAME	TYPE	DIRECTION	DISTANCE-A**	VALUE-A	DISTANCE-B**	VALUE-B
2	FORCE	LOCAL-2	0.0000	-0.1000	1.0000	−0.1000
3	FORCE	LOCAL-2	0.1667	-0.1000	0.7917	−0.2000
4	FORCE	LOCAL-2	0.0000	-0.0500	1.0000	−0.0500

FRAME SPAN POINT LOADS Load Case LOAD1

FRAME	TYPE	DIRECTION	DISTANCE◊	VALUE
1	FORCE	LOCAL-2▲	0.1111	−30.0000
1	FORCE	LOCAL-2	0.2222	−10.0000
4	MOMENT	LOCAL-3	0.5000	100.0000

*   SAP2000 considers beams as three-dimensional elements with six nodal coordinates at each joint.

**   Distance A and distance B are the relative distances measured from the left node of the beam element respectively, to the beginning and to the end of the distributed load in the span.

◊   Relative distance measured from the left node of the beam element.

▲   Local-1, Local-2 and Local-3 refer to the element local axes x, y and z, respectively with the axis 1 oriented along the beam, and axes 2 and 3, along the minor and major principal axes of the cross-sectional area.

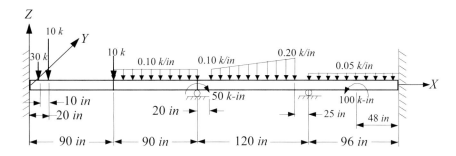

Fig. 3.1 Beam of Illustrative Example 3.1

Solution:

Begin: Load SAP2000 from the CD enclosed with this volume. Follow the installation instructions and the prompts in the software. After successful installation of the program, begin this example by opening the program.

Enter: Click OK to close the "Tip of the Day" screen.

       Hint: Maximize both screens for full views of all windows.

Units: In the lower right-hand corner of the window, use the drop-down list to select "kip-in".

Model: From the main menu select:
FILE > NEW MODEL FROM TEMPLATE.
Click the button displaying the drawing of a beam.
Change the number of spans to 4 and change the span length to 90.
Then OK.

Edit: Maximize the 2-D (plane XZ), which appears on the right side of the split screen.
Click on the PAN icon on the toolbar and drag the figure to the center of the screen.

Model: Translate the origin of the Global Coordinate System to the left end of the beam using the following commands from the main menu:
SELECT>SELECT>ALL. This will mark all members of the beam.
EDIT>MOVE.
In the "Move Selected Points" form, change the Delta X to 180.
Click OK.

Edit: Add grid lines to the last two supports of the beam with the following
commands from the main menu:
DRAW>EDIT GRID.
When the pop-up menu appears:
Highlight the value −180 and change it to 270.
Click the button labeled "Move Grid Line".
Highlight the value −90 and change it to 360.
Click "Move Grid Line".
The screen should now show values of 0, 90, 180, 270 and 360.
Click the check box labeled "glue joints to grid lines".
This will move the joints with the grid lines.
Click OK.

Edit: To change the length of the last two beam elements, edit the grid lines once
again. From the main menu select:
DRAW>EDIT GRID.
Highlight the value 270 and change it to 300.
Click "Move grid lines".
Highlight the value 360 and change it to 396.
Click "Move Grid Lines".
Click OK.

Material: From the main menu select:
DEFINE>MATERIALS.
Select STEEL from the list on the left side of the pop-up form.
Click MODIFY/SHOW MATERIAL.
Change the Modulus of Elasticity to $E = 29000$.
Click OK.

Section: Cross-sections: From the main menu select:
DEFINE>FRAME SECTIONS[2].
Select Import/Wide Flange from the window on the right side of the pop-
up screen.
Click OK.
Another pop-up window will appear with the label C:\SAP2000.
Select "sections.prop" then click OPEN.
Using the scroll bar, click the section "W14X82".
Click OK.
A pop-up window labeled "I/Wide Flange Section" will appear.
Click "Modification Factors". Another menu will pop-up labeled
"Analysis Property Modification Factors". In order not to include the

---

[2]SAP2000 refers to the beam elements as frames.

shear deformation in the calculation of the stiffness coefficients, change the Shear Area in the 2 direction to 0 (zero).
Click OK on all three forms.

Assign: Assign frame sections: From the main menu select: SELECT>SELECT>ALL. Then from the main menu select: ASSIGN>FRAME>SECTIONS.
In the Define Frame Sections form, select W14X82.
Click OK.

Labels: For viewing convenience, label the joints and elements of the beam by selecting from the Main Menu:
VIEW>SET ELEMENTS.
From the form, check the boxes for Joint Labels and Frame Labels.
Click OK.

Restraints: Click on the center of the "dot" on Joints 1 and 5. This command will mark the selected joints with an "X".
From the main menu select:
ASSIGN>JOINT>RESTRAINTS.
In the form labeled "Joint Restraints," select restraints in all directions.
Click OK.

Mark joint 2.
From the main menu select:
ASSIGN>JOINT>RESTRAINTS.
In the pop-up window, select no restraints with the symbol of a dot.
Click OK.

Mark joints 3 and 4.
From the main menu select:
ASSIGN>JOINT>RESTRAINTS.
Restrain only Translation-3.
Click OK.

Loads: From the main menu select:
DEFINE>STATIC LOAD CASES.
In the "Define Static Load Case Names" form, change the label DEAD to LIVE using the drop-down menu.
Set the Self-Weight Multiplier to 0 (zero).
Click the "Change Load" button.
Click OK.

Joint Loads: Mark joint 2. From the main menu select:
ASSIGN>JOINT STATIC LOAD>FORCES.
In the Joint Forces menu, change the force in the Global Z to -10.0.

Warning: Be certain to zero out any previous entries made on this form!
Click OK.

Mark joint 3. From the Main Menu select:
ASSIGN>JOINT STATIC LOAD>FORCES.
    Change the Moment Global YY to 50.
Click OK.

Frame Loads: Select Frame 1 by clicking on this frame element of the beam.
    Frame 1 will change from a continuous line to dashes.
    From the Main Menu select:
ASSIGN>FRAME STATIC LOADS>POINT AND UNIFORM LOADS.
    Select the button "forces" and direction Local 2 (Local axis y is
    vertical) using the drop-down menu.
    Click the "absolute distance" button.

Warning: Be certain to zero out any previous entries made on this form!

In the first two columns on the left, enter the following:
    Distance =   10.0   Distance =   20.0
    Load   = −30.0   Load   = −10.0
Click OK.

Select Frame 4. From the main menu select:
ASSIGN>FRAME STATIC LOADS>POINT AND UNIFORM LOADS.
    Select the "MOMENTS" button.
    Select "Local 3" in the drop-down menu.
    Click "absolute distance".

Warning: Be certain to zero out any previous entries made on this form!

In the left-hand column, enter the following:
    Distance = 48.0   Load[3] = 100.0
Click OK.

Select Frame 4. From the main menu select:
ASSIGN > FRAME STATIC LOADS > POINT AND UNIFORM LOADS
    Select the "Forces" button.
    Select Local 2 from the drop-down menu.

Warning: Be certain to zero out any previous entries made on this form!

In the box that is located in the left-hand lower corner of this window
enter:
    Uniform Load = −0.05.
Click OK.

---

[3] Positive moment in reference to the local coordinate axes for a beam element.

Select Frame 2. From the main menu select:
ASSIGN>FRAME STATIC LOAD>POINT AND UNIFORM LOAD.
  Select "Forces" and change the direction to "Local 2".
  Warning: Be certain to zero out any previous entries made on this form!
  Enter a uniform load of:
    Uniform Load = −0.10
  Click OK.

Select Frame 3. From the main menu select:
ASSIGN>FRAME STATIC LOADS>TRAPEZOIDAL.
  Select "Forces", and change the direction to "Local 2".
  Click the "Absolute Distance" button.
  Warning: Be certain to zero out any previous entries made on this form!
  Enter the following in the two left-hand columns:
    Distance =   20.0   Distance = 95.0
    Load   =   − 0.10   Load   = − 0.20
  Click OK.

Options: From the main menu select:
  ANALYZE>SET OPTIONS.
  In the box marked "Available DOF" make certain that the only boxes
  checked are UZ and RY.

Final Analysis: From the main menu select:
  ANALYZE>RUN.
    On the form requesting a name for the model, enter
    NEX 3.1".
    Click "SAVE".
    Click OK.

Observe: The deformed shape of the beam will be shown on the screen.
  From the main menu select:
  DISPLAY>SHOW DEFORMED SHAPE.
    Click the "Wire Shadow" option.
    Click the "XZ" icon on the menu bar to view the plot on that plane.
    Click OK.
    To view the displacement values at any node, right-click on the node
    and a pop-up window will show the nodal displacements.

Print Input Tables: From the main menu select:
  FILE>PRINT INPUT TABLES.
  Click Print to File.
  Click OK.

The previous Table 3.1 provides the edited input tables for Illustrative Example 3.1.

Print Output Tables: From the main menu select:
FILE>PRINT OUTPUT TABLES.
Click Print to File and to Append.
The following Table 3.2 provides the edited output tables for Illustrative Example 3.1.
Use a text editor, such as Note pad or WORD, to edit and to print the Input File and the Output File, which have been stored as a text file (NEX 3.1.txt).

Table 3.2 Edited[4] Output Tables for Illustrative Example 3.1 (Units: kips, inches)

JOINT DISPLACEMENTS

JOINT	LOAD	UX	UY	UZ	RX	RY	RZ
1	LOAD2	0.0000	0.0000	0.0000	0.0000	0.0000	0.0000
2	LOAD2	0.0000	0.0000	-0.0239	0.0000	4.548E-05	0.0000
3	LOAD2	0.0000	0.0000	0.0000	0.0000	-1.761E-04	0.0000
4	LOAD2	0.0000	0.0000	0.0000	0.0000	-7.099E-06	0.0000
5	LOAD2	0.0000	0.0000	0.0000	0.0000	0.0000	0.0000

JOINT REACTIONS

JNT	LOAD	F1	F2	F3	M1	M2	M3
1	LOAD2	0.0000	0.0000	46.9113	0.0000	-785.0463	0.0000
3	LOAD2	0.0000	0.0000	19.4328	0.0000	0.0000	0.0000
4	LOAD2	0.0000	0.0000	7.9866	0.0000	0.0000	0.0000
5	LOAD2	0.0000	0.0000	0.7193	0.0000	9.6171	0.0000

FRAME ELEMENT FORCES
[This section of Table 3.2 provides values for the Shear Force (V2) and the Bending Moment (M3) along selected locations of the beam.]

FRM	LOAD	LOC	P	V2	V3	T	M2	M3
1	LOAD2							
		0.00	0.00	-46.91	0.00	0.00	0.00	-785.05
		22.50	0.00	-6.91	0.00	0.00	0.00	-129.54
		45.00	0.00	-6.91	0.00	0.00	0.00	25.96
		67.50	0.00	-6.91	0.00	0.00	0.00	181.47
		90.00	0.00	-6.91	0.00	0.00	0.00	336.97
2	LOAD2							
		0.00	0.00	3.09	0.00	0.00	0.00	336.97
		22.50	0.00	5.34	0.00	0.00	0.00	242.16
		45.00	0.00	7.59	0.00	0.00	0.00	96.73

---

[4] To edit the input tables, select FILE > PRINT INTPUT TABLES and then "Print to File". Analogously, for the output tables, select FILE >PRINT OUTPUT TABLES and then "Print to File" and Append. Then save the new file containing the input and output tables. Use an editor (such as Word or Notepad) to open, edit and print this new file.

Table 3.2 (continued)

		67.50	0.00	9.84	0.00	0.00	0.00	−99.33
		90.00	0.00	12.09	0.00	0.00	0.00	−346.01
3	LOAD2							
		0.00	0.00	−7.34	0.00	0.00	0.00	−296.01
		30.00	0.00	−6.28	0.00	0.00	0.00	−98.24
		60.00	0.00	−2.28	0.00	0.00	0.00	33.08
		90.00	0.00	2.92	0.00	0.00	0.00	26.40
		120.00	0.00	3.91	0.00	0.00	0.00	−70.97
4	LOAD2							
		0.00	0.00	−4.08	0.00	0.00	0.00	−70.97
		24.00	0.00	−2.88	0.00	0.00	0.00	12.57
		48.00	0.00	−1.68	0.00	0.00	0.00	−32.69
		72.00	0.00	−0.48	0.00	0.00	0.00	−6.75
		96.00	0.00	0.72	0.00	0.00	0.00	−9.62

Plot Displacements: From the main menu select:
DISPLAY>SHOW DEFORMED SHAPE.
Click FILE>PRINT GRAPHICS.
(The deformed shape shown on the screen is reproduced in Figure 3.2.)

Fig. 3.2 Deformed shape for the beam of Illustrative Example 3.1

Plot Shear Forces: From the main menu select:
DISPLAY>SHOW ELEMENT FORCES / STRESSES>FRAMES.
Select Shear 2-2.
Click FILE>PRINT GRAPHICS.
(The shear force diagram shown on the screen is depicted in Figure 3.3.)

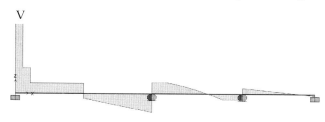

Fig. 3.3 Shear Force diagram for the beam of Illustrative Example 3.1

Plot the Bending Moment: From the main menu select:

DISPLAY>SHOW FORCES / STRESSES>FRAME.
Select Moment 3-3.
Click OK.
Click FILE>PRINT GRAPHICS.
(The bending moment diagram shown on the screen is depicted in Figure 3.4.)

Note: To plot the Moment Diagram on the compression side, from the main menu select OPTIONS and uncheck the "Moment Diagram on the Tension Side" option.

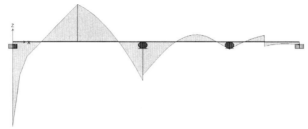

Fig 3.4   Bending Moment Diagram for the Beam of Illustrative Example 3.1

Note: To view plots of either the shear force or bending moment diagrams for any element of the beam, right-click on the element and a pop-up window will depict the diagram for the selected element. It should be observed that as the cursor is moved across the plot in this window, values of shear force or of the bending moment will be displayed.

Illustrative Example 3.2
Use SAP2000 to solve the beam of Illustrative Example 3.1, which is shown in Figure 3.1. Assume that, in addition to the loads of Illustrative Example 3.1, the supports 3 and 4 have undergone downward vertical displacements of 1.0 *in* and 2.0 *in*, respectively.

Solution:
The input data for this problem is the same as the input data for Illustrative Example 3.1, except for the additional data on the imposed displacements at nodes 3 and 4. The following data should be added to the file containing data for Illustrative Example 3.1.

Open File: From the main menu select:
          FILE > OPEN.
          Select File NEX3.1.
          Note: To use the file from Illustrative Example 3.1, it is necessary to unlock it by clicking the icon showing a padlock on the main menu. Then click OK.

Select: Click on node 3 to select it. Then from the main menu select:
        ASSIGN>JOINT STATIC LOADS>DISPLACEMENTS.
            Set translation Z = −1.0.
            Click OK.
        Select node 4 and from the main menu select:
        ASSIGN>JOINT STATIC LOADS>DISPLACEMENTS.
            Set translation Z = −2.0.
            Click OK.

Set Analysis: From the main menu select:
        ANALYZE>RUN.
            On the window requesting to name the model, enter its new name
            "NEX 3.2".
            Click SAVE.
            Click OK.

Observe: The deformed shape of the beam will display.
        On the main menu select:
        DISPLAY>SHOW DEFORMED SHAPE.
            Click on the "Wire Shadow" option.
            Click the "XZ" icon to set the plot to that plane.
            Click OK.

Results: To view the displacement values at any node, right-click on the node and
        a pop-up window will show the nodal displacements.

Print Input Tables: From the main menu select:
        FILE>PRINT INPUT TABLES.

        Table 3.3 contains the edited input tables for Illustrative Example 3.2.

Table 3.3 Edited Input Tables for Illustrative Example 3.2 (Units: kips, inches)

JOINT DATA

JNT	GLBL-X	GLBL-Y	GLBL-Z	RESTRAINTS
1	0.00000	0.00000	0.00000	1 1 1 1 1 1
2	90.00000	0.00000	0.00000	0 0 0 0 0 0
3	180.00000	0.00000	0.00000	0 0 1 0 0 0
4	300.00000	0.00000	0.00000	0 0 1 0 0 0
5	396.00000	0.00000	0.00000	1 1 1 1 1 1

FRAME ELEMENT DATA

FRM	JNT-1	JNT-2	SEC	ANGL	RELEASES	SEGTS	R1	R2	FACTR	LENGTH
1	1	2	W14X82	0.000	000000	4	0.000	0.000	1.000	90.000
2	2	3	W14X82	0.000	000000	4	0.000	0.000	1.000	90.000
3	3	4	W14X82	0.000	000000	4	0.000	0.000	1.000	120.000
4	4	5	W14X82	0.000	000000	4	0.000	0.000	1.000	96.000

Table 3.3 (continued)

J O I N T   F O R C E S   Load Case LOAD2

JNT	GLBL-X	GLBL-Y	GLBL-Z	GLBL-XX	GLBL-YY	GLBL-ZZ
3	0.000	0.000	0.000	0.000	50.000	0.000
2	0.000	0.000	-10.000	0.000	0.000	0.000

J O I N T   D I S P L A C E M E N T   L O A D S   Load Case LOAD2

JNT	GLBL-X	GLBL-Y	GLBL-Z	GLBL-XX	GLBL-YY	GLBL-ZZ
3	0.00000	0.00000	-1.00000	0.00000	0.00000	0.00000
4	0.00000	0.00000	-2.00000	0.00000	0.00000	0.00000

F R A M E   S P A N   D I S T R I B U T E D   L O A D S   Load Case LOAD2

FRM	TYPE	DIRECTION	DIST-A	VALUE-A	DIST-B	VALUE-B
4	FORCE	LOCAL-2	0.0000	-0.0500	1.0000	-0.0500
2	FORCE	LOCAL-2	0.0000	-0.1000	1.0000	-0.1000
3	FORCE	LOCAL-2	0.1667	-0.1000	0.7917	-0.2000

F R A M E   S P A N   P O I N T   L O A D S   Load Case LOAD2

FRM	TYPE	DIRECTION	DIST	VALUE
4	MOMENT	LOCAL-3	0.5000	100.0000
1	FORCE	LOCAL-2	0.1111	-30.0000
1	FORCE	LOCAL-2	0.2222	-10.0000

Print Output Tables: From the main menu select:
        FILE>PRINT OUTPUT TABLES.
        (Table 3.4 provides the output tables for Illustrative Example 3.2.)

Table 3.4   Edited Output tables for Illustrative Example 3.2 (Units: kips, inches)

J O I N T   D I S P L A C E M E N T S

JOINT	LOAD	UX	UY	UZ	RX	RY	RZ
1	LOAD2	0.0000	0.0000	0.0000	0.0000	0.0000	0.0000
2	LOAD2	0.0000	0.0000	-0.1774	0.0000	0.004.5	0.0000
3	LOAD2	0.0000	0.0000	-1.0000	0.0000	0.0152	0.0000
4	LOAD2	0.0000	0.0000	-2.0000	0.0000	-0.0152	0.0000
5	LOAD2	0.0000	0.0000	0.0000	0.0000	0.0000	0.0000

J O I N T   R E A C T I O N S

JOINT	LOAD	F1	F2	F3	M1	M2	M3
1	LOAD2	0.0000	0.0000	26.58	0.0000	-1144.52	0.0000
3	LOAD2	0.0000	0.0000	215.53	0.0000	0.00	0.0000
4	LOAD2	0.0000	0.0000	-608.05	0.0000	0.00	0.0000
5	LOAD2	0.0000	0.0000	440.98	0.0000	25199.59	0.0000

Table 3.4 (continued)

F R A M E   E L E M E N T   F O R C E S

FRAME	LOAD	LOC	P	V2	V3	T	M2	M3
1	LOAD2							
		0.00	0.00	−26.59	0.00	0.00	0.00	−1144.53
		22.50	0.00	13.41	0.00	0.00	0.00	−946.30
		45.00	0.00	13.41	0.00	0.00	0.00	−1248.07
		67.50	0.00	13.41	0.00	0.00	0.00	−1549.85
		90.00	0.00	13.41	0.00	0.00	0.00	−1851.62
2	LOAD2							
		0.00	0.00	23.41	0.00	0.00	0.00	−1851.62
		22.50	0.00	25.66	0.00	0.00	0.00	−2403.71
		45.00	0.00	27.91	0.00	0.00	0.00	−3006.42
		67.50	0.00	30.16	0.00	0.00	0.00	−3659.76
		90.00	0.00	32.41	0.00	0.00	0.00	−4363.72
3	LOAD2							
		0.00	0.00	−183.12	0.00	0.00	0.00	−4313.72
		30.00	0.00	−182.05	0.00	0.00	0.00	1157.29
		60.00	0.00	−178.05	0.00	0.00	0.00	6561.86
		90.00	0.00	−172.85	0.00	0.00	0.00	11828.42
		120.00	0.00	−171.87	0.00	0.00	0.00	17011.54
4	LOAD2							
		0.00	0.00	436.18	0.00	0.00	0.00	17004.29
		24.00	0.00	437.38	0.00	0.00	0.00	6521.52
		48.00	0.00	438.58	0.00	0.00	0.00	−4090.05
		72.00	0.00	439.78	0.00	0.00	0.00	−4630.42
		96.00	0.00	440.98	0.00	0.00	0.00	−5199.59

Plot Displacements: From the main menu select:
  DISPLAY>SHOW DEFORMED SHAPE.
    Select FILE>PRINT GRAPHICS.
  (The deformed shape shown on the screen is reproduced in Figure 3.5.)

Fig. 3.5 Deformed shape for the beam in Illustrative Example 3.2

Plot Shear Forces: From the main menu select:
  DISPLAY>SHOW ELEMENT FORCES / STRESSES>FRAMES.
    Select Shear 2-2.
    Select FILE>PRINT GRAPHICS.

  (The shear force diagram shown on the screen is depicted in Figure 3.6.)

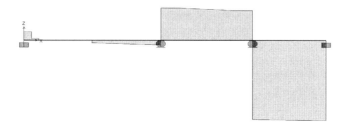

Fig. 3.6 Shear Force diagram for the beam of Illustrative Example 3.2

Plot the Bending Moment: From the main menu select:
        DISPLAY>SHOW FORCES / STRESSES>FRAME.
            Select Moment 3-3.
            Click OK.
            Select FILE>PRINT GRAPHICS.
    (The bending moment diagram shown on the screen is depicted in Figure 3.7.)

>   **Note:**   To plot the Moment Diagram on the compression side, from the
>   main menu select OPTIONS then uncheck the "Moment Diagram
>   on the Tension Side" option.

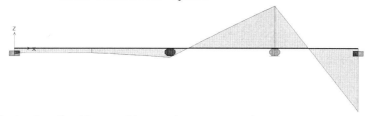

Fig 3.7 Bending Moment Diagram for the beam of Illustrative Example 3.2

Note: To view plots of either the shear force or bending moment diagrams for any
element of the beam, right-click on the element and a pop-up window will depict
the diagram for the selected element. It should be observed that as the cursor is
moved across the plot in this window, values of shear force or of the bending
moment will be displayed.

Illustrative Example 3.3
Use SAP2000 to solve the beam of Illustrative Example 3.1, which is shown in
Figure 3.1. In addition to the load specified in Illustrative Example 3.1, assume that
beam element 3 has a hinge (released for bending) at its left end.

Solution:

The input data for this example is the same as the input data for Illustrative Example 3.1, except for additional data to implement the release at the left end of element 3. The following data should be added to the Illustrative Example 3.1 file (NEX 3.1):

Select: Click on element 3. Then from the main menu select:
ASSIGN>FRAME>RELEASES.
Check Start Moment 3-3 (major axis of the cross-sectional area).
Click OK.

Analysis: From the main menu select:
ANALYZE>RUN.
On the window requesting to name the model, enter its new name:
"NEX 3.3".
Click "SAVE".

Observe: The deformed shape of the beam will be shown on the screen.
From the main menu select:
DISPLAY>SHOW DEFORMED SHAPE.
Click on the "Wire Shadow" option.
Click the "XZ" icon to view the plot on that plane.
Click OK.

Results: To view the displacement values at any node, right-click on the node and a pop-up window will show the nodal displacements.

Print Output Tables: From the main menu select:
FILE>PRINT OUTPUT TABLES.
(Table 3.5 contains the edited output tables for Illustrative Example 3.3.)

Table 3.5 Edited Output Tables for Illustrative Example 3.3 (Units: kips, inches)

JOINT DISPLACEMENTS

JOINT	LOAD	UX	UY	UZ	RX	RY	RZ
1	LOAD2	0.0000	0.0000	0.0000	0.0000	0.0000	0.0000
2	LOAD2	0.0000	0.0000	−0.0356	0.0000	$1.757E$-04	0.0000
3	LOAD2	0.0000	0.0000	0.0000	0.0000	−6.969E-04	0.0000
4	LOAD2	0.0000	0.0000	0.0000	0.0000	−9.390E-05	0.0000
5	LOAD2	0.0000	0.0000	0.0000	0.0000	0.0000	0.0000

JOINT REACTIONS

JOINT	LOAD	F1	F2	F3	M1	M2	M3
1	LOAD2	0.0000	0.0000	49.3781	0.0000	−933.0516	0.0000
3	LOAD2	0.0000	0.0000	13.7284	0.0000	0.0000	0.0000
4	LOAD2	0.0000	0.0000	12.6696	0.0000	0.0000	0.0000
5	LOAD2	0.0000	0.0000	−0.7261	0.0000	−36.6346	0.0000

Table 3.5 (continued)

FRAME ELEMENT FORCES

FRAME	LOAD	LOC	P	V2	V3	T	M2	M3
1	LOAD2							
		0.00	0.00	−49.38	0.00	0.00	0.00	−933.05
		22.50	0.00	−9.38	0.00	0.00	0.00	−222.05
		45.00	0.00	−9.38	0.00	0.00	0.00	−11.04
		67.50	0.00	−9.38	0.00	0.00	0.00	199.97
		90.00	0.00	−9.38	0.00	0.00	0.00	410.97
2	LOAD2							
		0.00	0.00	0.6.2	0.00	0.00	0.00	410.97
		22.50	0.00	2.87	0.00	0.00	0.00	371.67
		45.00	0.00	5.12	0.00	0.00	0.00	281.74
		67.50	0.00	7.37	0.00	0.00	0.00	141.18
		90.00	0.00	9.62	0.00	0.00	0.00	−50.00
3	LOAD2							
		0.00	0.00	−4.11	0.00	0.00	0.00	0.00
		30.00	0.00	−3.04	0.00	0.00	0.00	100.64
		60.00	0.00	0.9.6	0.00	0.00	0.00	134.83
		90.00	0.00	6.16	0.00	0.00	0.00	31.03
		120.00	0.00	7.14	0.00	0.00	0.00	−156.22
4	LOAD2							
		0.00	0.00	−5.53	0.00	0.00	0.00	−163.47
		24.00	0.00	−4.33	0.00	0.00	0.00	−45.24
		48.00	0.00	−3.13	0.00	0.00	0.00	−55.82
		72.00	0.00	−1.93	0.00	0.00	0.00	4.81
		96.00	0.00	−0.7.3	0.00	0.00	0.00	36.63

Plot Displacements: From the main menu select:
    DISPLAY>SHOW DEFORMED SHAPE.
        Select FILE>PRINT GRAPHICS.
    (The deformed shape shown on the screen is reproduced in Figure 3.8.)

Fig. 3.8 Deformed shape for the beam of Illustrative Example 3.3

Plot Shear Forces: From the main menu select:
    DISPLAY>SHOW ELEMENT FORCES / STRESSES>FRAMES.
        Select Shear 2-2.
        Click OK.
        To obtain a hard copy select FILE>PRINT GRAPHICS.
    (The shear force diagram shown on the screen is depicted in Figure 3.9.)

Fig. 3.9 Shear force diagram for the beam of Illustrative Example 3.3

Plot the Bending Moment: From the main menu select:
DISPLAY>SHOW FORCES / STRESSES>FRAME.
Select Moment 3-3.
Click OK.
Then to obtain a hard copy select FILE>PRINT GRAPHICS.
(The bending moment diagram shown on the screen is depicted in Figure 3.10.)

Note: To plot the Moment Diagram on the compression side, from the main menu select OPTIONS, then uncheck the "Moment Diagram on the Tension Side" option.

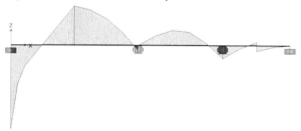

Fig 3.10 Bending Moment Diagram for the beam of Illustrative Example 3.3

Note: To view plots of either the shear force or bending moment diagrams for any element of the beam, right-click on the element and a pop-up window will depict the diagram for the selected element. It should be observed that as the cursor is moved across the plot in this window, values of shear force or of the bending moment will be displayed.

Illustrative Example 3.4
Use SAP2000 to analyze the loaded beam shown in Figure 3.1, assuming that beam element 3 is subjected to a linear change in temperature, with $T_2 = 60°F$ at the top and $T_1 = 120°F$ at the bottom. The coefficient of thermal expansion is $\alpha = 6.5xE-6 /°F$, and the height of the cross-section area is $h = 14.31$ *in*.

Solution:

The temperature gradient across the section of the beam is

$$\text{Temp. Gradiant} = \frac{T_2 - T_1}{h} = \frac{60 - 120}{14.31} = -4.19\,°F\!\!\Big/\!\!{}_{in}.$$

The beam and the load for this example are the same as in Illustrative Example 3.1, with the exception of the thermal load applied at element 3. Therefore, the stored data file for Illustrative Example 3.1 may be used by simply adding the additional equivalent load due to change in temperature in Element 3.

The following commands are then implemented in SAP2000:

Begin: From the main menu select:
FILE>OPEN.
Select the file for Illustrative Example 3.1 (NEX 3.1).

Unlock: Since the program has locked the data file used in the solution of Illustrative Example 3.1, it is necessary to unlock this file. To unlock, click the icon showing the symbol of a padlock and respond YES to the question that follows.

Thermal Load: Click on Element 3 of the beam.
 From the main menu select:
ASSIGN>FRAME STATIC LOADS>TEMPERATURE.
Click on Temperature Gradient 2-2.
Enter Temperature Gradient = − 4.19.
Click OK.

Analyze: From the main menu select:
 RUN.
Click OK.
The screen will show the deformed shape of the beam.

Print Input Tables: From the main menu select:
FILE>PRINT>INPUT TABLE.
(Table 3.6 contains the edited input tables for Illustrative Example 3.4.)

Table 3.6 Edited Input Table for Illustrative Example 3.4   (Units: kips, inches)

JOINT DATA

JOINT	GLBL-X	GLBL-Y	GLBL-Z	RESTRAINTS
1	0.00000	0.00000	0.00000	1 1 1 1 1 1
2	90.00000	0.00000	0.00000	0 0 0 0 0 0
3	180.00000	0.00000	0.00000	0 0 1 0 0 0
4	300.00000	0.00000	0.00000	0 0 1 0 0 0
5	396.00000	0.00000	0.00000	1 1 1 1 1 1

FRAME ELEMENT DATA

FRM	JNT-1	JNT-2	SEC	ANGL	RELEASES	SEG	R1	R2	FACTOR	LTH
1	1	2	W14X82	0.000	000000	4	0.000	0.000	1.000	90.000
2	2	3	W14X82	0.000	000000	4	0.000	0.000	1.000	90.000
3	3	4	W14X82	0.000	000000	4	0.000	0.000	1.000	120.000
4	4	5	W14X82	0.000	000000	4	0.000	0.000	1.000	96.000

JOINT FORCES Load Case LOAD2

JOINT	GLBL-X	GLBL-Y	GLBL-Z	GLBL-XX	GLBL-YY	GLBL-ZZ
3	0.000	0.000	0.000	0.000	50.000	0.000
2	0.000	0.000	-10.000	0.000	0.000	0.000

FRAME SPAN DISTRIBUTED LOADS Load Case LOAD2

FRAME	TYPE	DIRECTION	DIST-A	VALUE-A	DIST-B	VALUE-B
4	FORCE	LOCAL-2	0.0000	-0.0500	1.0000	-0.0500
2	FORCE	LOCAL-2	0.0000	-0.1000	1.0000	-0.1000
3	FORCE	LOCAL-2	0.1667	-0.1000	0.7917	-0.2000

FRAME SPAN POINT LOADS Load Case LOAD2

FRAME	TYPE	DIRECTION	DISTANCE	VALUE
4	MOMENT	LOCAL-3	0.5000	100.0000
1	FORCE	LOCAL-2	0.1111	-30.0000
1	FORCE	LOCAL-2	0.2222	-10.0000

FRAME THERMAL LOADS Load Case LOAD2

FRAME	TYPE	VALUE
3	GRAD 2-2	-4.1900

Print Output Tables: From the main menu select:
        FILE > PRINT > OUTPUT TABLE.

(Table 3.7 contains the edited output tables for Illustrative Example 3.4.)

Table 3.7 Output tables for Illustrative Example 3.4 (Units: kips-inches)

JOINT DISPLACEMENTS

JOINT	LOAD	UX	UY	UZ	RX	RY	RZ
1	LOAD2	0.0000	0.0000	0.0000	0.0000	0.0000	0.0000
2	LOAD2	0.0000	0.0000	−9.458E-03	0.0000	−1.150E-04	0.0000
3	LOAD2	0.0000	0.0000	0.0000	0.0000	4.658E-04	0.0000
4	LOAD2	0.0000	0.0000	0.0000	0.0000	−5.129E-04	0.0000
5	LOAD2	0.0000	0.0000	0.0000	0.0000	0.0000	0.0000

JOINT REACTIONS

JOINT	LOAD	F1	F2	F3	M1	M2	M3
1	LOAD2	0.0000	0.0000	43.8705	0.0000	−602.5990	0.0000
3	LOAD2	0.0000	0.0000	21.0223	0.0000	0.0000	0.0000
4	LOAD2	0.0000	0.0000	17.8605	0.0000	0.0000	0.0000
5	LOAD2	0.0000	0.0000	−7.7034	0.0000	−259.9073	0.0000

FRAME ELEMENT FORCES

FRAME	LOAD	LOC	P	V2	V3	T	M2	M3
1	LOAD2							
		0.00	0.00	−43.87	0.00	0.00	0.00	−602.60
		22.50	0.00	−3.87	0.00	0.00	0.00	−15.51
		45.00	0.00	−3.87	0.00	0.00	0.00	71.57
		67.50	0.00	−3.87	0.00	0.00	0.00	158.66
		90.00	0.00	−3.87	0.00	0.00	0.00	245.75
2	LOAD2							
		0.00	0.00	6.13	0.00	0.00	0.00	245.75
		22.50	0.00	8.38	0.00	0.00	0.00	82.52
		45.00	0.00	10.63	0.00	0.00	0.00	−131.33
		67.50	0.00	12.88	0.00	0.00	0.00	−395.80
		90.00	0.00	15.13	0.00	0.00	0.00	−710.91
3	LOAD2							
		0.00	0.00	−5.89	0.00	0.00	0.00	−660.91
		30.00	0.00	−4.83	0.00	0.00	0.00	−506.68
		60.00	0.00	−0.83	0.00	0.00	0.00	−418.89
		90.00	0.00	4.37	0.00	0.00	0.00	−469.11
		120.00	0.00	5.36	0.00	0.00	0.00	−602.76
4	LOAD2							
		0.00	0.00	−12.50	0.00	0.00	0.00	−610.01
		24.00	0.00	−11.30	0.00	0.00	0.00	−324.33
		48.00	0.00	−10.10	0.00	0.00	0.00	−167.45
		72.00	0.00	−8.90	0.00	0.00	0.00	60.63
		96.00	0.00	−7.70	0.00	0.00	0.00	259.91

Plot Displacements: From the main menu select:
  DISPLAY>SHOW DEFORMED SHAPE.
    Select FILE>PRINT GRAPHICS.

(The deformed shape shown on the screen is reproduced in Figure 3.11.)

Fig. 3.11 Deformed shape for the beam of Illustrative Example 3.4

Plot Shear Forces: From the main menu select:
DISPLAY>SHOW ELEMENT FORCES / STRESSES>FRAMES.
  Select Shear 2-2.
  Click OK.
  Select FILE>PRINT GRAPHICS.
(The shear force diagram shown on the screen is depicted in Figure 3.12.)

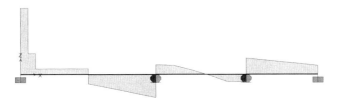

Fig. 3.12 Shear Force diagram for the beam of Illustrative Example 3.4

Plot the Bending Moment: From the main menu select:
DISPLAY>SHOW FORCES / STRESSES>FRAME
  Select Moment 3-3.
  Click OK.
  Select FILE>PRINT GRAPHICS
(The bending moment diagram on the screen is depicted in Figure 3.13.)

Fig 3.13 Bending Moment Diagram for the beam of Illustrative Example 3.4

Note: To plot the Moment Diagram on the compression side, on the main menu select OPTIONS and uncheck the "Moment Diagram on the Tension Side" option.

Note: To view plots of either the shear force or bending moment diagrams for any element of the beam, right-click on the element and a pop-up window will depict the diagram for the selected element. It should be observed that as the cursor is moved across the plot in this window, values of shear force or of the bending moment will be displayed.

Illustrative Example 3.5

Consider in Figure 3.14(a) a loaded concrete continuous beam of two spans and in Figure 3.14(b) the selected analytical model with 4 elements and 5 nodes. Use the program SAP2000 to perform the structural analysis to determine displacements and reactions, and to plot shear force and bending moment diagrams.

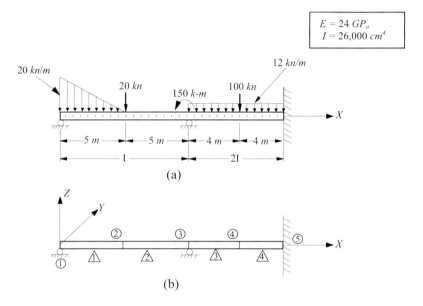

$$E = 24\ GP_a$$
$$I = 26{,}000\ cm^4$$

Fig. 3.14 (a) Concrete beam for Illustrative Example 3.5; (b) Analytical model

Table 3.8   Edited Input Data for Illustrative Example 3.5   (Units: KN, m)

JOINT DATA

JNT	GLBL-X	GLBL-Y	GLBL-Z	RESTRAINTS	AN-A	AN-B	AN-C
1	0.00000	0.00000	0.00000	0 0 1 0 0 0	0.000	0.000	0.000
2	5.00000	0.00000	0.00000	0 0 0 0 0 0	0.000	0.000	0.000
3	10.00000	0.00000	0.00000	0 0 1 0 0 0	0.000	0.000	0.000
4	14.00000	0.00000	0.00000	0 0 0 0 0 0	0.000	0.000	0.000
5	18.00000	0.00000	0.00000	1 1 1 1 1 1	0.000	0.000	0.000

Table 3.8 (continued)

```
F R A M E   E L E M E N T   D A T A
FRM JNT-1 JNT-2 SEC    ANGL RELEASES SGMNTS R1     R2    FACTOR LENGTH
```

FRM	JNT-1	JNT-2	SEC	ANGL	RELEASES	SGMNTS	R1	R2	FACTOR	LENGTH
1	1	2	FSEC3	0.000	000000	4	0.000	0.000	1.000	5.000
2	2	3	FSEC3	0.000	000000	4	0.000	0.000	1.000	5.000
3	3	4	FSEC2	0.000	000000	4	0.000	0.000	1.000	4.000
4	4	5	FSEC2	0.000	000000	4	0.000	0.000	1.000	4.000

```
J O I N T   F O R C E S   Load Case  LOAD1
```

JOINT	GLBL-X	GLBL-Y	GLBL-Z	GLBL-XX	GLBL-YY	GLBL-ZZ
3	0.000	0.000	0.000	0.000	-150.000	0.000
4	0.000	0.000	-100.000	0.000	0.000	0.000
2	0.000	0.000	-20.000	0.000	0.000	0.000

```
F R A M E   S P A N   D I S T R I B U T E D   L O A D S   Load Case  LOAD1
```

FRM	TYPE	DIRECTION	DIST-A	VALUE-A	DIST-B	VALUE-B
3	FORCE	GLOBAL-Z	0.0000	-12.0000	1.0000	-12.0000
4	FORCE	GLOBAL-Z	0.0000	-12.0000	1.0000	-12.0000
1	FORCE	GLOBAL-Z	0.0000	-20.0000	1.0000	0.0000

Solution:

Begin: Open SAP2000.

Hint: Maximize both screens for a full view of all windows and use the PAN icon to drag the plot to the center of the screen.

Units: Select kn-m (Kilo Newtons-Meters) in the drop-down list located in the lower right-hand corner of the window.

Model: From the main menu select:
FILE>NEW MODEL FROM TEMPLATE.
Select "Beam".
Change the number of spans to 4 and span length to 5.
Click OK.

Edit: Maximize the 2-D screen (X-Z screen). Then drag the figure to the center of the screen using the PAN icon on the toolbar.

Translate: From the main menu select:
SELECT>SELECT ALL.
From the main menu select:
EDIT>MOVE
Set Delta X = 10.
Click OK.
Use icon PAN to center figure.

Grid: From the main menu select:
DRAW>EDIT GRID.
  Check the "glue joints to grid lines" option.
  Highlight the –5 value and change it to 14.
  Click "Move Grid Line".
  Click OK.
  Highlight the –10 value and change it to 18.
  Click "Move grid line".
  Click OK.

Labels: From the main menu select:
VIEW>SET ELEMENTS.
  Check: Joint Labels.
  Check: Frame Labels.
  Click OK.

Restraints: Click (mark): Joints 1 and 3 and from the main menu select:
ASSIGN>JOINT>RESTRAINTS.
  Check: Restraint translation 3.
  Click OK.
Click Joint 5 and from the main menu select:
ASSIGN>JOINTS>RESTRAINTS.
  Check: Restrain for all directions.
  Click OK.

Material: From the main menu select:
DEFINE MATERIAL>CONC.
Click MODIFY/SHOW MATERIAL.
Set Modulus of Elasticity = $24E6$.
Click OK.

Section: From the main menu select:
DEFINE >FRAME SECTIONS.
Click Add/Wide Flange.
Select Add General (Section).
  Set Moment of Inertia about axis 3 to $2.6E\text{-}4$.
  Shears in 2 directions = 0.
  Click OK.
  Material: Scroll to CONC.
Click OK, OK.

Section: From the main menu select:
DEFINE >FRAME SECTIONS
Click Add/Wide Flange.
Select Add General (Section).

Set Moment of Inertia about axis 3 = 5.2E-4.
Shears in 2 direction = 0.
Click OK.
Material: Scroll to CONC.
Click OK, OK.

Assign Select Elements 1 and 2 and from the main menu select:
ASSIGN > FRAME > SECTIONS.
Click on FSEC2.
Click OK.

Assign Select Elements 3 and 4 and from the main menu select:
ASSIGN > FRAME > SECTIONS.
Click on FSEC3.
Click OK.

Loads: From the main menu select:
DEFINE>STATIC LOAD CASES.
Change DEAD to LIVE.
Set Self-Weight Multiplier = 0
Click Change Load.
Click OK.

Select Joint 2.
From the main menu select:
ASSIGN>JOINT STATIC LOAD>FORCES.
Check Forces Global Z = −20
Click OK.
Warning: Be certain to zero out the values of previous entries.
Select Joint 3.
From the main menu select:
ASSIGN>JOINT STATIC LOADS>FORCES.
Set Moment Global Y = −150
Click OK.

Select Joint 4.
From the main menu select:
ASSIGN>JOINT STATIC LOADS>FORCES.
Set Forces Global Z = −100
Click OK.

Assign Frame Loads: Select Frame 1 and from the main menu select:
ASSIGN>FRAME STATIC LOADS>TRAPEZOIDAL.
Verify: Checks on Forces and on Global Z.
Enter: Distance   0.0   1.0
           Load   −20   0

Click OK.
Select Frames 3 and 4.
From the main menu select:
ASSIGN>FRAME STATIC LOADS>POINT AND UNIFORM.
Set Uniform Load = −12.
Warning: Be certain to zero out the values of previous entries.
Click OK.

Options: From the main menu select:
ANALYZE>SET OPTIONS.
Check Available Degrees of Freedom UZ, RY.
Click OK.

Analyze: From the main menu select:
ANALYZE>RUN.
Input the Filename "NEX 3.5".
Click SAVE.
At the conclusion of the solution process, click OK.

Print Tables: From the main menu select:
FILE>PRINT INPUT TABLES.
Check: Joint Data: Coordinates.
        Element data: Frames.
        Static Loads: Joints, Frames.
Click OK.
(Table 3.8 reproduces the edited Input Data for Illustrative Example 3.5.)

From the main menu select:
FILE>PRINT OUTPUT TABLE.
Check: Displacements
        Reactions
        Frame Forces
Click OK.

(Table 3.9 provides the edited output table for Illustrative Example 3.5.)

Table 3.9 Edited Output tables for Illustrative Example 3.5 (Units: kN, m)

JOINT DISPLACEMENTS

JNT	LOAD	UX	UY	UZ	RX	RY	RZ
1	LOAD1	0.0000	0.0000	0.0000	0.0000	3.304E-03	0.0000
2	LOAD1	0.0000	0.0000	−8.007E-03	0.0000	−8.017E-04	0.0000
3	LOAD1	0.0000	0.0000	0.0000	0.0000	−7.648E-04	0.0000
4	LOAD1	0.0000	0.0000	−2.398E-03	0.0000	1.912E-04	0.0000
5	LOAD1	0.0000	0.0000	0.0000	0.0000	0.0000	0.0000

Table 3.9 Continued

JOINT REACTIONS

JNT	LOAD	F1	F2	F3	M1	M2	M3
1	LOAD1	0.0000	0.0000	45.4942	0.0000	0.0000	0.0000
3	LOAD1	0.0000	0.0000	131.4541	0.0000	0.0000	0.0000
5	LOAD1	0.0000	0.0000	89.0517	0.0000	140.1378	0.0000

FRAME ELEMENT FORCES

FRAME	LOAD	LOC	P	V2	V3	T	M2	M3
1	LOAD1	0.00	0.00	−45.49	0.00	0.00	0.00	0.00
		1.25	0.00	−23.62	0.00	0.00	0.00	42.54
		2.50	0.00	−7.99	0.00	0.00	0.00	61.65
		3.75	0.00	1.38	0.00	0.00	0.00	65.13
		5.00	0.00	4.51	0.00	0.00	0.00	60.80
2	LOAD1	0.00	0.00	24.51	0.00	0.00	0.00	60.80
		1.25	0.00	24.51	0.00	0.00	0.00	30.17
		2.50	0.00	24.51	0.00	0.00	0.00	−0.46
		3.75	0.00	24.51	0.00	0.00	0.00	−31.09
		5.00	0.00	24.51	0.00	0.00	0.00	−61.72
3	LOAD	0.00	0.00	−106.95	0.00	0.00	0.00	−211.72
		1.00	0.00	-94.95	0.00	0.00	0.00	−110.78
		2.00	0.00	−82.95	0.00	0.00	0.00	−21.83
		3.00	0.00	−70.95	0.00	0.00	0.00	55.12
		4.00	0.00	−58.95	0.00	0.00	0.00	120.07
4	LOAD1	0.00	0.00	41.05	0.00	0.00	0.00	120.07
		1.00	0.00	53.05	0.00	0.00	0.00	73.02
		2.00	0.00	65.05	0.00	0.00	0.00	13.97
		3.00	0.00	77.05	0.00	0.00	0.00	−57.09
		4.00	0.00	89.05	0.00	0.00	0.00	−140.14

Plot Output: From the main menu select:
  DISPLAY > SHOW DEFORMED SHAPE.
    Check the Wire Shadow option.
    Click OK.

  From the main menu select:
  FILE > PRINT GRAPHICS.
  [The deformed shape of the beam is reproduced in Figure 3.15(a).]

  From the main menu select:
  DISPLAY > SHOW ELEMENT FORCES/STRESSES > FRAMES.
    Click the Shear 3-2 option.
    Click OK.

  From the main menu select:
  FILE > PRINT GRAPHICS.
  [The shear force diagram is reproduced in Figure 3.15(b).]

  From the main menu select:

DISPLAY > SHOW ELEMENT FORCES/STRESSES > FRAMES.
   Click the Moment 3 – 3 option.
   Click OK.

From the main menu select:
FILE > PRINT GRAPHICS.
[The bending moment diagram is reproduced in Figure 3.15(c).]

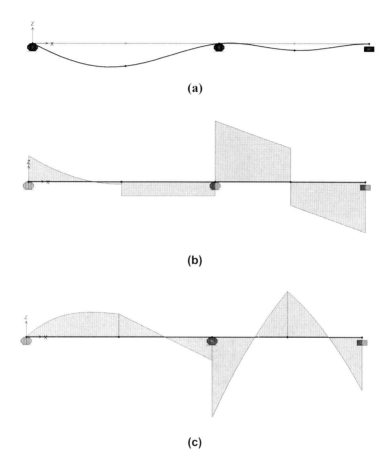

(a)

(b)

(c)

Fig. 3.15 (a) Deformed shape, (b) Shear Force diagram, and (c) Bending Moment
            Diagram for the beam of Illustrative Example 3.5

Illustrative Example 3.6
Use SAP2000 to analyze the beam shown in Figure 1.16 reproduced here for
convenience as Figure 3.16.

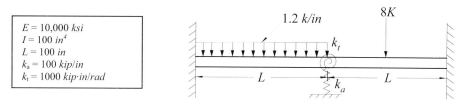

$E = 10,000 \; ksi$
$I = 100 \; in^4$
$L = 100 \; in$
$k_a = 100 \; kip/in$
$k_t = 1000 \; kip\cdot in/rad$

Fig. 3.16 Beam for Illustrative Example 3.6

Solution:

Begin: Open SAP2000.
Hint: Maximize both windows for a full view.

Units: Select "kip-in" in the drop-down list located in the lower right-hand corner of the window.

Model: From the main menu select:
FILE>NEW MODEL FROM TEMPLATE.
Select "Beam".
Change the number of spans to 2 and span length to 100.
Click OK.

Edit: Maximize the 2-D screen (X-Z screen). Then drag the figure to the center of the screen using the PAN icon on the toolbar.

Translate: From the main menu select:
SELECT>SELECT ALL.
From the main menu select:
EDIT>MOVE.
Set Delta X = 100.
Click OK.
Use the PAN icon to center figure.

Grid: From the main menu select:
DRAW>EDIT GRID.
Check "glue joints to grid lines".
Highlight the –100 value and change to 200.
Click "Move Grid Line".
Click OK.

From the main menu select:
VIEW.
Uncheck "Show Grid".

Labels: From the main menu select:
VIEW>SET ELEMENTS.
Check the Joint Labels option.
Check the Frame Labels option.
Click OK.

Restraints: Click (mark) Joints 1 and 3 and from the main menu select:
ASSIGN>JOINT>RESTRAINTS.
Check: Restrain all directions.
Click OK.

Click Joint 2.
From the main menu select:
ASSIGN>JOINT>RESTRAINTS.
        No restraints in all directions.
        Click OK.

Material: From the main menu select:
DEFINE MATERIAL>OTHER.
Click MODIFY/SHOW MATERIAL.
        Set: Modulus of Elasticity = $10E3$.
        Click OK, OK.

Sections: From the main menu select:
DEFINE>FRAME SECTIONS.
Click Add/Wide Flange.
Select: Add General (Section).
        Set: Moment of Inertia about axis 3 = 100.
        Shear area in 2 direction = 0.
        Scroll: Material to "Other".
        Click OK, OK.

Click (mark) Frames 1 and 2.
From the main menu select:
ASSIGN>FRAME>SECTIONS.
        Select FSFC2.
        Click OK.

Springs: Mark joint 2 and from the main menu select:
ASSIGN > JOINTS.
        Translation   3 = 100.
        Rotation   2 = 1000.
        Click OK.

Loads: From the main menu select:
DEFINE>STATIC LOAD CASES.
Change DEAD to LIVE.
Set Self-Weight Multiplier = 0.
Click Change Load.
Click OK.

Click on Frame 1.
From the main menu select:
ASSIGN>FRAME STATIC LOAD>POINT LOAD AND UNIFORM
LOADS
Warning: Be certain to zero out the values of previous entries.
Check Forces Z direction.
Uniform Load = −1.2.
Click OK.

Click on Frame 2.
From the main menu select:
ASSIGN>FRAME STATIC LOADS>POINT AND UNIFORM LOADS.
Warning: Be certain to zero out the values of previous entries.
Check Forces in Z Direction.
Enter Distance 0.5.
Load −8.0.
Click OK.

Options: From the main menu select:
ANALYZE>SET OPTIONS.
Check Available Degrees of Freedom: UZ, RY.
Click OK.

Analyze: From the main menu select:
ANALYZE>RUN.
Enter Filename "NEX 3.6".
Click SAVE.
At the conclusion of the solution process, click OK.

Print Tables: From the main menu select:
FILE>PRINT INPUT TABLES.
Check: Joint Data: Coordinates
Element data: Frames
Static Loads: Joints, Frames
Click OK.
(Table 3.10[5] reproduces the edited Input Data for Illustrative Example 3.6.)

---

[5] This table has been edited to conform to the formatting requirements of this text.

Table 3.10 Edited Input Data for Illustrative Example 3.6 (Units: kips, inches)

JOINT DATA

JOINT	GLOBAL-X	GLOBAL-Y	GLOBAL-Z	RESTRAINTS
1	0.00000	0.00000	0.00000	1 1 1 1 1 1
2	100.00000	0.00000	0.00000	0 0 0 0 0 0
3	200.00000	0.00000	0.00000	1 1 1 1 1 1

JOINT SPRING DATA

JOINT	K-U1	K-U2	K-U3	K-R1	K-R2	K-R3
2	0.000	0.000	100.000	0.000	1000.000	0.000

FRAME ELEMENT DATA

FRAME	JNT-1	JNT-2	SECTION	RELEASES	SEGMENTS	LENGTH
1	1	2	FSEC2	0.000	4	100.000
2	2	3	FSEC2	0.000	4	100.000

FRAME SPAN DISTRIBUTED LOADS Load Case LOAD1

FRAME	TYPE	DIRECTION	DISTANCE-A	VALUE-A	DISTANCE-B	VALUE-B
1	FORCE	GLOBAL-Z	0.0000	−1.2000	1.0000	−1.2000

FRAME SPAN POINT LOADS
Load Case LOAD1

FRAME	TYPE	DIRECTION	DISTANCE	VALUE
2	FORCE	GLOBAL-Z	0.5000	−8.0000

Enter: From the main menu select:
FILE>PRINT OUTPUT TABLE.
  Check: Displacements
      Reactions
      Frame Forces
      Spring Forces
  Click OK.

(Table 3.11 reproduces the edited Output Tables for Illustrative Example 3.6.)

Table 3.11[6] Edited Output Tables for Illustrative Example 3.6 (Units: kips, inches)

JOINT DISPLACEMENTS

JOINT	LOAD	UX	UY	UZ	RX	RY	RZ
1	LOAD1	0.0000	0.0000	0.0000	0.0000	0.0000	0.0000
2	LOAD1	0.0000	0.0000	-0.5161	0.0000	-0.0111	0.0000

JOINT REACTIONS

JOINT	LOAD	F1	F2	F3	M1	M2	M3
1	LOAD1	0.0000	0.0000	72.8602	0.0000	-1531.8997	0.0000
3	LOAD1	0.0000	0.0000	3.5269	0.0000	187.4552	0.0000

FRAME ELEMENT FORCES

FRAME	LOAD	LOC	P	V2	V3	T	M2	M3
1	LOAD1							
		0.00	0.00	-72.86	0.00	0.00	0.00	-1531.90
		25.00	0.00	-42.86	0.00	0.00	0.00	-85.39
		50.00	0.00	-12.86	0.00	0.00	0.00	611.11
		75.00	0.00	17.14	0.00	0.00	0.00	557.62
		100.00	0.00	47.14	0.00	0.00	0.00	-245.88
2	LOAD1							
		0.00	0.00	-4.47	0.00	0.00	0.00	-234.77
		25.00	0.00	-4.47	0.00	0.00	0.00	122.94
		50.00	0.00	3.53	0.00	0.00	0.00	-11.11
		75.00	0.00	3.53	0.00	0.00	0.00	-99.28
		100.00	0.00	3.53	0.00	0.00	0.00	-187.46

JOINT SPRING FORCES

JOINT	LOAD	F1	F2	F3	M1	M2	M3
2	LOAD1	0.0000	0.0000	51.6129	0.0000	11.1111	0.0000

Deformed Shape: From the main menu select:
DISPLAY>SHOW DEFORMED SHAPE.
Check the Wire Shadow option.
Click OK.

From the main menu select:
FILE>PRINT GRAPHICS.

(The deformed shape is reproduced in Figure 3.17.)

---

[6] This table has been edited to conform to the formatting requirements of this text.

Fig. 3.17   Deformed shape of the beam of Illustrative Example 3.6

Shear Force Diagram: From the main menu select:
      DISPLAY>SHOW ELEMENT FORCES/STRESSES>FRAMES.
         Check the Shear 2-2 option.
         Click OK.
      (The shear force diagram is reproduced in Figure 3.18.)

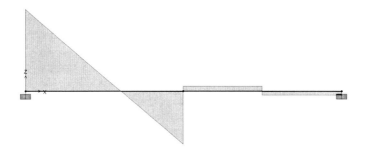

Fig. 3.18   Shear force diagram of the beam in Illustrative Example 3.6

Bending Moment Diagram: From the main menu select:
      DISPLAY>SHOW ELEMENT FORCES/STRESSES>FRAMES.
         Check the Moment 3-3 option.
         Click OK.
      (The Bending Moment diagram is reproduced in Figure 3.19.)

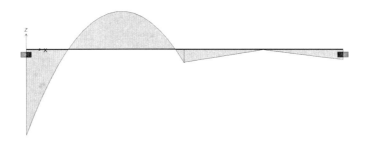

Fig. 3.19 Bending moment diagram of the beam in Illustrative Example 3.6

3.3   Applications in Structural Dynamics

Illustrative Example 3.7

For the cantilever beam shown in Figure 3.20, use SAP2000 to determine: (a) Natural frequencies and mode shapes; and (b) the response to the concentrated force of 100,000 N suddenly applied at the free end of the beam for 0.5 sec. Use time step of integration $\Delta t = 0.01\,(sec)$. (Neglect damping.)

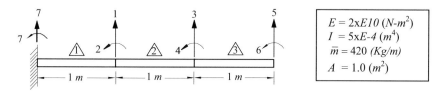

Fig. 3.20 Cantilever beam modeled for Illustrative Example 3.7

Solution:

Analytical Model: This beam is divided into three elements of 1.0 m length, as shown in Figure 3.20. The following commands are implemented in SAP2000.

Begin: Open SAP2000 and select "N-m" from the drop down menu in the lower right-hand corner of the window.
From the main menu select:
FILE>NEW MODEL FROM TEMPLATE.
Click the Beam button.
　　Enter: Number of Spans = 3.
　　　　　　Span length = 1.
Check Restraints and Gridlines.
Click OK.

Maximize the X-Z Plane window and center the drawing in the screen using the PAN icon.

Labels: From the main menu select:
VIEW>SET ELEMENTS.
　　Under Joints, check Labels and Restraints.
　　Under Frames, check Labels.
　　Accept all other default selections.
　　Click OK.
Use the PAN icon to center the drawing once again.

Restraints: Select Joints 1 by clicking on it and then from the main menu select:
ASSIGN>JOINT>RESTRAINTS.
Restrain in all directions.
Click OK.

Select Joints 2, 3 and 4 by clicking on them.
From the main menu select:
ASSIGN>JOINT>RESTRAINTS.
Select no restraints in any direction (free).
Click OK.

Materials: From the main menu select:
DEFINE>MATERIALS.
Under Materials, select STEEL.
Click Modify/Show Material.
Material Name = STEEL
Mass per Unit Volume = 420
Weight per Unit Volume = 0
Modulus of Elasticity = $2.0E10$
Accept all other default entries.
Click OK, OK.

Define: From the main menu select:
DEFINE>FRAME SECTIONS.
Click on the second drop-down menu, Add/Wide Flange.
Select ADD GENERAL (section).
Enter: Moment of Inertia about 3 axis = $5.0E-4$.
Shear Area in 2 Direction = 0.
Accept all other default entries.
Click OK.

In the General Section window, change the Section Name to GENERAL.
Under Material, use the drop-down menu to select STEEL.
Click OK, OK.

Assign: Select frame elements 1, 2 and 3 by clicking on them and from the main menu select:
ASSIGN>FRAME>SECTIONS.
In Frame Sections, select GENERAL.
Click OK.

Define: From the main menu select:
DEFINE>STATIC LOAD CASES
Load = LOAD1

Select Type = LIVE
Set Self-Weight Multiplier = 0
Click on "Change Load".
Click OK.

Time History Function: From the main menu select:
DEFINE>TIME HISTORY FUNCTIONS.
Click Add New Function.
Enter the following values and click "add" after each row is entered.

Time	Value
0	−100000
0.5	−100000

Click OK, OK.

Time History Cases: From the main menu select:
DEFINE>TIME HISTORY CASES.
Click to Add New History.
Accept HIST1 as history case name.
Enter: Analysis Type = Linear
Number of Time Steps = 50
Output Time Step Size = 0.01
Click on envelopes
Under Load Assignments
Load = LOAD1
Function = FUNC1
Scale Factor = 1
Arrival Time = 0
Then Click "ADD".
Click OK, OK.

Joint Load: Mark Joint 4 by clicking on it, and from the main menu select:
ASSIGN>JOINT STATIC LOADS>FORCES.
Load Case Name = LOAD1.
Force Global Z = 1, all others 0.
Click Add to existing loads.
Click OK.

Set Options: From the main menu select:
ANALYZE>SET OPTIONS.
Under Available DOFs (Degrees of Freedom), check only UZ and RY.
Click on "Dynamic Analysis" and on "Set Dynamic Parameters".
Enter Number of Modes = 3
Click OK, OK.

Analyze: From the main menu select:
  ANALYZE>RUN.
  Enter the filename NEXD 3.7.
  Click SAVE.
  When analysis is completed, check for errors and then click OK.
  Select X-Z plane to view deformed shape of beam.

Print Input Tables: From the main menu select:
  FILE>PRINT INPUT TABLES.
  Click "Print to File" and accept all other default values.
  Click OK.

(The edited Input Tables for Illustrative Example 3.7 are in Table 3.12.)

Table 3.12 Edited Input Data for Illustrative Example 3.7

STATIC LOAD CASES

STATIC CASE	CASE TYPE	SELF WT FACTOR
LOAD1	LIVE	0.0000

TIME HISTORY CASES

HISTORY CASE	HISTORY TYPE	NUMBER OF TIME STEPS	TIME STEP INCREMENT
HIST1	LINEAR	100	0.01000

JOINT DATA

JOINT	GLOBAL-X	GLOBAL-Y	GLOBAL-Z	RESTRAINTS
1	−1.50000	0.00000	0.00000	1 1 1 1 1 1
2	−0.50000	0.00000	0.00000	0 0 0 0 0 0
3	0.50000	0.00000	0.00000	0 0 0 0 0 0
4	1.50000	0.00000	0.00000	0 0 0 0 0 0

FRAME ELEMENT DATA

FRAME	JNT-1	JNT-2	SECTION	SEGMENTS	R1	R2	LENGTH
1	1	2	GENERAL	4	0.000	0.000	1.000
2	2	3	GENERAL	4	0.000	0.000	1.000
3	3	4	GENERAL	4	0.000	0.000	1.000

JOINT FORCES Load Case LOAD1

JOINT	GLOBAL-X	GLOBAL-Y	GLOBAL-Z	GLOBAL-XX	GLOBAL-YY	GLOBAL-ZZ
4	0.000	0.000	−1.000	0.000	0.000	0.000

Output Table Mode: From the main menu select:
  DISPLAY>OUTPUT TABLE MODE.
  Select LOAD1 Load Case and HIST1 History (hold down the CTRL key
  down to select both) then click OK.

a) Natural Periods and Frequencies:
Using Notepad or other word editor and open the file:
C:\SAP2000e\NEXD 3.7.OUT
The Modal Period and Frequency table may be found in this file.

(Table 3.13 provides the edited table of the Natural Period and Frequency values for Illustrative Example 3.7.)

Table 3.13 Natural Periods and Frequencies for Illustrative Example 3.7

MODAL PERIODS AND FREQUENCIES

MODE	PERIOD (SEC)	FREQUENCY (CYC/SEC)	FREQUENCY (RAD/TIME)	EIGENVALUE $(RAD/SEC)^2$
1	0.109537	9.129307	57.361125	3290.299
2	0.019405	51.533663	323.795552	104843.559
3	0.007793	128.325467	806.292689	650107.900

b) Mode Shapes:
Mode Shape 1:
From the main menu select:
DISPLAY>SHOW MODE SHAPE.
Mode Shape 1: Click the Wire Shadow option and the Cubic Curve
                           option.
Scale Factor = 1.
Click OK.

From the main menu select:
FILE>PRINT GRAPHICS.

(Figure 3.21 reproduces the plot for Mode Shape 1 for Illustrative Example 3.7.)

Fig. 3.21 *Mode shape 1 for Illustrative Example 3.7

*Mode Shape 2:
Click on the directional arrow in the lower right-hand corner of the screen to select Mode 2.
Scale = 1.
Click OK.

(Figure 3.22 reproduces plot of *Mode Shape 2 for Illustrative Example 3.7.)

Fig. 3.22 *Mode Shape 2 for Illustrative Example 3.7

Mode Shape 3:
Click on the directional arrow in the lower right-hand corner of the screen to select Mode 3
Scale Factor = 1.
Click OK.

(Figure 3.23 reproduces the plot of *Mode Shape 3 for Illustrative Example 3.7.)

Fig. 3.23 *Mode Shape 3 for Illustrative Example 3.7

*Note: Numerical values for a mode shape in the Z direction and rotation around the Y axis may be viewed on screen by right clicking on a node in the modal shape.

c)   **Plot Displacement Function:**
From the main menu select:
DISPLAY>SHOW TIME HISTORY TRACES.
Click Define Functions.
Click to "Add Joint Displs.-Forces".
     Joint ID = 4.
     Vector Type = Displacement.
     Component = UZ.
     Mode Number = Include All.
Click OK, OK.
Under List of Functions click on Joint 4.
Click ADD to move Joint 4 to the column "Plot Functions".
Axis Labels
     Horizontal = Time (*sec*)
     Vertical = Displacement Joint 4 (*m*)
Click on "Display" to view the Displacement function
     Max = 0
     Min = −1783 *m*

FILE>PRINT GRAPHICS to obtain a printed version of the plot.
(Figure 3.24 shows the plot of the displacement function of Joint 4.)

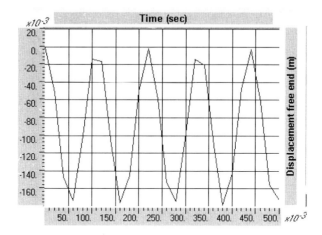

Fig. 3.24 Displacement function of joint 4 of Illustrative Example 3.7

Alternatively, a table of the displacement function may be obtained by
selecting:
FILE>PRINT TABLES.

d) Frame Element Force Function:
From the main menu select:
DISPLAY>SHOW TIME HISTORY TRACES.
Click "Define Functions".
Click to "Add Frame Element Forces"
   Element ID = 1
   Station = 1 (left end of the beam)
   Component = Moment 3-3 (reaction)
   Accept all other default settings.
   Click OK, OK.
Click once on "Frame1" and then ADD to move "Frame1" to the
column "Plot Functions"
Highlight "Joint4" and click to remove "Joint4" back to column "List
of Functions"
   Axis labels
   Horizontal = Time (*sec*)
   Vertical = Bending Moment at the support (*N-m*)
FILE>PRINT GRAPHICS
   Max = 0
   Min = $-1.1976E5$ (*N-m*)

Figure 3.25 shows the plot of the bending moment function at the support (reaction).

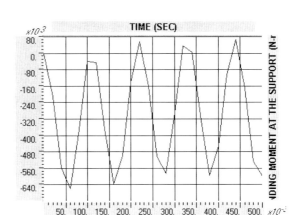

Fig. 3.25 Time-bending moment function at the support of the cantilever beam of Illustrative Example 3.7

Numerical values at any time of the bending moment function may be viewed on screen by moving the cursor along the plot line.

To obtain a numerical table of the function of the Time History Function shown on the screen, from the main menu select:
FILE>PRINT TABLES TO FILE.
The values of this function may also be plotted using Excel or similar software.

Illustrative Example 3.8

For the fixed beam shown in Figure 3.26, use SAP2000 to determine: (a) Natural frequencies and modal shapes, and (b) the response to the concentrated force of 10,000 *lb* suddenly applied at the center of the beam for 0.1 *sec* and removed linearly, as shown in Figure 3.27. Use time step of integration $\Delta t = 0.01\,sec$.

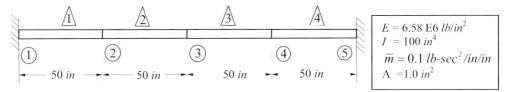

Fig. 3.26 Fixed beam modeled for Illustrative Example 3.8

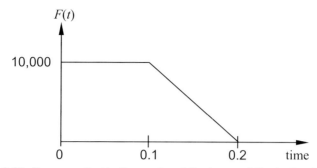

Fig. 3.27 Force applied to the center of the beam of Illustrative Example 3.8

Solution:

Analytical Model: This beam is divided into four segments of 50 *in* length, as shown in Figure 3.26. The following commands are implemented in SAP2000.

Begin: Open SAP2000 and select "lb-in" from the drop-down menu in the lower right-hand corner of the screen.
From the main menu select:
FILE>NEW MODEL FROM TEMPLATE.
Click on the Beam button and enter:
   Number of Spans = 4
   Span length = 50
   Check Restraints and Gridlines.
   Click OK.

Maximize the X-Z Plane window and center the drawing in the screen using the PAN icon.

Labels: From the main menu select:
VIEW>SET ELEMENTS.
Under Joints, check Labels and Restraints.
Under Frames, check Labels.
Accept all other default selections.
Click OK.
Use PAN icon to center drawing once again.

Restraints: Select Joints 1 and 1 by clicking on them and from the main menu select:
ASSIGN>JOINT>RESTRAINTS.
   Restrain in all directions.
   Click OK.

Select joints 2, 3 and 4 by clicking on them and from the main menu select:
ASSIGN>JOINT>RESTRAINTS.
> No restraints in any direction (free).
> Click OK.

Materials: From the main menu select:
DEFINE>MATERIALS.
Under Materials, select OTHER.
Click Modify/Show Material
> Material Name = OTHER
> Mass per Unit Volume = 0.1
> Weight per Unit Volume = 0
> Modulus of Elasticity = 6.58E6
> Accept all other default entries.
Click OK, OK.

Sections: From the main menu select:
DEFINE>FRAME SECTIONS.
Click on the second drop-down menu, Add/Wide Flange and select ADD.
IMPORT GENERAL.
Select Section properties.
Enter: Moment of Inertia about 3 axis = 100
> Shear Area in 2 Direction = 0
> Area = 1.0
> Accept all other default entries.
Click OK.

In the General Section window, change Section Name to GENERAL.
Under Material use drop-down menu to select OTHER.
Click OK, OK.

Assign: Select frame elements 1, 2, 3 and 4 by clicking on them and from the main menu select:
ASSIGN>FRAME>SECTIONS.
In Frame Sections, select GENERAL.
Click OK.

Define: From the main menu select:
DEFINE>STATIC LOAD CASES.
> Load = LOAD1
> Type = LIVE
> Self-Weight Multiplier = 0
Click "Change Load".
Click OK.

Time History Function: From the main menu select:
> DEFINE>TIME HISTORY FUNCTIONS.
> Click on Add New Function.
> Add the following values and click "Add" after each row has been entered.
>
> | Time | Value |
> |------|-------|
> | 0 | −10000 |
> | 0.1 | −10000 |
> | 0.2 | 0 |
> | 0.5 | 0 |
>
> Click OK, OK.

Time History Cases: From the main menu select:
> DEFINE>TIME HISTORY CASES.
> Click Add New History.
>> Accept HIST1 as history case name.
>> Enter: Analysis Type = Linear
>> Number of Time Steps = 50
>> Output Time Step Size = 0.01
>> Click on envelopes
>> Under Load Assignments
>>> Load = LOAD 1
>>> Function = FUNC1
>>> Scale Factor = 1
>>> Arrival Time = 0
>>> Then click "ADD".
> Click OK, OK.

Joint Load: Mark Joint 3 by clicking on it and from the main menu select:
> ASSIGN>JOINT STATIC LOADS>FORCES.
>> Load Case Name = LOAD1.
>> Force Global Z = 1, all others 0.
>> Click Add to existing loads.
> Click OK.

Set Option: From the main menu select:
> ANALYZE>SET OPTIONS.
> Under available DOFs (Degrees of Freedom), check only UZ and RY.
> Click on "Dynamic Analysis" and on "Set Dynamic Parameters".
> Enter: Number of Modes = 3.
> Click OK, OK.

Analyze: From the main menu select:
> ANALYZE>RUN.
> Enter filename EXD 3.8.

        Click SAVE.
        When analysis is completed, check for errors.
        Click OK.

        Select X-Z plane to view the deformed shape of the beam.

Output Table Mode: From the main menu select:
        DISPLAY>OUTPUT TABLE MODE.
        Select LOAD1 Load Case and HIST1 History (hold down CTRL key to
        select both).
        Click OK.

Print Input Tables: From the main menu select:
        FILE>PRINT INPUT TABLES.
        CLICK "Print to File" and accept all other default values.
        Click OK.
        Use an editor such as Notepad to retrieve and edit the Input Tables.

(The edited Input Tables of Illustrative Example 3.8 are in Table 3.14.)

Table 3.14 Edited Input Tables for Illustrative Example 3.8 (Units: lb, in, sec)

--STATIC LOAD CASES

STATIC	CASE	SELF-WT
CASE	TYPE	FACTOR
LOAD1	LIVE	0.0000

TIME HISTORY CASES

HISTORY	HISTORY	NUMBER OF	TIME STEP
CASE	TYPE	TIME STEPS	INCREMENT
HIST1	LINEAR	50	0.01000

JOINT DATA

JOINT	GLOBAL-X	GLOBAL-Y	GLOBAL-Z	RESTRAINTS
1	-100.00000	0.00000	0.00000	1 1 1 1 1 1
2	- 50.00000	0.00000	0.00000	0 0 0 0 0 0
3	0.00000	0.00000	0.00000	0 0 0 0 0 0
4	50.00000	0.00000	0.00000	0 0 0 0 0 0
5	100.00000	0.00000	0.00000	1 1 1 1 1 1

FRAME ELEMENT DATA

FRAME	JNT-1	JNT-2	SECTION	ANG	RELS	SEGMT	R1	R2	LENGTH
1	1	2	GENERAL	0.000	0000000	4	0.000	0.000	50.000
2	2	3	GENERAL	0.000	000000	4	0.000	0.000	50.000
3	3	4	GENERAL	0.000	000000	4	0.000	0.000	50.000
4	4	5	GENERAL	0.000	000000	4	0.000	0.000	50.000

Table 3.14 (continued)

J O I N T   F O R C E S   Load Case LOAD1

JOINT	GLOBAL-X	GLOBAL-Y	GLOBAL-Z	GLOBAL-XX	GLOBAL-YY	GLOBAL-ZZ
3	0.000	0.000	1.000	0.000	0.000	0.000

Plot Displacement Function: From the main menu select:
    DISPLAY>SHOW TIME HISTORY TRACES.
    Click Define Functions.
    Click to "Add Joint Disps/Forces".
        Joint ID = 3.
        Vector Type = Displacement.
        Component = UZ.
        Mode Number = Include All.
    Click OK, OK.

    Under List of Functions, click on Joint 3.
    Click ADD to move Joint3 to the Plot Functions column.
    Axis Labels
        Horizontal = Time (*sec*)
        Vertical = Displacement Joint 3 (*in*)

    Click "Display" to view the Displacement function
        Max = 1.241(*in*)
        Min = −0.45 (*in*)

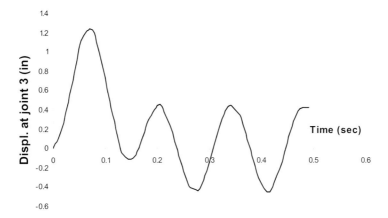

Fig. 3.28 Displacement function at joint 3 of Illustrative Example 3.8

Note: Displacement values may be viewed on the screen by moving the cursor along the curve shown on the screen.

Select FILE>PRINT GRAPHICS to obtain a printed version of the plot.

Alternatively, a table of displacement function may be obtained by selecting:
FILE>PRINT TABLES TO FILE.
The values in this table may be plotted using Excel or similar software.

Plot Frame Element Force Function: From the main menu select:
  DISPLAY>SHOW TIME HISTORY TRACES.
  Click "Define Functions".
  Click to "Add Frame Element Forces".
    Element ID = 2
    Station = 1
    Component = Moment 3-3
    Accept all other default settings.
    Click OK, OK.

Click once on "Frame2" and then ADD to move "Frame2" to the Plot Function column.
Highlight "Joint3" and click Remove to move "Joint3" back to List of Functions column.
Axis Labels
    Horizontal = Time (*sec*).
    Vertical = Bending Moment at left end of Element 2 *lb·in*.
Click Display to view the function plot.
    Max =   7.284E4 *lb·in*
    Min = −7.115E4 *lb·in*
FILE>PRINT GRAPHICS.
Click DONE.
(Figure 3.29 shows a plot of the bending moment time function for Element 3 of Illustrative Example 3.8.)

To obtain a table of the Time History Function, select
FILE>PRINT TABLES TO FILE.
Alternatively, the values of this function may be plotted using Excel or similar software.

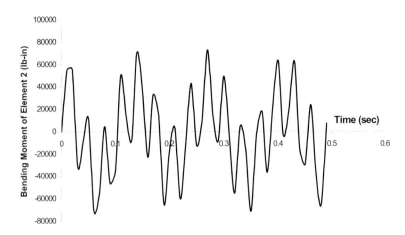

Fig. 3.29 Bending moment function left end Element 2 of Illustrative Example 3.8

Natural Periods and Frequencies:
Using Notepad or other word editor and open the file:
C:/SAP2000/EXD 3.8.OUT
The Modal Period and Frequency table may be found in this file.

(Table 3.15 provides the edited table of the Natural Period and Frequency values for Illustrative Example 3.8.)

Table 3.15 Natural Periods and Frequencies for Illustrative Example 3.8

MODAL PERIODS AND FREQUENCIES

MODE	PERIOD (SEC)	FREQUENCY (CYC/SEC)	FREQUENCY (RAD/SEC)	EIGENVALUE $(RAD/SEC)^2$
1	0.138924	7.198200	45.227623	2045.538
2	0.052290	19.124041	120.159893	14438.400
3	0.031811	31.436088	197.518764	39013.662

[2]Note: Numerical values at any time of the bending moment time function may be viewed on screen by moving the cursor along the plot line shown on the screen.

Mode Shapes: From the main menu select:
DISPLAY>SHOW MODE SHAPE
Modal Shape 1: Click the "Wire Shadow" option and the "Cubic Curve" option.
Scale Factor = 1.
Click OK.

FILE>PRINT GRAPHICS

(Figure 3.30 reproduces the plot of Modal Shape 1 for Illustrative Example 3.8.)

Mode 1*

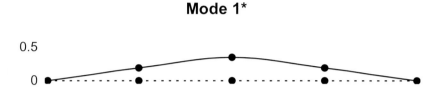

Fig. 3.30 Modal Shape 1 of Illustrative Example 3.8

Mode Shape 2: Click on the directional arrow in the lower right-hand corner of the screen to select Mode 2.
Scale Factor = 1.
Click OK.

(Figure 3.31 reproduces plot of Modal Shape 2 for Illustrative Example 3.8.)

Mode 2*

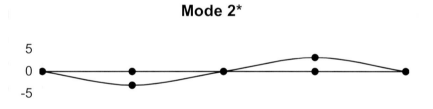

Fig. 3.31 Modal Shape 2 of Illustrative Example 3.8

Modal Shape 3: Click on the directional arrow in the lower right-hand corner of the screen to select Mode 3.
Scale Factor = 1.
Click OK.

(Figure 3.32 reproduces plot of Modal Shape 3 for Illustrative Example 3.8.)

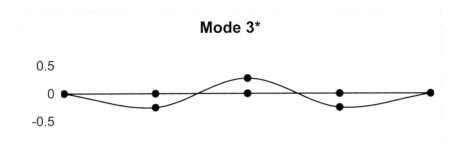

Fig. 3.32 Modal Shape 3 of Illustrative Example 3.8

* Numerical values for a mode shape in the Z direction and rotation around the Y axis may be viewed on screen by right clicking on a node in the modal shape.

3.4   Problems in Structural Analysis

Use SAP2000 to solve the problems in Chapter 1, repeated here for convenience.

Problem 3.1
For the beam shown in Figure P3.1, determine:

(a.) Displacements at nodes 1 and 2
(b.) Reactions at the support

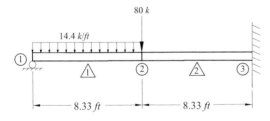

Fig. P3.1

Problem 3.2
For the beam shown in Figure P3.2, determine:

(a.) Displacements at node 2
(b.) End forces on the beam elements
(c.) Reactions at the supports

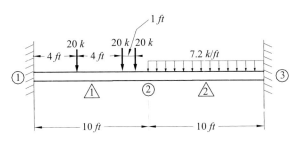

Fig P3.2

Problem 3.3

For the beam shown in Figure P3.3, determine:

(a.) Displacement at nodes 2 and 3
(b.) End forces on the beam elements
(c.) Reactions at the supports

$$E = 200 \ GP_a$$
$$I = 40,000 \ cm^4$$

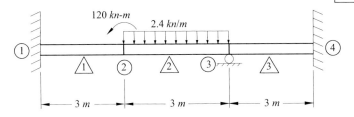

Fig. P3.3

Problem 3.4

For the beam shown in Figure P3.4, determine:

(a.) Displacement at node 2
(b.) End forces on the beam elements
(c.) Reactions at the supports

$$E = 4500 \ ksi$$
$$I = 1200 \ in^4$$

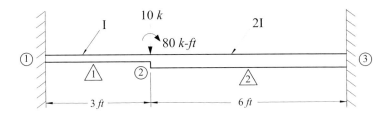

Fig. P3.4

Problem 3.5

For the beam shown in Figure P3.5, determine:

(a.) Displacements at nodes 2 and 3
(b.) End forces on the beam elements
(c.) Reactions at the supports.

$$E = 200 \ GP_a$$
$$I = 40,000 \ cm^4$$

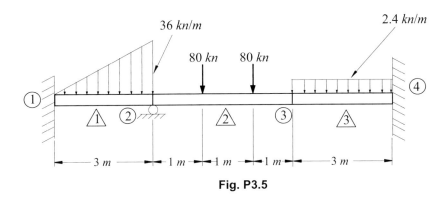

Fig. P3.5

Problem 3.6

For the beam shown in Figure P3.6, determine:

(a.) Displacements at nodes 1, 2, and 3
(b.) End forces on the beam elements
(c.) Reactions at the supports.

$$E = 30,000 \ ksi$$
$$I = 120 \ in^4$$

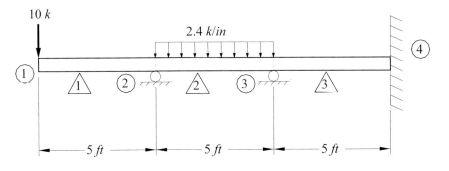

Fig. P3.6

Problem 3.7
Solve Problem 3.1 assuming that a hinge exists at the right end of beam element 1.

Problem 3.8
Solve Problem 3.2 assuming a hinge at the left end of beam element 2.

Problem 3.9
Solve Problem 3.3 assuming a hinge at the right end of beam element 2.

Problem 3.10
Solve Problem 3.4 assuming a hinge at node 2 (right end of beam element 1).

Problem 3.11
Solve Problem 3.5 assuming a hinge at the left end of beam element 2.

Problem 3.12
Solve Problem 3.6 assuming hinges at the two ends of beam element 2.

Problem 3.13
Solve Problem 3.1 assuming that the support at node 2 underwent a downward displacement of 1.0 *in* and rotational displacement of 5 degrees in the clockwise direction.

Problem 3.14
Solve Problem 3.5 assuming that the support at node 2 underwent a downward displacement of 2.0 *in*.

Problem 3.15
Solve Problem 3.6 assuming that the support at node 3 underwent a downward displacement of 0.5 *in* and clockwise rotation of 10 degrees.

Problem 3.16
Solve Problem 3.5 assuming that the temperature at the top side of beam element 2 is $30°F$ and at the bottom side is $80°F$.

Problem 3.17
Solve Problem 3.6 knowing that the temperature along the top of the beam is $40°F$ and at its bottom side is $100°F$.

Problem 3.18
For the beam shown in Figure P3.18 determine:

(a.) Displacements at nodes 1 and 2
(b.) The force and the moment of the spring at node 2

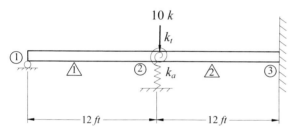

Fig. P3.18

Problem 3.19
For the beam shown in Figure P3.19 determine:

(a.) Displacements at nodes 2 and 3
(b.) Force in the spring at node 2
(c.) Reactions at the supports

$E = 30,000\ ksi$
$I = 400\ in^4$
$k_a = 10\ kip/in$
$L = 100\ in$

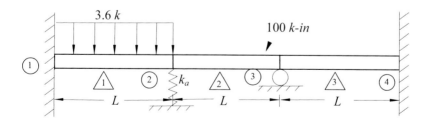

Fig. P3.19

3.5   Problems in Structural Dynamics

Problem 3.20

A uniform beam shown in Figure 3.20(a) of flexural stiffness $EI = 10^9$ lb in^2 and length 300 in supporting an impulsive load depicted in Figure 3.20(b) applied at nodal coordinate 5. The beam has one end fixed and the other simply supported. Model this beam with six elements (use the command "Divide Frames") and determine:

(a.) Lowest six natural frequencies
(b.) Plot the lowest three mode shapes
(c.) Time-response plot at nodal coordinate 5
(d.) Time-bending moment plot at the center between nodal coordinates 4 and 5
    shown in Figure 3.20(a)

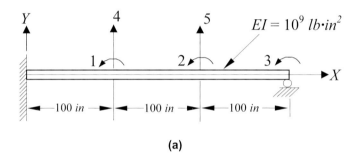

(a)

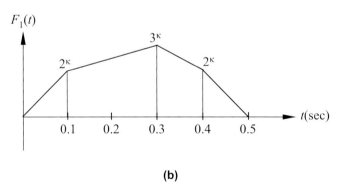

(b)

Fig. P3.20

Problem 3.21

The beam shown in Figure P3.21 is subjected to a force $F = 10$ kip suddenly applied at the node 2 as shown in the figure.

Determine

(a.) Lowest three natural frequencies
(b.) Plot the lowest three mode shapes
(c.) Time-displacement plot at the center of the beam
(d.) Time-bending moment plot at the center of the beam

$$E = 29,000 \ ksi$$
$$I = 404 \ in^4$$

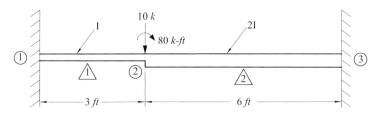

Fig. P3.21

Problem 3.22
Solve Problem 3.21 for a harmonic force $F(t) = F_0 \sin \bar{\omega} t \ kip$ at node 2 in which $F_0 = 5.0 \ kip$ and select $\bar{\omega}$ with a value 5% greater than the lowest natural frequency.

Problem 3.23
Determine the steady-state response for the beam shown in Figure P3.23 which is acted upon by a harmonic force $F(t) = 5000 \sin 30t \ lb$ applied as shown in the figure. Assume 5% damping in all the modes.

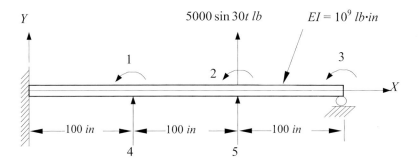

Fig. P3.23

4 Plane Frames

4.1   Introduction

Structural analysis using the matrix stiffness method for structures modeled as beams has been presented in the preceding three chapters. This method of analysis when applied to structures modeled as *plane frames* requires the inclusion of the axial effect in the element stiffness matrix; hence, the inclusion of axial effect in the system stiffness matrix. It also requires a coordinate transformation of element end forces and displacements from *element or local coordinate axes* to the *system or global coordinate axes*. Except for the consideration of the axial effect and the need to transform the element end forces and displacements, the matrix stiffness method when applied to plane frames is identical to the analysis of beams presented in the preceding chapters.

4.2   Stiffness Coefficients for Axial Forces

The inclusion of axial forces and axial deformations in the stiffness matrix of a flexural beam element requires the determination of the stiffness coefficients for axial loads. To derive the stiffness matrix for an axially loaded member, consider in Figure 4.1 a beam element acted on by the axial forces P_1 and P_2, producing axial displacements δ_1 and δ_2 at the nodes of the element. For a prismatic and uniform beam segment of length L and cross-sectional A, it is relatively simple to obtain the stiffness coefficients for axial effects by applying Hooke's law. In relation to the beam shown in Figure 4.1, the displacements δ_1 produced by the force P_1 acting at node 1 while node 2 is maintained fixed ($\delta_2 = 0$) is given by

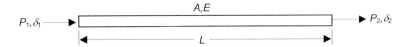

Fig. 4.1   Beam element with nodal axial loads P_1 and P_2, and corresponding nodal displacements δ_1 and δ_2

$$\delta_1 = \frac{P_1 L}{AE}. \tag{4.1}$$

From eq. (4.1) and the definition of the stiffness coefficient k_{11} (force at node 1 to produce a unit displacement $\delta_1 = 1.0$), we obtain

$$k_{11} = \frac{P_1}{\delta_1} = \frac{AE}{L}. \tag{4.2a}$$

The equilibrium of the beam element acted upon by the force k_{11} requires an opposite force k_{21} at the other end, namely

$$k_{21} = k_{11} = -\frac{AE}{L}. \tag{4.2b}$$

Analogously, the other stiffness coefficients are

$$k_{22} = \frac{AE}{L} \tag{4.2c}$$

and

$$k_{12} = -\frac{AE}{L}. \tag{4.2d}$$

The stiffness coefficients given by eqs. (4.2) are conveniently arranged in the stiffness matrix relating axial forces $\{P\}$ and displacements $\{\delta\}$ for a uniform prismatic beam element, namely

$$\begin{Bmatrix} P_1 \\ P_2 \end{Bmatrix} = \frac{AE}{L} \begin{bmatrix} 1 & -1 \\ -1 & 1 \end{bmatrix} \begin{Bmatrix} \delta_1 \\ \delta_2 \end{Bmatrix} \tag{4.3}$$

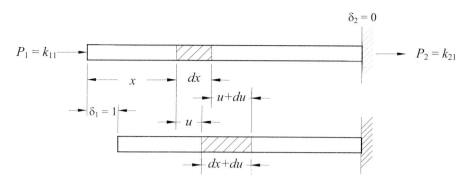

Fig. 4.2 Beam element showing deformation due to a unit displacement $\delta_1 = 1$

4.3   Displacement Functions for an Axially Loaded Beam.

Consider in Figure 4.2 a beam element undergoing a unit displacement $\delta_1 = 1$ at its left end. If $u = u(x)$ is the displacement at section x, the displacement at section $x + dx$ will be $u + du$. The element dx in this new position has changed in length by the amount du, and thus, the strain is du/dx. By Hooke's law, the strain is equal to the stress (P/A) divided by the modulus of elasticity E, namely

$$\frac{du(x)}{dx} = \frac{P}{AE}. \tag{4.4}$$

Integration of eq. (4.4) with respect to x yields

$$u(x) = \frac{P}{AE}x + C$$

in which C is a constant of integration. Introducing the boundary conditions $u = 1$ at $x = 0$ and $u = 0$ at $x = L$, we obtain the displacement function $u_1 = u_1(x)$ corresponding to $\delta_1 = 1$ as

$$u_1(x) = 1 - \frac{x}{L}. \tag{4.5}$$

Analogously, the displacement function $u_2(x)$ corresponding to a unit displacement $\delta_2 = 1$ is

$$u_2(x) = \frac{x}{L}. \tag{4.6}$$

The general expression for the stiffness coefficient k_{ij} for axial effect obtained by application of the Principle for Virtual Work is (see Analytical Problem 4.1, Section 4.15):

$$k_{ij} = \int_0^L AE\, u_i'(x)\ u_j'(x)\, dx \tag{4.7}$$

where u_i' or u_j' is the derivative with respect to x of the displacement function $u_1(x)$ or $u_2(x)$ given by eq. (4.5) or (4.6).

Using eq. (4.7) the reader may verify the results obtained in eqs. (4.2) for a uniform beam element. However, eq. (4.7) could as well be used for non-uniform elements in which, in general, AE would be a function of x. In practice, the same displacement functions $u_1(x)$ and $u_2(x)$ obtained for a uniform beam element also are used in eq. (4.7) for a non-uniform element.

4.4   Element Stiffness Matrix for Plane Frame Elements

The stiffness matrix corresponding to the local coordinates for an element of a plane frame (Figure 4.3) is obtained by combining in a single matrix the stiffness matrix for axial effects [eq. (4.3)] and the stiffness matrix for flexural effects [eq. (1.11)]. The resulting matrix relates the forces P_i and the displacements δ_i at the nodal coordinates indicated in Fig 4.3 as

$$\begin{Bmatrix} P_1 \\ P_2 \\ P_3 \\ P_4 \\ P_5 \\ P_6 \end{Bmatrix} = \frac{EI}{L^3} \begin{bmatrix} AL^2/I & 0 & 0 & -AL^2/I & 0 & 0 \\ 0 & 12 & 6L & 0 & -12 & 6L \\ 0 & 6L & 4L^2 & 0 & -6L & 2L^2 \\ -AL^2/I & 0 & 0 & AL^2/I & 0 & 0 \\ 0 & -12 & -6L & 0 & 12 & -6L \\ 0 & 6L & 2L^2 & 0 & -6L & 4L^2 \end{bmatrix} \begin{Bmatrix} \delta_1 \\ \delta_2 \\ \delta_3 \\ \delta_4 \\ \delta_5 \\ \delta_6 \end{Bmatrix} \tag{4.8}$$

or, in concise notation,

$$\{P\} = [k]\{\delta\} \tag{4.9}$$

in which $[k]$ is the stiffness matrix in reference to local coordinate axes of an element of a plane frame and $\{P\}$ and $\{\delta\}$ are, respectively, the nodal force and the nodal displacement vectors of the element.

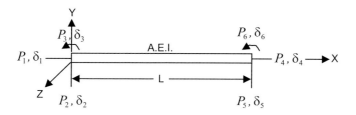

Fig. 4.3   Beam element showing flexural and axial nodal forces and displacements

4.5   Coordinate Transformation

The stiffness matrix for the beam element of a plane frame in eq. (4.8) refers to local coordinates defined by coordinate axes fixed on the beam element. These axes (x, y, and z) are called *local* or *element axes*, while the coordinate axes (X, Y, Z) for the whole structure are known as *global* or *system axes*. Figure 4.4 shows a beam element of a plane frame with nodal forces (P_1, P_2,...P_6) in reference to the local coordinate axes x, y, z, and ($\bar{P}_1, \bar{P}_2,...\bar{P}_6$), referring to the global coordinate axes X, Y, Z. The objective is to transform the element stiffness matrix from the reference of local coordinate axes to the global coordinate axes. This transformation is required in order that the matrices for all the elements refer to the same set of coordinates; hence, the matrices become compatible for assemblage into the system matrix for the structure. We begin by expressing the forces (P_1, P_2, and $P3$) in terms of the forces ($\bar{P}_1, \bar{P}_2, \bar{P}_3$). Since these two sets of forces are equivalent, we obtain from Figure 4.4 the following relationships:

$$P_1 = \bar{P}_1\cos\theta + \bar{P}_2\sin\theta$$
$$P_2 = -\bar{P}_1\sin\theta + \bar{P}_2\cos\theta \qquad (4.10)$$
$$P_3 = \bar{P}_3$$

Analogously, we obtain for the forces P_4, P_5 and P_6 on the other node, the relationships:

$$P_4 = \bar{P}_4\cos\theta + \bar{P}_5\sin\theta$$
$$P_5 = -\bar{P}_4\sin\theta + \bar{P}_5\cos\theta \qquad (4.11)$$
$$P_6 = \bar{P}_6$$

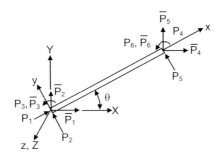

Fig. 4.4   Beam element showing nodal forces P_i in local (x, y, z) and nodal forces $\overline{P}_i$ in global coordinate axes (X, Y, Z)

Equations (4.10) and (4.11) may conveniently be arranged in matrix form as

$$\begin{Bmatrix} P_1 \\ P_2 \\ P_3 \\ P_4 \\ P_5 \\ P_6 \end{Bmatrix} = \begin{bmatrix} \cos\theta & \sin\theta & 0 & 0 & 0 & 0 \\ -\sin\theta & \cos\theta & 0 & 0 & 0 & 0 \\ 0 & 0 & 1 & 0 & 0 & 0 \\ 0 & 0 & 0 & \cos\theta & \sin\theta & 0 \\ 0 & 0 & 0 & -\sin\theta & \cos\theta & 0 \\ 0 & 0 & 0 & 0 & 0 & 1 \end{bmatrix} \begin{Bmatrix} \overline{P}_1 \\ \overline{P}_2 \\ \overline{P}_3 \\ \overline{P}_4 \\ \overline{P}_5 \\ \overline{P}_6 \end{Bmatrix} \qquad (4.12)$$

or in condensed notation

$$\{P\} = [T]\{\overline{P}\} \qquad (4.13)$$

in which $\{P\}$ and $\{\overline{P}\}$ are the vectors of the element nodal forces, respectively, in local and in global coordinates and $[T]$ is the transformation matrix given by the square matrix in eq. (4.12). Repeating the same procedure, we obtain the relationship between the nodal displacements $(\delta_1, \delta_2, ...\delta_6)$ in the local coordinate system and the nodal displacements $(\overline{\delta}_1, \overline{\delta}_2, ...\overline{\delta}_6)$ in the global coordinate system, namely

$$\begin{Bmatrix} \delta_1 \\ \delta_2 \\ \delta_3 \\ \delta_4 \\ \delta_5 \\ \delta_6 \end{Bmatrix} = \begin{bmatrix} \cos\theta & \sin\theta & 0 & 0 & 0 & 0 \\ -\sin\theta & \cos\theta & 0 & 0 & 0 & 0 \\ 0 & 0 & 1 & 0 & 0 & 0 \\ 0 & 0 & 0 & \cos\theta & \sin\theta & 0 \\ 0 & 0 & 0 & -\sin\theta & \cos\theta & 0 \\ 0 & 0 & 0 & 0 & 0 & 1 \end{bmatrix} \begin{Bmatrix} \overline{\delta}_1 \\ \overline{\delta}_2 \\ \overline{\delta}_3 \\ \overline{\delta}_4 \\ \overline{\delta}_5 \\ \overline{\delta}_6 \end{Bmatrix} \qquad (4.14)$$

or

$$\{\delta\} = [T]\{\overline{\delta}\} \qquad (4.15)$$

Now, the substitution of $\{P\}$ from eq. (4.13) and of $\{\delta\}$ from eq. (4.15) into the stiffness equation referred to local axes, $\{P\} = [k]\{\delta\}$, results in

$$[T]\{\overline{P}\} = [k][T]\{\overline{\delta}\}$$

or

$$\{\overline{P}\} = [T]^{-1}[k][T]\{\overline{\delta}\}$$

where $[T]^{-1}$ is the inverse of matrix $[T]$. The reader may verify that the transformation matrix $[T]$ in eq. (4.12) is an orthogonal matrix, that is, $[T]^{-1} = [T]^{T}$.

Hence:

$$\{\overline{P}\} = [T]^{T}[k][T]\{\overline{\delta}\} \qquad (4.16)$$

or, in a more convenient notation,

$$\{\overline{P}\} = [\overline{k}]\{\overline{\delta}\} \qquad (4.17)$$

in which

$$[\overline{k}] = [T]^{T}[k][T] \qquad (4.18)$$

is the stiffness matrix of an element of a plane frame in reference to the global system of coordinates.

4.6   Equivalent Nodal Forces for an Axially Loaded Beam

The application of the Principle of Virtual Work (see Analytical Problem 4.2, Section 4.15) provides the general expression to calculate the equivalent nodal forces for an axially loaded beam element as

$$Q_i = \int_0^L p(x)\, u_i(x)\, dx \qquad (4.19)$$

in which $p(x)$ is the applied distributed axial force along the beam element and $u_i(x)$ $(i = 1, 2)$ is the displacement function given by eq. (4.5) or (4.6). Appendix I Case (e) or Case (g) provides, respectively, the expression for the equivalent nodal forces for a beam element loaded axially with a concentrated force or with a uniform distributed force.

4.7   Element End Forces

The element end forces in reference to local coordinates are calculated as in the case of beams [eq. (1.23)] by

$$\{P\} = [k]\{\delta\} - \{Q\} \qquad (4.20)$$

in which

$$\{P\} = \text{element end forces}$$

$$[k] = \text{element stiffness matrix}$$

$$\{\delta\} = \text{element nodal displacements}$$

$$\{Q\} = \text{equivalent forces at the nodal coordinates for the loads applied on the beam element.}$$

Illustrative Example 4.1
For the loaded plane frame shown in Figure 4.5 determine:

(a)   Displacements at node 2
(b)   End forces on the elements
(c)   Reactions at the supports

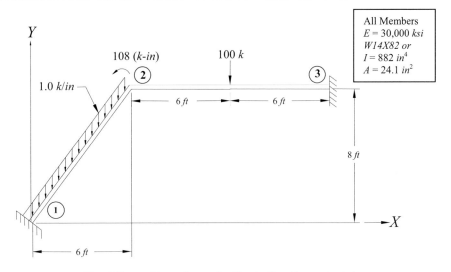

Fig. 4.5   Plane frame for Illustrative Example 4.1

Solution:

1. Modeling the structure.

    The plane frame shown in Figure 4.5 is modeled, as shown in Figure 4.6, into two beam elements, three nodes and nine system nodal coordinates. The first three system nodal coordinates correspond to the free nodal coordinates and the last six to the fixed nodal coordinates, as shown in Figure 4.6.

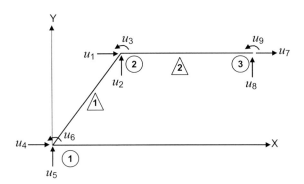

Fig. 4.6   Modeled plane frame for Illustrative Example 4.1 showing the system nodal coordinates u_1 through u_9

2. Element stiffness matrices (local coordinates).

ELEMENT 1

Substituting numerical values into the stiffness matrix in eq. (4.8) results in

$$[k]_1 = \frac{30x10^3 x\,882}{120^3} \begin{bmatrix} 241x120^2/882 & 0 & 0 & -241x120^2/882 & 0 & 0 \\ 0 & 12 & 6\text{ x }120 & 0 & -12 & 6\text{ x }120 \\ 0 & 6\text{ x }120 & 4\text{ x }120^2 & 0 & -6\text{ x }120 & 2\text{ x }120^2 \\ -241x120^2/882 & 0 & 0 & 241x120^2/882 & 0 & 0 \\ 0 & -12 & -6\text{ x }120 & 0 & 12 & -6\text{ x }120 \\ 0 & 6\text{ x }120 & 2\text{ x }120^2 & 0 & -6\text{ x }120 & 4\text{ x }120^2 \end{bmatrix}$$

or

$$[k]_1 = \begin{bmatrix} 6.025E3 & 0 & 0 & -6.025E6 & 0 & 0 \\ 0 & 1.838E2 & 1.103E4 & 0 & -1.838E2 & 1.103E4 \\ 0 & 1.103E4 & 8.820E5 & 0 & -1.103E4 & 4.410E5 \\ -6.025E6 & 0 & 0 & 6.025E3 & 0 & 0 \\ 0 & -1.838E2 & -1.103E4 & 0 & 1.838E2 & -1.103E4 \\ 0 & 1.103E4 & 4.410E5 & 0 & -1.103E4 & 8.820E5 \end{bmatrix} \quad \text{(a)}$$

ELEMENT 2

Analogously,

$$[k]_2 = \begin{array}{c} \\ \\ \\ \\ \\ \\ \end{array} \begin{array}{cccccc} 1 & 2 & 3 & 7 & 8 & 9 \\ \begin{bmatrix} 5.021E3 & 0 & 0 & -5.021E3 & 0 & 0 \\ 0 & 1.063E2 & 7.656E3 & 0 & -1.063E2 & 7.656E3 \\ 0 & 7.656E3 & 7.350E5 & 0 & -7.656E3 & 3.67E5 \\ -5.021E3 & 0 & 0 & 5.021E3 & 0 & 0 \\ 0 & -1.063E2 & -7.656E3 & 0 & 1.063E2 & -7.656E3 \\ 0 & 7.656E3 & 3.675E5 & 0 & -7.656E3 & 7.350E5 \end{bmatrix} & \begin{array}{c} 1 \\ 2 \\ 3 \\ 7 \\ 8 \\ 9 \end{array} \end{array} \quad \text{(b)}$$

3. Transformation matrix:

ELEMENT 1

From eq. (4.12) with $\cos\theta = 0.60$ and $\sin\theta = 0.80$

$$[T]_1 = \begin{bmatrix} 0.60 & 0.80 & 0 & 0 & 0 & 0 \\ -0.80 & 0.60 & 0 & 0 & 0 & 0 \\ 0 & 0 & 1 & 0 & 0 & 0 \\ 0 & 0 & 0 & 0.60 & 0.80 & 0 \\ 0 & 0 & 0 & -0.80 & 0.60 & 0 \\ 0 & 0 & 0 & 0 & 0 & 1 \end{bmatrix} \tag{c}$$

ELEMENT 2

The transformation matrix for element 2 is the unitary matrix $[T]_2 = [I]$ since $\theta_2 = 0$.

4. Element stiffness matrices (global coordinates).

ELEMENT 1

The element stiffness matrix in reference to the global system of coordinates is obtained by substituting into eq. (4.18) the element stiffness matrix and the element transformation matrix, respectively, from eqs. (a) and (c). After performing the multiplication of the matrices as indicated in eq. (4.18), we obtain

$$\left[\bar{k}\right]_1 = \begin{bmatrix} & 4 & 5 & 6 & 1 & 2 & 3 & \\ 2.287E3 & 2.804E3 & -8.820E3 & -2.287E3 & -2.804E3 & -8.820E3 & 4 \\ 2.804E3 & 3.922E3 & 6.615E3 & -2.804E3 & -3.922E3 & 6.665E3 & 5 \\ -8.820E3 & 6.615E3 & 8.820E3 & 8.820E3 & -6.615E3 & 4.410E3 & 6 \\ -2.287E3 & -2.804E3 & 8.820E3 & 2.287E3 & 2.804E3 & 8.820E5 & 1 \\ -2.804E3 & -3.922E3 & -6.615E3 & 2.804E3 & 3.922E3 & -6.615E3 & 2 \\ -8.820E3 & 6.615E3 & 4.410E3 & 8.820E5 & -6.615E3 & 8.820E5 & 3 \end{bmatrix} \tag{d}$$

ELEMENT 2

Since the transformation matrix for this element is the unitary matrix $[I]$, its stiffness matrix in global coordinates is identical to the stiffness matrix in the local coordinate as given by eq. (b).

5. Assemblage of the reduced system matrix.

To assemble the reduced system stiffness matrix for the structure (considering only the free nodal coordinates), we proceed to transfer each coefficient from the two element matrices, eqs. (d) and (b), into the system stiffness matrix. To this objective, we write at the top and on the right side of these two matrices the nodal coordinates assigned for these elements in global nodal coordinates as labeled in Figure 4.6.

Proceeding systematically to transfer the coefficients in the element stiffness matrices, eqs. (d) and (b), according to the rows and columns indicated on the right and at the top of these two matrices results in the reduced system stiffness matrix:

$$[K]_R = \begin{bmatrix} 2.287E3 + 5.021E3 & 2804E3 + 0 & 8.820E5 + 0 \\ 2.804E3 + 0 & 3.922E3 + 1.063E2 & -6.615E3 + 7.656E3 \\ 8.820E3 + 0 & -6.615E3 + 7.656E3 & 8.820E5 + 7.350E5 \end{bmatrix}$$

or

$$[K]_R = \begin{bmatrix} 0.7307E4 & 0.2804E4 & 0.8820E6 \\ 0.2804E4 & 0.4028E4 & 0.1041E4 \\ 0.8820E6 & 0.1041E4 & 0.1617E7 \end{bmatrix} \qquad (e)$$

6. Equivalent element nodal forces (local coordinates).

ELEMENT 1

Figure 4.7 shows the vertically distributed load on element 1 equated to the superposition of the normal and the axial components of this load. This figure also shows the equivalent nodal forces for these two loads. The equivalent nodal forces for element 1 are calculated using the formulas in Appendix I as follows (L = 120 *in*):

$$Q_1 = \frac{-0.8L}{2} = -48 \; kip \qquad Q_2 = \frac{-0.6L}{2} = -36 \; kip \qquad Q_3 = \frac{-0.6L^2}{12} = -720 \; kip{\cdot}in$$

$$Q_4 = \frac{-0.8L}{2} = -48 \; kip \qquad Q_5 = \frac{-0.6L}{2} = -36 \; kip \qquad Q_6 = \frac{0.6L^2}{12} = 720 \; kip{\cdot}in$$

or in vector notation as

$$\{Q\}_1 = \begin{Bmatrix} -48 \\ -36 \\ -720 \\ -48 \\ -36 \\ 720 \end{Bmatrix} \qquad (f)$$

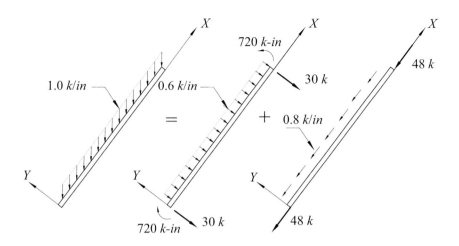

Fig. 4.7   Element 1 showing its load equated to the superposition of the normal and the axial components and also showing the equivalent forces at the nodal coordinates

ELEMENT 2

The equivalent nodal forces are calculated using the formulas for Case (a) of Appendix I to obtain

$$Q_1 = 0 \qquad Q_2 = \frac{W}{2} = -50 \; kip \qquad Q_3 = \frac{WL}{8} = -\frac{100 x 144}{8} = -1800 \; kip \cdot in$$

$$Q_4 = 0 \qquad Q_5 = \frac{W}{2} = -50 \; kip \qquad Q_6 = -\frac{WL}{8} = 1800 \; kip \cdot in$$

Or in vector notation

$$\{Q\}_2 = \begin{Bmatrix} 0 \\ -50 \\ -1800 \\ 0 \\ -50 \\ 1800 \end{Bmatrix} \tag{g}$$

7. Element equivalent nodal forces (global coordinates).

The equivalent nodal forces $\{\overline{Q}\}$ in global coordinates are calculated using the transformation of forces obtained from eq. (4.13):

$$\{\overline{Q}\} = [T]^T \{Q\} \tag{h}$$

ELEMENT 1

The substitution into eq. (h) of the transpose of matrix $[T]_1$ from eq. (c) and $\{Q\}_1$ from eq. (f) results in

$$\{\overline{Q}\}_1 = \begin{bmatrix} 0.60 & -0.80 & 0 & 0 & 0 & 0 \\ 0.80 & 0.60 & 0 & 0 & 0 & 0 \\ 0 & 0 & 1 & 0 & 0 & 0 \\ 0 & 0 & 0 & 0.60 & -0.80 & 0 \\ 0 & 0 & 0 & 0.80 & 0.60 & 0 \\ 0 & 0 & 0 & 0 & 0 & 1 \end{bmatrix} \begin{Bmatrix} -48 \\ -36 \\ -720 \\ -48 \\ -36 \\ 720 \end{Bmatrix} = \begin{Bmatrix} 0 \\ -60 \\ -720 \\ 0 \\ -60 \\ 720 \end{Bmatrix}$$

$$\{\overline{Q}\}_1 = \begin{Bmatrix} 0 \\ -60 \\ -720 \\ 0 \\ -60 \\ 720 \end{Bmatrix} \begin{matrix} 4 \\ 5 \\ 6 \\ 1 \\ 2 \\ 3 \end{matrix} \tag{i}$$

ELEMENT 2 (no transformation is needed since $\theta = 0$). From eq. (g):

$$\{\overline{Q}\}_2 = \{Q\}_2 = \begin{Bmatrix} 0 \\ -50 \\ -1800 \\ 0 \\ -50 \\ 1800 \end{Bmatrix} \begin{matrix} 1 \\ 2 \\ 3 \\ 7 \\ 8 \\ 9 \end{matrix} \tag{j}$$

8. Assemblage of the reduced system force vector.

The reduced system force vector $\{F\}_R$, corresponding to the free nodal coordinates (1, 2, 3), is assembled by transferring the coefficients in the force vectors in eqs. (i) and (j) to the locations indicated on the right side of these vectors and adding the force applied directly to a nodal coordinate, namely,

$$\{F\}_R = \begin{Bmatrix} 0+0 \\ -60-50 \\ 720-1800 \end{Bmatrix} + \begin{Bmatrix} 0 \\ 0 \\ 108 \end{Bmatrix} = \begin{Bmatrix} 0 \\ -110 \\ -972 \end{Bmatrix} \tag{k}$$

9. System stiffness equation.

Substitution of the reduced stiffness matrix [eq. (e)] and of the reduced force vector [eq. (k)] into the reduced system stiffness equation, $\{F\}_R = [K]_R\{u\}$, results in:

$$\begin{Bmatrix} 0 \\ -110 \\ -972 \end{Bmatrix} = \begin{bmatrix} 0.7307E4 & 0.2804E4 & 0.8820E6 \\ 0.2804E4 & 0.4028E4 & 0.1041E4 \\ 0.8820E6 & 0.1041E4 & 0.1617E7 \end{bmatrix} \begin{Bmatrix} u_1 \\ u_2 \\ u_3 \end{Bmatrix} \tag{l}$$

10. Solution of the unknown displacements.

$$u_1 = 0.0153 \; in$$
$$u_2 = -0.0378 \; in \tag{m}$$
$$u_3 = -6.602E\text{--}4 \; rad$$

11. Element nodal displacement vectors (global coordinates).

The displacements at the nodal coordinates of elements 1 and 2 are identified from the global coordinates assigned in the analytical model shown in Figure 4.6.

$$\{\bar{\delta}\}_1 = \begin{Bmatrix} 0 \\ 0 \\ 0 \\ 0.0153 \\ -0.0378 \\ -6.602E\text{--}4 \end{Bmatrix} \qquad \{\bar{\delta}\}_2 = \begin{Bmatrix} 0.0153 \\ -0.0378 \\ -6.602E\text{--}4 \\ 0 \\ 0 \\ 0 \end{Bmatrix} \tag{n}$$

12. Element nodal displacement vectors (local coordinates).

The transformation of the element nodal coordinates is given by eq. (4.15) as

$$\{\delta\} = [T]\{\bar{\delta}\} \qquad \text{(4.15) repeated}$$

ELEMENT 1

The substitution into eq. (4.15) of $\{\bar{\delta}\}_1$, and $[T]_1$, respectively, from eqs. (n) and (c) results in

$$\{\delta\}_1 = \begin{bmatrix} 0.6 & 0.8 & 0 & 0 & 0 & 0 \\ -0.8 & 0.6 & 0 & 0 & 0 & 0 \\ 0 & 0 & 1 & 0 & 0 & 0 \\ 0 & 0 & 0 & 0.6 & 0.8 & 0 \\ 0 & 0 & 0 & -0.8 & 0.6 & 0 \\ 0 & 0 & 0 & 0 & 0 & 1 \end{bmatrix} \begin{Bmatrix} 0 \\ 0 \\ 0 \\ 0.0153 \\ -0.0378 \\ -0.00066 \end{Bmatrix} = \begin{Bmatrix} 0 \\ 0 \\ 0 \\ -0.211 \\ -0.035 \\ -0.00066 \end{Bmatrix}$$

ELEMENT 2

For element 2, $\{\delta\}_2 = \{\bar{\delta}\}_2$ since $[T]_2 = [I]$.

13. Element end forces (local coordinates).

Element end forces are calculated by eq. (4.20):

$$\{P\} = [k]\{\delta\} - \{Q\} \qquad \text{(4.20) repeated}$$

ELEMENT 1

$$\begin{Bmatrix} P_1 \\ P_2 \\ P_3 \\ P_4 \\ P_5 \\ P_6 \end{Bmatrix} = \begin{bmatrix} 6.025E3 & 0 & 0 & -6.025E3 & 0 & 0 \\ 0 & 1.838E2 & 1.103E4 & 0 & -1.838E2 & 1.103E4 \\ 0 & 1.103E4 & 8.820E5 & 0 & -1.103E4 & 4.410E5 \\ -6.025E3 & 0 & 0 & 6.025E3 & 0 & 0 \\ 0 & -1.838E2 & -1.103E4 & 0 & 1.838E2 & -1.103E4 \\ 0 & 1.103E4 & 4.410E5 & 0 & -1.103E4 & 8.820E5 \end{bmatrix} \begin{Bmatrix} 0 \\ 0 \\ 0 \\ -0.211 \\ -0.035 \\ -0.00066 \end{Bmatrix} \begin{Bmatrix} -48 \\ -36 \\ -720 \\ -48 \\ -36 \\ 720 \end{Bmatrix} \begin{Bmatrix} 174.81 \\ 35.13 \\ 813.63 \\ -78.81 \\ 36.87 \\ -917.52 \end{Bmatrix}$$

ELEMENT 2

$$\begin{Bmatrix} P_1 \\ P_2 \\ P_3 \\ P_4 \\ P_5 \\ P_6 \end{Bmatrix} = \begin{bmatrix} 5.02E3 & 0 & 0 & -5.02E3 & 0 & 0 \\ 0 & 1.063E2 & 7.656E3 & 0 & -1.063E2 & 7.656E3 \\ 0 & 7.656E3 & 7.350E5 & 0 & -7.656E3 & 3.675E5 \\ -5.02E3 & 0 & 0 & 5.02E3 & 0 & 0 \\ 0 & -1.063E2 & -7.656E3 & 0 & 1.063E2 & -7.656E3 \\ 0 & 7.656E3 & 3.675E5 & 0 & -7.656E3 & 7.350E5 \end{bmatrix} \begin{Bmatrix} 0.0153 \\ -0.0378 \\ -0.00066 \\ 0 \\ 0 \\ 0 \end{Bmatrix} \begin{Bmatrix} 0 \\ -50 \\ -1800 \\ 0 \\ -50 \\ 1800 \end{Bmatrix} = \begin{Bmatrix} 76.78 \\ 40.93 \\ 1025 \\ -76.78 \\ 59.07 \\ -2331.0 \end{Bmatrix}$$

Figure 4.8 shows the end forces on the two elements of the plane frame in Illustrative Example 4.1.

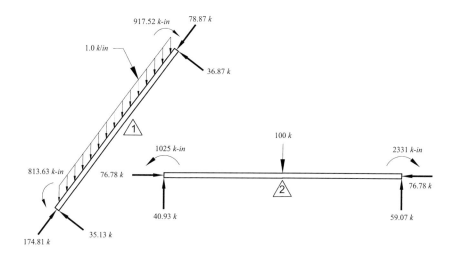

Fig. 4.8   End forces for Elements 1 and 2

14. Support reactions (global coordinates).

The support reactions in reference to the global coordinates are identified from the components of the calculated element end forces at the supports as follows:

Joint 1

$$\overline{R}_4 = 0.6\,P_{11} - 0.8\,P_{21}$$
$$\overline{R}_4 = 0.6x174.81 - 0.8x35.13 = 76.792\ kip$$
$$\overline{R}_5 = 0.8\,P_{11} + 0.6\,P_{21}$$
$$\overline{R}_5 = 0.8x174.81 + 0.6x35.13 = 160.926\ kip$$
$$\overline{R}_6 = P_{31} = 813.63\ kip{\cdot}in$$

Joint 3

$$\overline{R}_7 = P_{42} = -76.78\ kip$$
$$\overline{R}_8 = P_{52} = 59.07\ kip$$
$$\overline{R}_9 = P_{62} = -2331.86\ kip{\cdot}in$$

Note: P_{ij} is the force (in reference to local coordinates) at nodal coordinate i for element j.

4.8 Inclined Roller Supports

The structures considered thus far have been supported such that fixed joint forces (reactions) are in the direction of the global coordinate axes oriented in the horizontal and vertical directions. Occasionally, situations may exist in which restrained displacements are inclined with respect to direction of the global axes. Figure 4.9 shows an example of a plane frame having an inclined roller support. This frame is analyzed later in this chapter using SAP2000 (Illustrative Example 4.6). A simple example of a one span beam with an inclined roller support is presented now to illustrate the necessary calculations required for the analysis of structures having inclined roller supports.

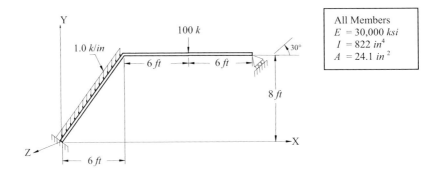

Fig. 4.9   Example of a plane frame having an inclined roller support

Illustrative Example 4.2
Consider in Figure 4.10(a) a beam having on the right end an inclined roller support. Use matrix structural analysis to determine:

(d)   Displacements on the right end of the beam.
(e)   End forces on the beam element.
(f)   Reactions at the supports.

1.   Analytical Model.

Figure 4.10(b) shows the analytical model in which the nodal coordinates in reference to global axes, u_1 through u_6 are indicated. This figure also shows the nodal coordinates u_4' and u_5' oriented along the auxiliary axis (x', y', z') with axes x' and y' oriented along the direction of the inclined support and normal to it, respectively.

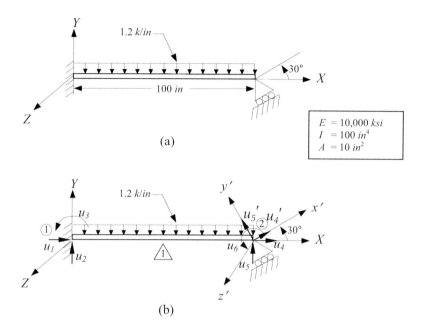

Fig. 4.10 (a) Beam having an inclined roller support for Illustrative Example 4.2
(b) Analytical model showing nodal coordinates in global axes (X, Y, Z)
and in the auxiliary axes (x', y', z')

2. Stiffness matrix $(X, Y, Z$ axes).

The element stiffness matrix in local coordinates is given in eq. (4.8). For this example, the global coordinates coincide with the element local coordinates. The substitution of numerical values in eq. (4.8) yields:

$$[k] = \begin{bmatrix} 1000 & 0 & 0 & -1000 & 0 & 0 \\ 0 & 12 & 600 & 0 & -12 & 600 \\ 0 & 600 & 40000 & 0 & -600 & 20000 \\ -1000 & 0 & 0 & 1000 & 0 & 0 \\ 0 & -12 & -600 & 0 & 12 & -600 \\ 0 & 600 & 20000 & 0 & -600 & 40000 \end{bmatrix}$$ (a)

3. Stiffness matrix in reference to the auxiliary coordinate axes (x', y', z').

The relationship at node 2 [(Figure 4.10(b)] between the nodal displacements u_4, u_5 along the axes X, Y, and nodal displacements u'_4 and u'_5 along the auxiliary axes x', y' is given, in general, for an inclined support in an angle α by

$$u'_4 = u_4 \cos\alpha + u_5 \sin\alpha$$

$$u'_5 = -u_4 \sin\alpha + u_5 \cos\alpha$$

(b)

Solving for u_4 and u_5 and using matrix notation yields

$$\begin{Bmatrix} u_4 \\ u_5 \end{Bmatrix} = \begin{bmatrix} \cos\alpha & -\sin\alpha \\ \sin\alpha & \cos\alpha \end{bmatrix} \begin{Bmatrix} u'_4 \\ u'_5 \end{Bmatrix}$$

(c)

We may write the following relationship to include all of the nodal displacements shown in Figure 4.10(b):

$$\begin{Bmatrix} u_1 \\ u_2 \\ u_3 \\ u_4 \\ u_5 \\ u_6 \end{Bmatrix} = \begin{bmatrix} 1 & 0 & 0 & 0 & 0 & 0 \\ 0 & 1 & 0 & 0 & 0 & 0 \\ 0 & 0 & 1 & 0 & 0 & 0 \\ 0 & 0 & 0 & \cos\alpha & -\sin\alpha & 0 \\ 0 & 0 & 0 & \sin\alpha & \cos\alpha & 0 \\ 0 & 0 & 0 & 0 & 0 & 1 \end{bmatrix} \begin{Bmatrix} u_1 \\ u_2 \\ u_3 \\ u'_4 \\ u'_5 \\ u_6 \end{Bmatrix}$$

(d)

or in condensed notation

$$\{u\} = [T']\{u'\}$$

(e)

in which the transformation matrix $[T']$ defined in eq. (d) relates the displacement vector $\{u\}$ with the displacement vector $\{u'\}$. These displacement vectors are in reference to the global axes and to the auxiliary coordinate axes, respectively. The same transformation matrix $[T']$ also relates force vector $\{F\}$ with the force vector $\{F'\}$ in reference to these two systems of coordinates, namely

$$\begin{Bmatrix} F_1 \\ F_2 \\ F_3 \\ F_4 \\ F_5 \\ F_6 \end{Bmatrix} = \begin{bmatrix} 1 & 0 & 0 & 0 & 0 & 0 \\ 0 & 1 & 0 & 0 & 0 & 0 \\ 0 & 0 & 1 & 0 & 0 & 0 \\ 0 & 0 & 0 & \cos\alpha & -\sin\alpha & 0 \\ 0 & 0 & 0 & \sin\alpha & \cos\alpha & 0 \\ 0 & 0 & 0 & 0 & 0 & 1 \end{bmatrix} \begin{Bmatrix} F_1' \\ F_2' \\ F_3' \\ F_4' \\ F_5' \\ F_6' \end{Bmatrix} \qquad (f)$$

or in condensed notation

$$\{F\} = [T']\{F'\} \quad \text{and} \quad \{F'\} = [T']^T \{F\} \qquad (g)$$

since $[T']$ is an orthogonal matrix $\left([T']^{-1} = [T']^T\right)$.

The substitution of $\{u\}$ from eq. (e) and of $\{F\}$ from eq. (g) into the stiffness equation $\{F\} = [k]\{u\}$ results in

$$[T']\{F'\} = [k][T']\{u'\} .$$

Then, since $[T']$ is an orthogonal matrix $([T']^{-1} = [T']^T)$

$$\{F'\} = [T']^T [k][T']\{u'\}$$

or

$$\{F'\} = [k']\{u'\} \qquad (h)$$

where

$$[k'] = [T']^T [k][T'] \qquad (i)$$

in which $[T']$, from eq. (f), is:

$$[T'] = \begin{bmatrix} 1 & 0 & 0 & 0 & 0 & 0 \\ 0 & 1 & 0 & 0 & 0 & 0 \\ 0 & 0 & 1 & 0 & 0 & 0 \\ 0 & 0 & 0 & 0.866 & -0.5 & 0 \\ 0 & 0 & 0 & 0.5 & 0.866 & 0 \\ 0 & 0 & 0 & 0 & 0 & 1 \end{bmatrix} \qquad (j)$$

Substituting into eqs. (k) and (i), $[T']$ and its transpose from eq. (j), $[k]$ from eq. (a) and performing the multiplications indicated in eq. (i) results in

$$\begin{Bmatrix} F_1 \\ F_2 \\ F_3 \\ F_4' \\ F_5' \\ F_6 \end{Bmatrix} = \begin{matrix} & 1 & 2 & 3 & 4' & 5 & 6 & \\ & \begin{bmatrix} 1 & 0 & 0 & -866 & 500 & 0 \\ 0 & 12 & 600 & -6 & -10.39 & 600 \\ 0 & 600 & 40000 & -300 & -519.6 & 20000 \\ -866 & -6 & -300 & 753 & -428 & -300 \\ 500 & -10.39 & -519.6 & -428 & -259 & -519.6 \\ 0 & 600 & 20000 & -300 & -519.6 & 40000 \end{bmatrix} & \begin{matrix} 1 \\ 2 \\ 3 \\ 4' \\ 5 \\ 6 \end{matrix} \end{matrix} \begin{Bmatrix} u_1 \\ u_2 \\ u_3 \\ u_4' \\ u_5' \\ u_6 \end{Bmatrix} \quad \text{(k)}$$

4.   Equivalent nodal forces vector $\{Q\}$ (X, Y, Z axes):

The equivalent nodal force vector for the distributed load on the span of this beam is obtained by substituting numerical values in Case (c) of Appendix I:

$$\{Q\} = \begin{Bmatrix} 0 \\ wL/2 \\ wL^2/12 \\ 0 \\ wL/2 \\ -wL^2/12 \end{Bmatrix} = \begin{Bmatrix} 0 \\ -60 \\ -1000 \\ 0 \\ -60 \\ 1000 \end{Bmatrix} \quad \text{(l)}$$

in which $w = -1.2$ k/in and $L = 100$ in.

5.   Equivalent nodal force vector $\{Q'\}$ (auxiliary coordinates):

The equivalent nodal force vector in the auxiliary coordinates is given as in eq. (g) by

$$\{Q'\} = [T']^T \{Q\} . \quad \text{(m)}$$

Substituting into eq. (m) $[T']^T$ from eq. (j) and $\{Q\}$ from eq. (l) results in

$$\{Q'\} = \begin{Bmatrix} Q_1 \\ Q_2 \\ Q_3 \\ Q_4' \\ Q_5' \\ Q_6 \end{Bmatrix} = \begin{Bmatrix} 0 \\ -60 \\ -1000 \\ -30 \\ -52 \\ 1000 \end{Bmatrix} \begin{matrix} 1 \\ 2 \\ 3 \\ 4' \\ 5 \\ 6 \end{matrix} \qquad (n)$$

6. Reduced stiffness matrix equation (auxiliary system).

    The reduced stiffness matrix equation is assembled by transferring from
    the matrix in eq. (j) the coefficients corresponding to the free coordinates
    u_4 and u_6. Likewise, the reduced system force vector is obtained from
    eq. (n) by transferring the coefficients in the force vector $\{Q'\}$
    corresponding to these two free coordinates, to obtain

$$\begin{Bmatrix} -30 \\ 1000 \end{Bmatrix} = \begin{Bmatrix} 753 & -300 \\ -300 & 40000 \end{Bmatrix} \begin{Bmatrix} u_4' \\ u_6 \end{Bmatrix} \qquad (o)$$

7. Nodal displacements (auxiliary system):

    The solution of eq. (o) gives the displacements at the free nodal
    coordinates as

$$u_4' = -0.030 \ in$$
$$u_6 = 0.0248 \ rad$$

8. Element end forces:

    The element end forces are given from eq. (4.20) as

$$\{P\} = [k]\{\delta\} - \{Q\} \qquad \text{(4.20) repeated}$$

    in which the displacement vector $\{\delta\}$ in reference to the coordinate
    system (X, Y, Z) is identified from Figure 4.10(b) as

$$\{\delta\} = \begin{Bmatrix} u_1 \\ u_2 \\ u_3 \\ u_4 \\ u_5 \\ u_6 \end{Bmatrix} = \begin{Bmatrix} 0 \\ 0 \\ 0 \\ u_4' \cos\alpha \\ u_4' \sin\alpha \\ u_6 \end{Bmatrix} = \begin{Bmatrix} 0 \\ 0 \\ 0 \\ -0.026 \\ -0.015 \\ 0.0248 \end{Bmatrix} \tag{p}$$

where $\alpha = 30°$.

Then substituting into eq. (4.20) of the stiffness matrix $[k]$ from eq. (a), $\{\delta\}$ from eq. (p) and $\{Q\}$ from eq. (l) results in

$$\begin{Bmatrix} P_1 \\ P_2 \\ P_3 \\ P_4 \\ P_5 \\ P_6 \end{Bmatrix} = \begin{bmatrix} 1000 & 0 & 0 & -1000 & 0 & 0 \\ 0 & 12 & 600 & 0 & -12 & 600 \\ 0 & 600 & 40000 & 0 & -600 & 20000 \\ -1000 & 0 & 0 & 1000 & 0 & 0 \\ 0 & -12 & -600 & 0 & 12 & -600 \\ 0 & 600 & 20000 & 0 & -600 & 40000 \end{bmatrix} \begin{Bmatrix} 0 \\ 0 \\ 0 \\ -0.026 \\ -0.015 \\ 0.0248 \end{Bmatrix} - \begin{Bmatrix} 0 \\ -60 \\ -1000 \\ 0 \\ -60 \\ 1000 \end{Bmatrix}$$

or

$$\begin{Bmatrix} P_1 \\ P_2 \\ P_3 \\ P_4 \\ P_5 \\ P_6 \end{Bmatrix} = \begin{Bmatrix} 26 \\ 75.04 \\ 1504.5 \\ -26 \\ 45.04 \\ 0 \end{Bmatrix} \tag{q}$$

9. Reaction at the supports.

The reactions at the supports are calculated using the end forces given in eq. (q):

At the left support:

$$R_1 = P_1 = 26 \ kip$$
$$R_2 = P_2 = 75.04 \ kip$$
$$R_3 = P_3 = 1504.5 \ kip \cdot in$$

At the right support:

$$R_5' = -P_4 \sin 30° + P_5 \cos 30°$$

$$R_5' = -(-26)(0.5) + (45.04)(0.866)$$

$$R_5' = 52.00 \; kip$$

4.9   Analysis of Plane Frames Using SAP2000

Illustrative Example 4.3

Use SAP2000 to analyze the plane frame in Illustrative Example 4.1. For convenience, the plane frame for this example is reproduced in Figure 4.11 using global axis Z along the vertical direction as adopted by SAP2000. The input data is given in Table 4.1.

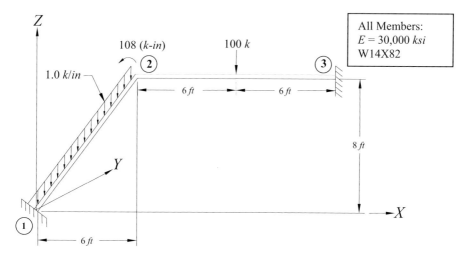

Fig 4.11   Plane Frame for Illustrative Example 4.3

Table 4. 1 Edited Input Data for Illustrative Example 4.3 (units: kips, inches)

```
JOINT  DATA
  JOINT     GLOBAL-X     GLOBAL-Y     GLOBAL-Z     RESTRAINTS
    1        0.00000      0.00000      0.00000      1 1 1 1 1 1
    2       72.00000      0.00000     96.00000      0 0 0 0 0 0
    3      216.00000      0.00000     96.00000      1 1 1 1 1 1
```

Table 4.1 continued

```
FRAME  ELEMENT  DATA

FRAME  JNT-1  JNT-2  SECTION  RELEASES  SEGMENTS   LENGTH
  1      1      2     W14X82   000000      2        120.000
  2      2      3     W14X82   000000      4        144.000

JOINT  FORCES  Load Case LOAD1
 JOINT  GLOBAL-X  GLOBAL-Y  GLOBAL-Z  GLOBAL-XX  GLOBAL-YY  GLOBAL-ZZ
   2     0.000     0.000     0.000     0.000     -108.000     0.000

FRAME  SPAN  DISTRIBUTED  LOADS  Load Case  LOAD1
 FRAME    TYPE  DIRECTION  DISTANCE-A   VALUE-A  DISTANCE-B   VALUE-B
   1     FORCE  GLOBAL-Z    0.0000     -1.0000    1.0000     -1.0000

FRAME  SPAN  POINT  LOADS  Load Case  LOAD1
 FRAME    TYPE  DIRECTION   DISTANCE     VALUE
   2     FORCE  GLOBAL-Z     0.5000    -100.0000
```

Solution:

The following commands are implemented in SAP2000:

Begin: Open SAP2000.

Enter: Click "OK" to close the "Tip of the Day" form.

        Hint: Maximize both screens for full views of all windows.

Units: In the lower right-hand corner of the window, use the drop-down list to select "kip-in".

Model: From the main menu select:
FILE>NEW MODEL.
In the Coordinate System Definition:
Number of Divisions:
    $x = 10$
    $y = 0$
    $z = 10$
Grid Spacing:
    $x = 24$
    $y = 1$
    $z = 24$
Accept the other default values.

View: Click the XZ icon on the toolbar to set the view to the XZ Plane.

Edit: Maximize the X-Z window. Click on the icon on the toolbar with the symbol of a hand (the PAN icon) and drag the plot to the center of the window.

Draw: From the main menu select:
DRAW>DRAW FRAME ELEMENT.
Click in the lower left grid intersection point and drag the cursor to a point 3 grid lines to the right and 4 grid lines upward. Then click again at this location.

Drag again horizontally 6 grid lines and double-click at this location. Then press Enter to disable the command to draw frame elements.

Axes: Translate the origin of the global coordinate system to the left lower node of the frame using the following commands:
From the main menu select:
SELECT>SELECT ALL. This command will mark all the elements of the frame.
From the main menu select:
EDIT>MOVE.
In the "Move Selected Points" form, change $x = 240$.
Click OK.

Beam Material: From the main menu select:
DEFINE >MATERIALS.
Select STEEL.
Click MODIFY / SHOW MATERIALS.
    Change the value of the Modulus of elasticity to 30000.
Click Ok, OK.

Sections: From the main menu select:
DEFINE>FRAME SECTIONS.
Select Import / Wide Flange.
Click OK.
In the next window, open the file "section.prop".
Scroll down the Section labels and select "W14 X 82".
Click OK.
    Click Modification Factors.
    Change shear area in the y direction to 0 (to exclude shear deformation in the stiffness matrix of the elements).
    Click OK, OK, OK.

Assign Frame Sections: From the main menu select:
SELECT>SELECT ALL.
From the main menu select:

ASSIGN>FRAME>SECTIONS.
In the Define Frame Section form, select "W 14 X 82".
Click OK.

Label: For viewing convenience, label joints and elements,
From the main menu select:
VIEW>SET ELEMENTS.
Click on the boxes labeled "Joint Labels" and "Frame Labels".
Click OK.

Boundary: Restraint Setting: Click on Joints 1 and 3. This command will mark
these joints with an "X".
From the main menu select:
ASSIGN>JOINTS>RESTRAINTS.
In the "Joint Restraints" form, select restraints in all directions.
Click OK.

Assign Loads: From the main menu select:
DEFINE>STATIC LOAD CASES.
In the "Define Static Load Case Names" form, change the label DEAD to
LIVE.
Set the Self-Weight Multiplier to 0 (zero).
Click on "Change Load".
Click OK.

Assign: Select Joint 2. Then from the main menu select:
ASSIGN>JOINT STATIC LOAD>FORCES.
Change the value of the moment YY to –108.
Click OK.

Assign: Select Frame 1 by clicking on it. Frame 1 will change from a continuous
line to dashes. From the main menu select:
ASSIGN>FRAME STATIC LOADS>POINT AND UNIFORM LOADS.
Select Forces and Global Direction Z.
Then enter Uniform Load = –1.0.
Click OK.

Assign: Select Frame 2. From the main menu select:
ASSIGN>FRAME STATIC LOADS>POINT AND UNIFORM LOAD.

Warning: Make certain to zero out any previous entries made on this form!

Select Forces and Global Direction Z.
Enter: Distance = 0.5 and Load = –100.

Options: From the main menu select:

ANALYZE>SET OPTIONS.
In the "Available DOF," check only UX, UZ and RY.
Alternatively, click on the picture labeled "Plane Frame".
Click OK.

Analyze: From the main menu select:
ANALYZE>RUN.
Name the model "Example 4.3".
Click "SAVE". The calculations will scroll on the screen. When the analysis is completed, a message will appear on the bottom of the pop-up window "The analysis is complete".
Click OK.

Observe: From the main menu select:
DISPLAY>SHOW DEFORMED SHAPE.
Click on the "Wire Shadow" option.
Also click on the "XZ" icon to view the plot on that plane.
Click OK.
To display the displacement values at any node, right-click on the node and a pop-up window will show the nodal displacements.

Print Input Tables: From the main menu select:
FILE>PRINT INPUT TABLES.
(The previous Table 4.1 provides the edited input tables for Illustrative Example 4.3.)

Print Output Tables: From the main menu select:
FILE>PRINT OUTPUT TABLES.
(Table 4.2 provides the edited output tables for Illustrative Example 4.3.)

Table 4.2 Edited Output Tables for Illustrative Example 4.3 (Units: kips, inches)

JOINT DISPLACEMENTS

JOINT	LOAD	UX	UY	UZ	RX	RY	RZ
1	LOAD1	0.0000	0.0000	0.0000	0.0000	0.0000	0.0000
2	LOAD1	0.0153	0.0000	−0.0378	0.0000	6.602E-04	0.0000
3	LOAD1	0.0000	0.0000	0.0000	0.0000	0.0000	0.0000

JOINT REACTIONS

JOINT	LOAD	F1	F2	F3	M1	M2	M3
1	LOAD1	76.7783	0.0000	160.9282	0.0000	−813.6295	0.0000
3	LOAD1	−76.7783	0.0000	59.0718	0.0000	2331.8601	0.0000

Table 4.2 (continued)

FRAME ELEMENT FORCES

FRAME	LOAD	LOC	P	V2	V3	T	M2	M3
1	LOAD1							
		0.00	-174.81	-35.13	0.00	0.00	0.00	-813.63
		60.00	-126.81	8.657E-01	0.00	0.00	0.00	214.43
		120.00	-78.81	36.87	0.00	0.00	0.00	-917.52
2	LOAD1							
		0.00	-76.78	-40.93	0.00	0.00	0.00	-1025.52
		36.00	-76.78	-40.93	0.00	0.00	0.00	447.90
		72.00	-76.78	59.07	0.00	0.00	0.00	1921.31
		108.00	-76.78	59.07	0.00	0.00	0.00	-205.27
		144.00	-76.78	59.07	0.00	0.00	0.00	-2331.86

Plot Displacements: From the main menu select:
　　　　DISPLAY>SHOW DEFORMED SHAPE.
　　　　Select FILE>PRINT GRAPHICS.
　　　　(The deformed shape shown on the screen is reproduced in Figure 4.12.)

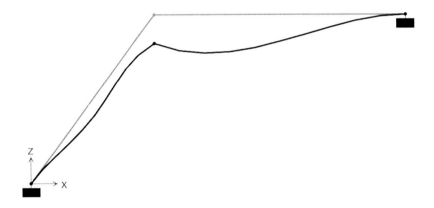

Fig. 4.12　Deformed shape for the Plane Frame in Illustrative Example 4.3

Plot Shear Forces: From the main menu select:
　　　　DISPLAY>SHOW ELEMENT FORCES / STRESSES>FRAMES.
　　　　Select Shear　-22.
　　　　Select FILE>PRINT GRAPHICS.
　　　　(The shear force diagram shown on the screen is depicted in Figure 4.13.)

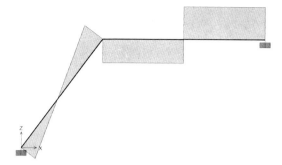

Fig. 4.13   Shear force diagram for the plane frame in Illustrative Example 4.3

Plot the Bending Moment: From the main menu select:
        DISPLAY>SHOW FORCES / STRESSES>FRAME.
        Select Moment 3-3 and OK.
        Select FILE>PRINT GRAPHICS.
        (The bending moment diagram shown on the screen is depicted in Figure
        4.14.)

Note: To plot the Moment Diagram on the compression side, from the main menu
select OPTIONS; then uncheck the "Moment Diagram on the Tension Side" option.

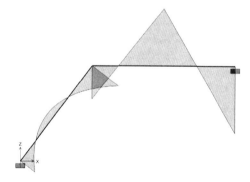

Fig. 4.14   Bending Moment Diagram for the plane frame of Illustrative Example 4.3

Note: To view plots of the shear force or bending moment diagrams for any
element of the frame, right-click on the element and a pop-up window will depict
the diagram for the selected element. It should be observed that as the cursor is
moved in this window across the plot, values of shear force or of the bending
moment will be displayed.

Illustrative Example 4.4

Use SAP2000 to analyze the plane frame shown in Figure 4.15. The cross bracings are to be modeled as rod elements resisting only axial force. Input data tables for this Illustrative Example are given in Table 4.3.

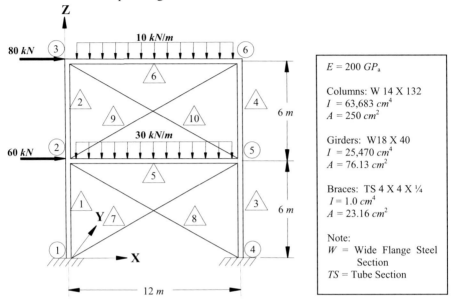

Fig. 4.15 Plane Frame of Illustrative Example 4.4

Table 4.3 Edited Input data tables for Illustrative Example 4.4 (Units: KN, m)

```
JOINT  DATA
  JOINT    GLOBAL-X    GLOBAL-Y    GLOBAL-Z    RESTRAINTS
    1       0.00000     0.00000     0.00000     1 1 1 1 1 1
    2       0.00000     0.00000     6.00000     0 0 0 0 0 0
    3       0.00000     0.00000    12.00000     0 0 0 0 0 0
    4      12.00000     0.00000     0.00000     1 1 1 1 1 1
    5      12.00000     0.00000     6.00000     0 0 0 0 0 0
    6      12.00000     0.00000    12.00000     0 0 0 0 0 0
```

```
FRAME  ELEMENT  DATA
  FRAME  JNT-1  JNT-2   SECTION    RELEASES  SEGMENTS    LENGTH
    1      1      2    W14X132     000000       2        6.000
    2      2      3    W14X132     000000       2        6.000
    4      5      6    W14X132     000000       2        6.000
    3      4      5    W14X132     000000       2        6.000
    5      2      5    W18X40      000000       4       12.000
    6      3      6    W18X40      000000       4       12.000
    7      2      6    TS4X4X1/4   000000       2       13.416
```

Table 4.3 (continued)

8	3	5	TS4X4X1/4	000000	2	13.416
9	1	5	TS4X4X1/4	000000	2	13.416
10	2	4	TS4X4X1/4	000000	2	13.416

JOINT FORCES Load Case LOAD1

JOINT	GLOBAL-X	GLOBAL-Y	GLOBAL-Z	GLOBAL-XX	GLOBAL-YY	GLOBAL-ZZ
2	60.000	0.000	0.000	0.000	0.000	0.000
3	80.000	0.000	0.000	0.000	0.000	0.000

FRAME SPAN DISTRIBUTED LOADS Load Case LOAD1

FRAME	TYPE	DIRECTION	DISTANCE-A	VALUE-A	DISTANCE-B	VALUE-B
6	FORCE	GLOBAL-Z	0.0000	−10.0000	1.0000	−10.0000
5	FORCE	GLOBAL-Z	0.0000	−30.0000	1.0000	−30.0000

Solution:

Table 4.4 provides the edited output tables for Illustrative Example 4.4. The deformed shape, the axial force, the shear force diagram and the bending moment diagrams are reproduced in Figs. 4.16(a) through (d).

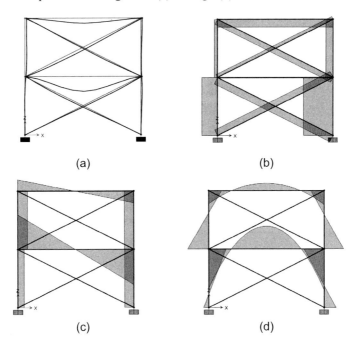

(a)   (b)

(c)   (d)

Fig. 4.16 Plots for the plane frame of Illustrative Example 4.4:
(a) Deformed shape, (b) Axial force diagram, (c) Shear force diagram and (d) Bending moment diagram

Table 4.4 Edited Output tables for Illustrative Example 4.4 (Units: KN, m)

JOINT DISPLACEMENTS

JOINT	LOAD	UX	UY	UZ	RX	RY	RZ
1	LOAD1	0.0000	0.0000	0.0000	0.0000	0.0000	0.0000
2	LOAD1	2.317E–03	0.0000	–1.966E–04	0.0000	2.333E–03	0.0000
3	LOAD1	4.167E–03	0.0000	–2.374E–04	0.0000	6.219E–04	0.0000
4	LOAD1	0.0000	0.0000	0.0000	0.0000	0.0000	0.0000
5	LOAD1	2.262E–03	0.0000	–3.731E–04	0.0000	–1.558E–03	0.0000
6	LOAD1	3.545E–03	0.0000	–4.638E–04	0.0000	–3.081E–04	0.0000

JOINT REACTIONS

JOINT	LOAD	F1	F2	F3	M1	M2	M3
1	LOAD1	–24.2079	0.0000	135.3580	0.0000	49.8529	0.0000
4	LOAD1	–115.7921	0.0000	344.6420	0.0000	–114.1489	0.0000

FRAME ELEMENT FORCES

FRAME	LOAD	LOC	P	V2	V3	T	M2	M3
1	LOAD1							
		0.00	–164.02	–33.12	0.00	0.00	0.00	–49.85
		3.00	–164.02	–33.12	0.00	0.00	0.00	49.52
		6.00	–164.02	–33.12	0.00	0.00	0.00	148.90
2	LOAD1							
		0.00	–34.06	–49.64	0.00	0.00	0.00	–185.23
		3.00	–34.06	–49.64	0.00	0.00	0.00	–36.32
		6.00	–34.06	–49.64	0.00	0.00	0.00	112.59
3	LOAD1							
		0.00	–311.28	49.07	0.00	0.00	0.00	114.15
		3.00	–311.28	49.07	0.00	0.00	0.00	–33.06
		6.00	–311.28	49.07	0.00	0.00	0.00	–180.28
4	LOAD1							
		0.00	–75.69	48.68	0.00	0.00	0.00	172.56
		3.00	–75.69	48.68	0.00	0.00	0.00	26.52
		6.00	–75.69	48.68	0.00	0.00	0.00	–119.52
5	LOAD1							
		0.00	–6.99	–178.44	0.00	0.00	0.00	–334.12
		3.00	–6.99	–88.44	0.00	0.00	0.00	66.20
		6.00	–6.99	1.56	0.00	0.00	0.00	196.52
		9.00	–6.99	91.56	0.00	0.00	0.00	56.84
		12.00	–6.99	181.56	0.00	0.00	0.00	–352.84
6	LOAD1							
		0.00	–78.90	–59.42	0.00	0.00	0.00	–112.59
		3.00	–78.90	–29.42	0.00	0.00	0.00	20.68
		6.00	–78.90	0.58	0.00	0.00	0.00	63.95
		9.00	–78.90	30.58	0.00	0.00	0.00	17.22
		12.00	–78.90	60.58	0.00	0.00	0.00	–119.52
7	LOAD1							
		0.00	33.79	0.00	0.00	0.00	0.00	0.00
		6.71	33.79	0.00	0.00	0.00	0.00	0.00
		13.42	33.79	0.00	0.00	0.00	0.00	0.00

Table 4.4 Continued

8	LOAD1							
		0.00	−56.72	0.00	0.00	0.00	0.00	0.00
		6.71	−56.72	0.00	0.00	0.00	0.00	0.00
		13.42	−56.72	0.00	0.00	0.00	0.00	0.00
9	LOAD1							
		0.00	64.10	0.00	0.00	0.00	0.00	0.00
		6.71	64.10	0.00	0.00	0.00	0.00	0.00
		13.42	64.10	0.00	0.00	0.00	0.00	0.00
10	LOAD1							
		0.00	−74.60	0.00	0.00	0.00	0.00	0.00
		6.71	−74.60	0.00	0.00	0.00	0.00	0.00
		13.42	−74.60	0.00	0.00	0.00	0.00	0.00

Illustrative Example 4.5

Use SAP2000 to solve Illustrative Example 4.2, which consists of a beam having an inclined roller support, as shown in Figure 4.17.

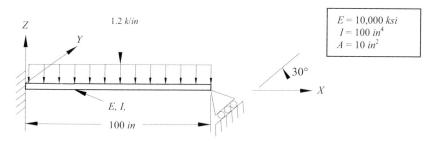

Fig. 4.17 Beam for Illustrative Example 4.5 having an inclined roller support

Solution:

The following commands are implemented in SAP2000:

Begin: Open SAP2000.
Hint: Maximize both windows for a full view.

Units: Select "kip-in" in the drop-down menu located on the lower right-hand corner of the window.

Model: From the main menu select:
FILE>NEW MODEL
Set: Number $x = 2$, $y = 0$ and $z = 1$
Distance $x = 100$, $y = 48$ and $z = 48$
Click OK.

View: Maximize the 2-D window (X-Z View). Then drag the figure to the center of the window using the PAN icon on the toolbar.

Draw: From the main menu select:
DRAW>DRAW FRAME MEMBER.
Click at the origin of XZ coordinates and drag the cursor horizontally to the next grid line on the right. Then double-click and press the Enter key.

Labels: From the main menu select:
VIEW>SET ELEMENTS.
Check Joints-Labels and Frames-Labels.
Click OK.

Boundaries: Click on Joint 1 and from the main menu select:
ASSIGN>JOINT>RESTRAINTS.
Check restraints in all directions.
Click OK.

Click on joint 2.
From the main menu select:
ASSIGN>JOINT>LOCAL AXES.
Enter: Rotation in degrees of local axis y about global axis $Y = -30$
Click OK.

Click on joint 2.
From the main menu select:
ASSIGN>JOINTS>RESTRAINTS.
Check (only) Joint restraint Translation 3.
Click OK.

Material: From the main menu select:
DEFINE >MATERIALS.
Select OTHER.
Click Modify/Show Materials
     Set the Modulus of Elasticity = 10000.
Click OK, OK.

Section: From the main menu select:
DEFINE>FRAME>SECTIONS.
Click Add/Wide Flange and scroll to Add General Section
        Enter: Cross-section area = 10.
        Moment of Inertia about 3 axis = 100.
        Shear area in 2 direction = 0.
        Click OK.

Select material = OTHER.
Click OK, OK.

Assign: Click on the beam element and from the main menu select:
ASSIGN>FRAME>SECTIONS.
Click on FSEC2.
Click OK.

Loads: From the main menu select:
DEFINE>STATIC LOAD CASES.
Change DEAD load to LIVE load.
Set self-weight multiplier = 0.
Then click on "Change Load".
Click OK.

Click on beam element 1; then from the main menu select:
ASSIGN>FRAME STATIC LOADS>POINTS>AND UNIFORM.
Enter: Uniform load = −1.2.
Click OK.

Analyze: From the main menu select:
ANALYZE>SET OPTIONS.
Click on XZ plane.
Click OK.

From the main menu select:
ANALYZE>RUN.
Enter File Name "Example 4.5".
Click SAVE.
At the conclusion of the calculation, click OK.

Plots: Click the XZ view button on the toolbar.
Use the PAN icon to center the deformed structure in the window.
From the main menu select:
DISPLAY>SHOW DEFORMED SHAPE.
FILE>PRINT GRAPHICS.

From the main menu select:
DISPLAY> SHOW ELEMENT FORCES/STRESSES>FRAMES.
Click Axial force.
Click OK.
FILE>PRINT GRAPHICS.

From the main menu select:
DISPLAY>SHOW ELEMENT FORCES/STRESSES>FRAMES.

Click Show force 2-2.
Click OK.

From the main menu select:
DISPLAY>SHOW ELEMENT FORCES/STRESSES>FRAMES.
Click Moment 3-3.
Click OK.
FILE>PRINT GRAPHICS.

The deformed plot and the axial force diagrams are reproduced in Figure 4.18 (a) and (b). The shear force and the bending moment diagram are reproduced in Figure 4.19 (a) and (b).

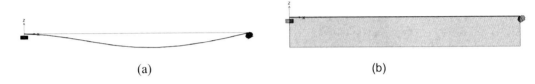

(a)   (b)

Fig. 4.18   (a) Deformed shape, and (b) Axial force diagram for the beam of Illustrative Example 4.5.

(a)   (b)

Fig. 4.19   (a) Shear force and (b) Bending moment diagrams for the beam of Illustrative Example 4.5

Tables: From the main menu select:
FILE>PRINT INPUT TABLES.
Click Print to File.
Click OK.
From the main menu select:
FILE>PRINT OUTPUT TABLES.

Click "Print to File" and "Append".
Click OK.

(The edited Input Tables and Output Tables for Illustrative Example 4.5 are shown in Tables 4.5 and 4.6, respectively.)

Table 4.5 Edited Input Tables for Illustrative Example 4.5 (Units: Kip-in)

JOINT DATA

JOINT	GLOBAL-X	GLOBAL-Y	GLOBAL-Z	RESTRAINTS	ANG-A	ANG-B	ANG-C
1	0.00000	48.00000	0.00000	1 1 1 1 1 1	0.000	0.000	0.000
2	100.00000	48.00000	0.00000	0 0 1 0 0 0	0.000	-30.000	0.000

FRAME ELEMENT DATA

FRAME	JNT-1	JNT-2	SCTN	ANG	RLS	SGMNTS	R1	R2	FCTR	LENGTH
1	1	2	FSEC2	0.000	000000	4	0.000	0.000	1.000	100.000

FRAME SPAN DISTRIBUTED LOADS Load Case LOAD1

FRAME	TYPE	DIRECTION	DISTANCE-A	VALUE-A	DISTANCE-B	VALUE-B
1	FORCE	GLOBAL-Z	0.0000	-1.2000	1.0000	-1.2000

Table 4.6 Output tables for Illustrative Example 4.5 (Units: Kip-in)

JOINT DISPLACEMENTS (Global System of Coordinates)

JOINT	LOAD	UX	UY	UZ	RX	RY	RZ
1	LOAD1	0.0000	0.0000	0.0000	0.0000	0.0000	0.0000
2	LOAD1	-0.0260	0.0000	-0.0150	0.0000	-0.0248	0.0000

JOINT REACTIONS

JOINT	LOAD	F1	F2	F3	M1	M2	M3
1	LOAD1	25.9548	0.0000	75.0450	0.0000	-1504.4955	0.0000
2	LOAD1	0.0000	0.0000	51.9096	0.0000	0.0000	0.0000

FRAME ELEMENT FORCES

FRAME	LOAD	LOC	P	V2	V3	T	M2	M3
1	LOAD1							
		0.00	-25.95	-75.04	0.00	0.00	0.00	-1504.50
		25.00	-25.95	-45.04	0.00	0.00	0.00	-3.37
		50.00	-25.95	-15.04	0.00	0.00	0.00	747.75
		75.00	-25.95	14.96	0.00	0.00	0.00	748.88
		100.00	-25.95	44.96	0.00	0.00	0.00	0.00

Illustrative Example 4.6

Consider the plane frame in Figure 4.20 having an inclined roller support. Use SAP2000 to determine:

(a) Displacements at the joints
(b) End forces on the members
(c) Reactions at the supports

$E = 30,000\ ksi$
$I = 882\ in^4$
$A = 24.1\ in^2$

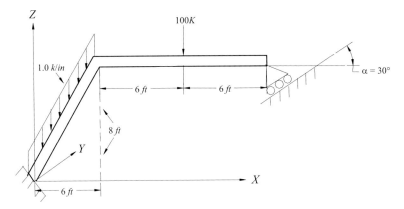

Fig. 4.20   Plane frame having an inclined roller support for Illustrative Example 4.6

Solution:

The necessary commands for SAP2000 can easily be implemented by analogy with detailed solutions presented in Illustrative Example 4.3 and 4.5. In particular, Illustrative Example 4.5 provides the necessary commands in SAP2000 to implement the condition of an inclined roller support.

Tables containing the edited input data and the edited output results for Illustrative Example 4.6 are reproduced as Tables 4.7 and 4.8, respectively. The deformed shape for the frame of Illustrative Example 4.6 is reproduced in Figure 4.21.

Table 4.7 Edited Input tables for Illustrative Example 4.6 (Units: kips, in)

JOINT DATA

JOINT	GLOBAL-X	GLOBAL-Y	GLOBAL-Z	RESTRAINTS	ANGLE-A	ANGLE-B	ANGLE-C
1	0.00000	30.00000	0.00000	1 1 1 1 1 1	0.000	0.000	0.000
2	80.00000	30.00000	60.00000	0 0 0 0 0 0	0.000	0.000	0.000
3	180.00000	30.00000	60.00000	0 0 1 0 0 0	0.000	-30.000	0.000

Table 4.7 (continued)

F R A M E   E L E M E N T   D A T A

FRAME	J1	J2	SECTION	ANGLE	RELEASES	SEGS	R1	R2	FACTOR	LENGTH
1	1	2	FSEC2	0.000	000000	2	0.000	0.000	1.000	100.000
2	2	3	FSEC2	0.000	000000	4	0.000	0.000	1.000	100.000

J O I N T   F O R C E S   Load Case LOAD1

JOINT	GLOBAL-X	GLOBAL-Y	GLOBAL-Z	GLOBAL-XX	GLOBAL-YY	GLOBAL-ZZ
2	0.000	0.000	−10.000	0.000	0.000	0.000

Table 4.8 Edited Output tables for Illustrative Example 4.6 (Units: kips, in)

J O I N T   D I S P L A C E M E N T S

JOINT	LOAD	UX	UY	UZ	RX	RY	RZ
1	LOAD1	0.0000	0.0000	0.0000	0.0000	0.0000	0.0000
2	LOAD1	0.1473	0.0000	−0.1980	0.0000	9.029E-04	0.0000
3	LOAD1	0.1471	0.0000	0.0849	0.0000	−4.696E-03	0.0000

J O I N T   R E A C T I O N S

JOINT	LOAD	F1	F2	F3	M1	M2	M3
1	LOAD1	1.3964	0.0000	7.5814	0.0000	−280.8675	0.0000
3	LOAD1	0.0000	0.0000	2.7928	0.0000	0.0000	0.0000

F R A M E   E L E M E N T   F O R C E S

FRAME	LOAD	LOC	P	V2	V3	T	M2	M3
1	LOAD1							
		0.00	−5.67	−5.23	0.00	0.00	0.00	−280.87
		50.00	−5.67	−5.23	0.00	0.00	0.00	−19.50
		100.00	−5.67	−5.23	0.00	0.00	0.00	241.86
2	LOAD1							
		0.00	−1.40	2.42	0.00	0.00	0.00	241.86
		25.00	−1.40	2.42	0.00	0.00	0.00	181.40
		50.00	−1.40	2.42	0.00	0.00	0.00	120.93
		75.00	−1.40	2.42	0.00	0.00	0.00	60.47
		100.00	−1.40	2.42	0.00	0.00	0.00	0.00

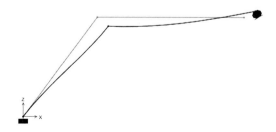

Fig. 4.21 The deformed shape for the plane frame having an inclined roller support of Illustrative Example 4.6

4.10 Dynamic Analysis of Plane Frames

The stiffness matrix analysis of structures subjected to dynamic loading is based on the determination of the stiffness and mass matrices for an element in reference to the local coordinate system followed by the transformation of these matrices to the global system of coordinates. The development of the stiffness matrix for an element of a plane frame as well as the transformation matrix were presented in Sections 4.4 and 4.5, thus it remains only to develop the mass matrix.

4.11 Axial Effects: Lumped Mass Matrix

The determination of mass influence coefficients for axial effects of an element of a plane frame may be carried out by either of the two methods presented in Chapter 2 for a beam element: (1) the lumped mass method and (2) the consistent mass method. In the lumped mass method, the mass allocation to the nodes of the beam element is found from static considerations, which for a uniform beam is equal to one-half of the total mass of the beam element allocated at each node. Then for a prismatic beam element, the relationship between nodal axial forces and corresponding nodal accelerations, as shown in Figure 4.22, is given by

$$\begin{Bmatrix} P_1 \\ P_4 \end{Bmatrix} = \frac{\bar{m}L}{2} \begin{bmatrix} 1 & 0 \\ 0 & 1 \end{bmatrix} \begin{Bmatrix} \ddot{\delta}_1 \\ \ddot{\delta}_4 \end{Bmatrix} \qquad (4.21)$$

where $\bar{m}$ is the mass per unit of length and L the element length.

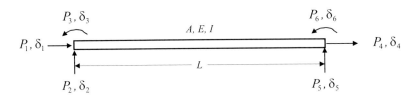

Fig. 4.22 Beam element showing flexural and axial nodal forces and displacements

The combination of the flexural lumped mass coefficient and the axial mass coefficients, in reference to the nodal coordinates in Figure 4.22, results in the following diagonal matrix, establishing the relationship between the nodal inertial forces and corresponding nodal accelerations:

$$\begin{Bmatrix} P_1 \\ P_2 \\ P_3 \\ P_4 \\ P_5 \\ P_6 \end{Bmatrix} = \frac{\overline{m}L}{2} \begin{bmatrix} 1 & & & & & \\ & 1 & & & & \\ & & 0 & & & \\ & & & 1 & & \\ & & & & 1 & \\ & & & & & 0 \end{bmatrix} \begin{Bmatrix} \ddot{\delta}_1 \\ \ddot{\delta}_2 \\ \ddot{\delta}_3 \\ \ddot{\delta}_4 \\ \ddot{\delta}_5 \\ \ddot{\delta}_6 \end{Bmatrix}$$ (4.22)

4.12   Axial Effects: Consistent Mass Matrix

The coefficients for the consistent mass matrix, m_{ij} (force at nodal coordinate i due to a unit axial acceleration at coordinate j) may be determined from the following expression (see Analytical Problem 4.3, Section 4.15):

$$m_{ij} = \int_0^L \overline{m}(x)u_i(x)u_j(x)dx$$ (4.23)

in which $\overline{m}(x)$ is the mass per unit of length, and u_i and u_j displacement functions given by eqs. (4.5) and (4.6).

The application of eq. (4.23) to the special case of a uniform beam element results in

$$m_{11} = \int_0^L \overline{m}\left(1-\frac{x}{L}\right)^2 dx = \frac{\overline{m}L}{3}$$ (4.24)

Similarly,

$$m_{22} = \frac{\overline{m}L}{3}$$

and

$$m_{12} = m_{21} = \int_0^L \overline{m}\left(1-\frac{x}{L}\right)\left(\frac{x}{L}\right)dx = \frac{\overline{m}L}{6}$$ (4.25)

Equations (4.24) and (4.25) establishing, for a uniform beam element, the consistent mass relationship of the axial inertial force with the axial acceleration may be written in matrix notation as

$$\begin{Bmatrix} P_1 \\ P_4 \end{Bmatrix} = \frac{\overline{m}L}{6} \begin{bmatrix} 2 & 1 \\ 1 & 2 \end{bmatrix} \begin{Bmatrix} \ddot{\delta}_1 \\ \ddot{\delta}_4 \end{Bmatrix}$$ (4.26)

Finally, combining the mass matrix in eq. (2.6) for flexural effects with the eq. (4.26) for the axial effects results in the consistent mass matrix for a uniform element of a plane frame in reference to the local coordinates shown in Figure 4.22:

$$
\begin{Bmatrix} P_1 \\ P_2 \\ P_3 \\ P_4 \\ P_5 \\ P_6 \end{Bmatrix} = \frac{\bar{m}L}{420}
\begin{bmatrix}
140 & 0 & 0 & 70 & 0 & 0 \\
0 & 156 & 22L & 0 & 54 & -13L \\
0 & 22L & 4L^2 & 0 & 13L & -3L^2 \\
70 & 0 & 0 & 140 & 0 & 0 \\
0 & 54 & 13L & 0 & 156 & -22L \\
0 & -13L & -3L^2 & 0 & -22L & 4L^2
\end{bmatrix}
\begin{Bmatrix} \ddot{\delta}_1 \\ \ddot{\delta}_2 \\ \ddot{\delta}_3 \\ \ddot{\delta}_4 \\ \ddot{\delta}_5 \\ \ddot{\delta}_6 \end{Bmatrix}
\qquad (4.27)
$$

or, in condensed notation,

$$\{P\} = [m]_C \{\ddot{\delta}\}$$

in which $[m]_C$ is the consistent mass matrix for an element of a plane frame.

4.13   Coordinate Transformation

The mass matrix in eq. (4.27) is referred to nodal coordinates defined by local coordinate axes fixed on the beam element. As explained in Section 4.5, a transformation is required in order that the mass matrices for all the elements of the structure refer to the same set of coordinates; hence, the matrices become compatible for assemblage into the system mass matrix. The transformation developed in Section 4.5 to transform the stiffness matrix from the local coordinate system to the global also serves to transform the mass matrix. Then by analogy to eq. (4.17), we obtain, in reference to the global coordinates, the following relationship between inertial forces and the nodal acceleration:

$$\{\bar{P}\} = [\bar{m}] \{\ddot{\bar{\delta}}\}$$

in which

$$\{\bar{m}\} = [T]^T [M] [T] \qquad (4.28)$$

where $[\bar{m}]$ is the mass matrix for an element of a plane frame in reference to the global system of coordinates, and $[T]$ is the transformation matrix given by the square matrix in eq. (4.12).

Illustrative Example 4.7

Consider in Figure 4.23 a plane frame having two prismatic beam elements modeled with three degrees of freedom at joint 2 (translations along the X and Y directions and rotation around the Z axis). Determine the natural frequencies and corresponding normalized mode shapes. Use the lumped mass method. (Neglect damping.)

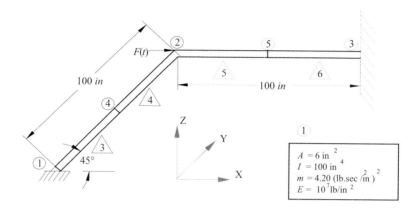

Fig. 4.23 Plane Frame of Illustrative Example 4.7

Solution

The stiffness matrix for element 1 or 2 in local coordinates is given by the matrix in eq. (4.8). After introducing numerical values, we obtain:

$$[k]_1 = [k]_2 = 10^3 \begin{bmatrix} 600 & 0 & 0 & -600 & 0 & 0 \\ 0 & 12 & 600 & 0 & -12 & 600 \\ 0 & 600 & 40000 & 0 & -600 & 20000 \\ -600 & 0 & 0 & 600 & 0 & 0 \\ 0 & -12 & -600 & 0 & 12 & -600 \\ 0 & 600 & 20000 & 0 & -600 & 40000 \end{bmatrix} \quad (a)$$

The transformation matrix for element 1 given by eq. (4.18) with $\theta = 45°$ results in:

$$
[T]_1 = \frac{1}{\sqrt{2}}
\begin{bmatrix}
1 & 1 & 0 & 0 & 0 & 0 \\
-1 & 1 & 0 & 0 & 0 & 0 \\
0 & 0 & \sqrt{2} & 0 & 0 & 0 \\
0 & 0 & 0 & 1 & 1 & 0 \\
0 & 0 & 0 & -1 & 1 & 0 \\
0 & 0 & 0 & 0 & 0 & \sqrt{2}
\end{bmatrix}
\tag{b}
$$

and for element 2 with $\theta = 0°$, the transformation matrix is the identity matrix (having ones in the main diagonal and zeros for the rest of the coefficients). Hence:

$$[T]_2 = [I].$$

The element stiffness matrices in reference to the global system of coordinates are calculated by eqs. (4.18) to yield:

Element 1

$$
[\bar{k}]_1 = 10^3
\begin{array}{cccccc}
4 & 5 & 6 & 1 & 2 & 3 \\
\begin{bmatrix}
306 & 294 & -424 & -306 & -294 & -424 \\
294 & 306 & 424 & -294 & -306 & 424 \\
-424 & 424 & 40{,}000 & 424 & -424 & 20{,}000 \\
-306 & -294 & 424 & 306 & 294 & 424 \\
-294 & -306 & -424 & 294 & 306 & -424 \\
-424 & 424 & 20{,}000 & 424 & -424 & 40{,}000
\end{bmatrix}
\begin{array}{c}
4 \\ 5 \\ 6 \\ 1 \\ 2 \\ 3
\end{array}
\end{array}
\tag{c}
$$

Element 2:

$$
[\bar{k}]_2 = 10^3
\begin{array}{cccccc}
1 & 2 & 3 & 7 & 8 & 9 \\
\begin{bmatrix}
600 & 0 & 0 & -600 & 0 & 0 \\
0 & 12 & 600 & 0 & -12 & 600 \\
0 & 600 & 40{,}000 & 0 & -600 & 20{,}000 \\
-600 & 0 & 0 & 600 & 0 & 0 \\
0 & -12 & -600 & 0 & 12 & -600 \\
0 & 600 & 20{,}000 & 0 & -600 & 40{,}000
\end{bmatrix}
\begin{array}{c}
1 \\ 2 \\ 3 \\ 7 \\ 8 \\ 9
\end{array}
\end{array}
\tag{d}
$$

The system stiffness matrix is then assembled by transferring the coefficients in matrices (c) and (d) to the system stiffness matrix. As stated in the previous chapters, it is expedient for the process assembling the system stiffness to indicate the corresponding system nodal coordinates at the top and on right of each matrix of the element. The transfer of the coefficients from eqs. (c) and (d) considering only the three free nodal coordinates with labels 1, 2, and 3 results in the reduced system stiffness matrix:

$$[K]_R = 10^3 \begin{bmatrix} 906 & 294 & 424 \\ 294 & 318 & 176 \\ 424 & 176 & 80{,}000 \end{bmatrix} \tag{e}$$

The system lumped mass matrix is readily assembled by considering the distributed mass in the beam elements lumped at the joints. Thus, at joint 2 the contributions of one-half of the beam elements intersecting at this joint result in 420 ($lb\text{-}sec^2/in$). Thus, the reduced system mass matrix results in

$$[M]_R = \begin{bmatrix} 420 & 0 & 0 \\ 0 & 420 & 0 \\ 0 & 0 & 0 \end{bmatrix} \tag{f}$$

Because the mass matrix in eq. (f) contains a zero in its main diagonal, it is necessary to condense the stiffness matrix to eliminate the nodal coordinate 3, which has an allocated zero mass value.

The application of the condensation method presented in Section 1.6 of Chapter 1 results in the following condensed system stiffness matrix:

$$[\overline{K}] = 10^3 \begin{bmatrix} 906.41 & 295.51 \\ 295.51 & 320.37 \end{bmatrix} \tag{g}$$

and transformation matrix:

$$[\overline{T}] = \begin{bmatrix} 0.005162 & 0.002138 \end{bmatrix}. \tag{h}$$

The mass matrix is readily condensed by simply eliminating the last column and the last row in eq. (f.) to obtain:

$$[\overline{M}] = \begin{bmatrix} 420 & 0 \\ 0 & 420 \end{bmatrix} \tag{i}$$

The natural frequencies are then found as the roots of the characteristic eq. (2.12):

$$|\,[\overline{K}] - \omega^2 [\overline{M}]\,| = 0. \tag{2.12 repeated}$$

Then, substituting into eq. (2.12), $[\overline{K}]$ from eq. (g) and $[\overline{M}]$ from eq. (i) results in:

$$\begin{vmatrix} 906E3 - 420\lambda & 296E3 \\ 296E3 & 320E3 - 420\lambda \end{vmatrix} = 0 \qquad (j)$$

in which, conveniently, it has been set $\lambda = \omega^2$.

The expansion of the determinant in eq. (j) results in the following quadratic equation:

$$\lambda^2 - 2.92 \times 10^3 + 1.147 \times 10^6 = 0. \qquad (k)$$

The roots of eq. (k) are then found to be:

$$\lambda_1 = \omega_1{}^2 = 468.0 \qquad\qquad \lambda_2 = \omega_2{}^2 = 2453.0.$$

Then the natural frequencies are

$$\omega_1 = 21.63 \; rad/sec \qquad\qquad \omega_2 = 49.5 \; rad/sec \qquad (l)$$

or in cps $(f = \omega/2\pi)$

$$f_1 = 3.44 \; cps \qquad\qquad f_2 = 7.88 \; cps. \qquad\qquad \text{Ans. (m)}$$

The mode shapes are determined as the nontrivial solution of eq. (2.11),

$$\left[[\overline{K}] - \omega^2 [\overline{M}] \right] \{a\} = \{0\}. \qquad (2.11) \text{ repeated}$$

Substituting into eq. (2.11) $\omega_1{}^2 = 468$ and setting $a_{11} = 1.0$, we obtain the first mode shape as

$$\{a\}_1 = \begin{Bmatrix} a_{11} \\ a_{21} \end{Bmatrix} = \begin{Bmatrix} 1.0000 \\ -2.3968 \end{Bmatrix}_1 \qquad (n)$$

which is conveniently normalized by dividing its components by the factor given by eq. (IV-5b) in Appendix IV:

$$\sqrt{\sum_{i=1}^{2} m_i a_{ij}^2} = \sqrt{420 x 1.000^2 + 420 x (-2.3968)^2} = 53.22.$$

The normalized first mode shape results in:

$$\{\phi\}_1 = \begin{Bmatrix} \phi_{11} \\ \phi_{21} \end{Bmatrix} = \begin{Bmatrix} 0.021879 \\ -0.04503 \end{Bmatrix}_1 \tag{o}$$

Analogously, the second mode shape results in

$$\{\phi\}_2 = \begin{Bmatrix} \phi_{12} \\ \phi_{22} \end{Bmatrix} = \begin{Bmatrix} 0.04499 \\ 0.002024 \end{Bmatrix}_2 \tag{p}$$

The components of the secondary condensed mode shape, ϕ_{31} or ϕ_{32} are then calculated by introducing into eq. (1.17b), $[\overline{T}]$ from eqs. (h), and $\{\phi\}_1$ from eq. (o) or $\{\phi\}_2$ from eq. (p) as follows:

$$\phi_{31} = [0.005162 \quad 0.002138] \begin{Bmatrix} 0.021879 \\ -0.04503 \end{Bmatrix}_1 = 0$$

and

$$\phi_{32} = [0.005162 \quad 0.002138] \begin{Bmatrix} 0.04499 \\ 0.002024 \end{Bmatrix}_2 = 2.43E{-}4. \tag{q}$$

Finally, from eqs. (0), (p), and (q), the three components of the first and of the second modal vectors are arranged in the columns of the modal matrix:

$$[\Phi] = \begin{bmatrix} 0.021879 & 0.04499 \\ -0.04503 & 0.002024 \\ 0 & 2.43E{-}4 \end{bmatrix} \tag{r}$$

Illustrative Example 4.8
Determine the maximum displacement at the nodal coordinates of the frame in Figure 4.23 when a force of magnitude 100,000 *lb* is suddenly applied at joint 2 along the X-direction. Use the lumped mass method and neglect damping.

Solution:
From Illustrative Example 4.7, the natural frequencies are $\omega_1 = 21.63$ *rad/sec* and $\omega_2 = 49.50$ *rad/sec*, and the modal matrix is

$$[\Phi] = \begin{bmatrix} 0.021879 & 0.04499 \\ -0.04503 & 0.002024 \\ 0 & 2.43E\text{--}4 \end{bmatrix} \qquad \text{(a)}$$

The modal equations are then given by eqs. (2.18) and (2.19):

$$\ddot{z}_j + \omega_j^2 z_j = P_j \qquad \text{(2.18) repeated}$$

where

$$P_j = \sum_{i=1}^{N} \phi_{ij} F_i. \qquad \text{(2.19) repeated}$$

In this example, the nodal applied forces are

$$F_1 = 100{,}000 \; lb, \qquad F_2 = 0, \qquad F_3 = 0.$$

We thus obtain, after substituting numerical values into eqs. (2.18) and (2.19), the modal equations as

$$\ddot{z}_1 + 468 z_1 = 2188$$
$$\ddot{z}_2 + 2453 z_2 = 4499. \qquad \text{(b)}$$

The general solution of equations (2.18) is of the form [see eq. (b) of Illustrative Example 2.3]:

$$z_j = \frac{P_j}{\omega_j^2}(1 - \cos \omega_j t) \qquad (j = 1, \, 2). \qquad \text{(c)}$$

Substitution of P_j from eq. (2.19) and ω_j yields

$$z_1 = 4.675(1 - \cos 21.63t)$$
$$z_2 = 1.834(1 - \cos 49.50t). \qquad \text{(d)}$$

The nodal displacements are then obtained from eq. (2.16):

$$\{u\} = [\Phi]\{z\} \qquad \text{(2.16) repeated}$$

which upon substitution of $[\Phi]$ from eq. (a) and $\{z\}$ from eq. (d) results in

$$u_1 = 0.1848 - 0.1023 \cos 21.63t - 0.0825 \cos 49.50t \quad in$$
$$u_2 = -0.2068 + 0.2105 \cos 21.63t - 0.0037 \cos 49.50t \quad in$$
$$u_3 = 0.00045 - 0.00045 \cos 4950t \; rad \qquad \text{(e)}$$

The maximum possible displacements at the nodal coordinates may then be estimated as the summation of the absolute values of the coefficients in the preceding expressions (cosine $= \pm 1$). Hence:

$$u_{1max} = 0.370 \; in \qquad u_{2max} = 0.421 \; in \qquad u_{3max} = 0.00090 \; rad \qquad\qquad \text{Ans.}$$

4.14   Dynamic Analysis of Plane Frames Using SAP2000

Illustrative Example 4.9
Use SAP2000 to solve Illustrative Examples 4.7 and 4.8.

Solution:
The following commands are implemented in SAP2000:

Begin: Open SAP2000
            Hint: Maximize both windows for a full view.

Units: In the lower right-hand corner of the screen select "lb-in".

Model: From the main menu select:
            FILE>NEW MODEL
            Number of Grid Spaces:
                 X direction = 2
                 Y direction = 0
                 Z direction = 1
            Grid Spaces
                 X direction = 70.7
                 Y direction = 1
                 Z direction = 70.7
            Click OK.

View: Click on the XZ icon on the main menu.
            Maximize the XZ window and use the PAN icon in the toolbar to drag the figure to the center of the screen.

Edit: From the main menu select:
            DRAW>EDIT GRID.
            Click on Direction X.
            In "X Location" highlight the entry 70.7 and change it to 100.
            Click on "Move Grid Line".
            Accept all other default entries.
            Click OK.

Draw: From the main menu select:
DRAW>DRAW FRAME ELEMENT.
Click on (−70.7, 0, 0) and drag cursor to (0, 0, 70.7), located at intersections of the grid lines. (One frame section completed.)
Click on (0, 0, 70.7) and drag cursor to (100, 0, 70.7).
(Second frame element completed.)
Click OK.

Label: From the main menu select:
VIEW>SET ELEMENTS.
Click on Joint Labels and Frame Labels. Then OK.
Use the PAN icon on the toolbar to center the figure on the screen.

Material: From the main menu select:
DEFINE>MATERIALS.
Under "Materials", select OTHER.
Click Modify/Show Material.
Enter:
    Material Name = OTHER
    Mass per Unit Volume = 0.7
    Weight per Unit Volume = 0
    Modulus of Elasticity = $10E6$
    Accept remaining default values.
    Click OK, OK.

Sections: From the main menu select:
DEFINE>FRAME SECTIONS.
Click Add I/Wide Flange and scroll down to click on "Add General".
Enter:
    Cross Sectional Area = 6
    Moment of Inertia about 3 axis = 100
    Shear area in directions 2 and 3 = 0
    Accept remaining default values.
    Material select OTHER.
    Click OK, OK, OK.

Assign: Select all the elements of the frame by clicking on them.
From the main menu select:
ASSIGN>FRAME>SECTIONS.
Select FSEC2.
Click OK.

Boundary: Select joints 1 and 3 by clicking on them and enter:
From the main menu select:

ASSIGN>JOINT>RESTRAINTS.
Select restraints in all directions.
Click OK.

Load: From the main menu select:
DEFINE>STATIC LOAD CASES.
Enter:
> Load = LOAD1
> Type = LIVE
> Self Weight Multiplier = 0
> Click on "Add New Load".
> Click OK.

Load Function: From the main menu select:
DEFINE>TIME HISTORY FUNCTIONS.
Click on Add New Function and accept default name FUNC1.
Enter: Time = 0
> Value = 100000

Click ADD.
Enter: Time = 0.5
> Value = 100000

Click ADD.
Click OK, OK.

Load Case: From the main menu select:
DEFINE>TIME HISTORY CASES.
Enter:
> Add New History
> Analysis Type = Linear
> Number of Output Time Steps = 50
> Time Steps = 0.01.

Click on "Envelopes".
Load Assignments:
> Load = LOAD1
> Function = FUNC1
> Accept all other default values.

Click ADD.
Click OK, OK.

Assign Loads: Click on Joint 2 and from the main menu select:
ASSIGN>JOINT STATIC LOADS>FORCES.
For LOAD1: Enter Load, Force Global X = 1.
Click OK.

Analyze: From the main menu select:
ANALYZE>SET OPTIONS.
Check only UX, RY, UZ in Available DOFs (Degrees of Freedom).
[Alternatively, click on picture of Plane Frame (Quick DOFs).]
Check Dynamic Analysis and then click Set Dynamic Parameters.
Enter: Number of Modes = 6.
Click OK, OK.

From the main menu select:
ANALYZE>RUN.
Enter filename: EXD 4.9.
Click SAVE.
At the conclusion of the calculations, after checking for errors, click OK.
Select X-Z view.

Print Input Tables: From the main menu select:
FILE>PRINT INPUT TABLES.
CLICK "Print to File" and accept all other default values.
Click OK.
Note: Using Notepad or similar word editor, the file EXD 4.9.txt may be
       opened, edited and printed.

(The edited Input Tables of Illustrative Example 4.9 are given in Table 4.8.)

Table 4.8   Edited Input Tables for Illustrative Example 4.9

STATIC LOAD CASES

STATIC CASE	CASE TYPE	SELF WT FACTOR
LOAD1	LIVE	0.0000

TIME HISTORY CASES

HISTORY CASE	HISTORY TYPE	NUMBER OF TIME STEPS	TIME STEP INCREMENT
HIST1	LINEAR	51	0.01000

JOINT DATA

JOINT	GLOBAL-X	GLOBAL-Y	GLOBAL-Z	RESTRAINTS
1	−70.70000	0.00000	0.00000	1 1 1 1 1 1
2	0.00000	0.00000	70.70000	0 0 0 0 0 0
3	100.00000	0.00000	70.70000	1 1 1 1 1 1

FRAME ELEMENT DATA

FRAME	JNT1	JNT2	SECT	ANGLE	RELEASES	SEG	R1	R2	FACTOR	LENGTH
1	1	4	FSEC2	0.000	000000	2	0.000	0.000	1.000	100.00
2	4	2	FSEC2	0.000	000000	2	0.000	0.000	1.000	100.00

Table 4.8 (continued)

JOINT	GLOBAL-X	GLOBAL-Y	GLOBAL-Z	GLOBAL-XX	GLOBAL-YY	GLOBAL-ZZ
2	1.000	0.000	0.000	0.000	0.000	0.000

J O I N T   F O R C E S   Load Case LOAD1

Plot Displacement Function: From the main menu select:
      DISPLAY>SHOW TIME HISTORY TRACES.
      Click Define Functions.
      Click to "Add Joint Disps/Forces".
          Joint ID = 2
          Vector Type = Displacement
          Component = UX
          Mode Number = Include All
          Click OK.

      Click On Define Functions
          Vector Type = Displacement
          Click to "Add Joint Disps/Forces"
          Joint ID = 2
          Component = UZ
          Mode Number = Include All
          Click OK, OK.

      Under List of Functions click once on "Joint2".
      Click ADD to move "Joint2" to Plot Functions column.

      Under List of Functions click once on "Joint2-1".
      Click ADD to move "Joint2-1" to Plot Functions column.

      Axis Labels
          Horizontal = Time (sec)
          Vertical = Displ. X, Z at Joint 2 (in)

      Click "Display" to view Displacement functions.
      FILE>PRINT GRAPHICS to obtain a printed version of these plots.

      (Figure 4.24 reproduces the plot for the displacement components X and Z at Joint 2 of the plane frame of Illustrative Example 4.9.)

      Alternatively, a table of the displacement component functions in the X and Z directions of joint 2 may be obtained by selecting from the main menu:
      FILE>PRINT TABLES TO FILE.

Then, using Notepad or similar word processor, the file EXD4.9.txt may be accessed and edited to obtain a table of displacement functions. An abbreviated table of the displacement component functions in the X and Z directions for joint 2 is reproduced in Table 4.10. The values in this table may be plotted using Excel or similar software.

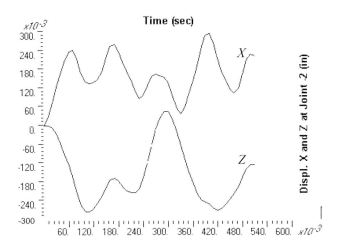

Fig 4.24   Displacements UX and UZ at node 2 for the
plane frame of Illustrative Example 4.9

Note: Displacement values may be viewed on screen by moving the cursor along the curves. Table 4.9 provides the maximum displacement components in the X and Z directions at joint 2 for the plane frame of Illustrative Example 4.9.

Table 4.9 Maximum Displacements at node 2 in X and Z directions for the plane frame of Illustrative Example 4.9

Direction	Maximum Displacement (in)
X	0.296
Z	0.274

Table 4.10 Displacement Component Functions at Joint 2 in the X and Z directions for the plane frame of Illustrative Example 4.9

Time (sec)	Displ. Components at Joint 2	
	X Direction (in)	Z Direction (in)
0.00	0	0
0.01	0.02218	−0.00005
0.02	0.07267	−0.00675
0.03	0.12346	−0.02569
0.04	0.16516	−0.05606
0.05	0.20431	−0.09073
----------------------	----------------------	----------------------
----------------------	----------------------	----------------------
0.49	0.05393	0.02457
0.50	0.06737	0.04241

Natural Periods and Frequencies:

Using Notepad or other word editor, open the following file:
C:/SAP2000e/EXD 4.9.OUT.
The Natural Period and Frequency table may be found in this file.

(Table 4.11 provides the edited Natural Period and Frequency values for the plane frame of Illustrative Example 4.9.)

Table 4.11 Natural period and frequencies for the plane frame of Illustrative Example 4.9

NATURAL PERIODS AND FREQUENCIES

MODE	PERIOD (SEC)	FREQUENCY (CYC/SEC)	FREQUENCY (RAD/SEC)	EIGENVALUE $(RAD/SEC)^2$
1	0.291	3.440	21.616	467.24
2	0.127	7.863	49.406	2441.00

Illustrative Example 4.10

Solve Illustrative Example 4.9 after modeling the plane frame of that example into 4, 8, and 16 frame elements using the command "Divide Frames".

Open the file EXD 4.9 in SAP2000

Enter the following commands:

Unlock file EXD 4.9: Click the padlock icon on the toolbar and then "Yes".

Divide Frames:
Select all the frame elements by clicking on each frame, or alternatively, by clicking on any point outside the plot and dragging the pointer to wrap the plane frame. Then from the main menu select:
EDIT > DIVIDE FRAMES.
Accept the default "Divide frame in 2".
Click OK.

Analyze: From the main menu select:
ANALYZE > RUN.
At the conclusion of the calculations, check for errors; then click OK.

Natural Periods and Frequencies:
Using Notepad or other word editor, open the following file:
C:/SAP2000E/EXD 5.9.OUT.
The Natural Period and Frequency table may be found in this file.

(Table 4.12 provides the edited table of Natural Period and Frequency values for the plane frame of Illustrative Example 4.10.)

Table 4.12   Natural Period and Frequencies for the Plane Frame of Illustrative Example 4.10

NATURAL PERIODS AND FREQUENCIES

MODE	PERIOD (SEC)	FREQUENCY (CYC/SEC)	FREQUENCY (RAD/SEC)	EIGENVALUE $(RAD/SEC)^2$
1	0.288	3.471	21.806	475.502
2	0.275	3.632	22.819	520.722
3	0.154	6.486	40.754	1660.873
4	0.113	8.849	55.603	3091.686
5	0.056	17.702	111.222	12370.304
6	0.0467	21.422	134.596	18116.000

Displacements: From the main menu select:
DISPLAY>SHOW TIME HISTORY TRACES.
Click Define Functions.
Click to "Add Joint Disps/Forces".
    Joint ID = 2
    Vector Type = Displacement
    Component = UX
    Mode Number = Include All
    (UX displacement function is now stored in "Joint2".)
    Click OK.

Click Define Functions.
Click to "Add Joint Disps/Forces"

Joint ID = 2
Vector Type = Displacement
Component = UZ
Mode Number = Include All
(UZ displacement function is now stored in "Joint2-1")
Click OK.

Under List of Functions, click once on "Joint2".
Click ADD to move "Joint2" to Plot Functions column.
Axis Labels
    Horizontal = Time (*sec*)
    Vertical = Displ. X at Joint 2 (*in*)
To view the plotted displacement function, click Display.
To obtain a printed version of this plot enter:
FILE>PRINT GRAPHICS.

Max. and Min. Displacement values are shown on the right side of the
window.
(Maximum absolute value of the displacement UX = 0.30 *in.*)
Click OK.

Click on Joint2 to remove it from the plot function column.
Under List of Functions, click once on "Joint2-1".
Click ADD to move "Joint2-1" to the Plot Functions column.
Axis Labels ﹅
    Horizontal = Time (*sec*)
    Vertical = Displacement Component UZ at Joint 2 (*in*)
Click Display to view the plot Displacement function.
To obtain a printed version of this plot select:
FILE>PRINT GRAPHICS.

Max. and Min. Displacement values are shown on the right side of the
window.
(Maximum absolute value of the displacement UZ = 0.34 *in.*)
Click OK.
Click DONE.

To analyze the plane frame of this example with 8 and then with 16 frame
elements, repeat the commands listed previously for this Illustrative
Example 4.10.

Table 4.13 provides the natural frequencies obtained in the analysis of the
plane frame of Illustrative Example 4.10 modeled using 2, 4, 8, and 16
frame elements.

Table 4.14 provides the maximum values for the displacements of joint 2 of the plane frame of Illustrative Example 4.10 modeled using 2, 4, 8, and 16 frame elements.

Table 4.13 Natural frequencies (*cps*) for the grid frame of Illustrative Example 4.10 modeled using 2, 4, 8, and 16 frame elements

	Number of Frame Elements in the Model			
Mode #	2 frames	4 frames	8 frames	16 frames
1	3.44	3.47	3.53	3.53
2	7.86	3.63	3.77	3.77
3		6.48	7.71	7.75
4		8.85	9.15	9.21
5		17.70	12.13	12.40
6		21.42	16.09	16.70

Table 4.14 Maximum displacements (*in*) for the grid frame of Illustrative Example 4.10 modeled using 2, 4, 8, and 16 frame elements

	Number of Frame Elements in the Model			
Component	2 frames	4 frames	8 frames	16 frames
UX	0.31	0.30	0.26	0.25
UZ	0.35	0.34	0.33	0.33

4.15 Analytical Problems

Analytical Problem 4.1
Demonstrate that the stiffness coefficients of a beam element, k_{ij}, for axial effect may be calculated by eq. (4.7), repeated here for convenience.

$$k_{ij} = \int_0^L AE\, u_i'(x)\, u_j'\, dx \qquad \text{(4.7) repeated}$$

in which

    E is the modulus of elasticity

    A is the cross-sectional area

    L is the length of the beam element, and

    u_i', u_j' are derivatives with respect to x of shape functions given by

$$u_1(x) = 1 - \frac{x}{L} \qquad \text{(for } \delta_1 = 1.0 \text{)} \qquad \text{(4.5) repeated}$$

and

$$u_2(x) = \frac{x}{L} \qquad \text{(for } \delta_2 = 1.0 \text{)} \qquad \text{(4.6) repeated}$$

Consider in Figure P4.1 a beam element to which a unit displacement $\delta_1 = 1$ has been applied at node 1. The axial force required to produce this displacement is given by eq. (4.4) as

$$P = AE \frac{du}{dx} = AEu_1'. \qquad (a)$$

To this displaced beam, we superimpose an axial deformation by giving a unit displacement ($\delta_2 = 1$) at node 2 that results in deformation du_2 of a differential element dx of the beam, which may be expressed as

$$du_2 = u_2'(x)dx. \qquad (b)$$

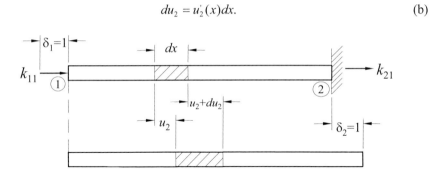

Fig. P4.1   Axial displacement given at node 2 ($\delta^2 = 1$) of a beam element previously subjected to a unitary axial displacement at node 1 ($\delta_1 = 1$)

The Principle of Virtual Work, which states that during the virtual displacement the work of external forces is equal to the work performed by the internal force, is applied. The external work W_E is calculated as the product of the external force k_{21} times the displacement $\delta_2 = 1.0$, that is

$$W_E = k_{21}. \qquad (c)$$

The internal work dW_I on the differential element dx is then calculated as the product of the axial force P times the differential displacement du_2:

$$dW_I = Pdu_2$$

or using eqs. (a) and (b)

$$dW_I = AEu_1' u_2' dx$$

and for the entire beam element as

$$W_I = \int_0^L AE\, u_1'\, u_2'\, dx. \tag{d}$$

Equating the external virtual work, eq. (c), with the internal virtual work, eq. (d), results in

$$k_{21} = \int_0^L AE\, u_i'\, u_2'\, dx \tag{e}$$

and in general

$$k_{ij} = \int_1^L AE\, u_i'\, u_j'\, dx. \tag{Q.E.D}$$

Analytical Problem 4.2
Demonstrate that the equivalent nodal forces, Q_i, for a beam element axially loaded by a force function $p(x)$ may be calculated by

$$Q_i = \int_0^L p(x)\, u_i(x)dx \tag{4.19) repeated}$$

in which $u_i(x)$ is the shape function [eq. (4.5) or (4.6)] resulting from a unit nodal displacement.

Solution:
Consider in Figure P4.2(a) a beam element showing a general axially distributed force $p(x)$ and in Figure P4.2(b) the beam element showing the nodal equivalent axial forces Q_1 and Q_2. We then give to both beams a unit virtual displacement at nodal coordinate ($\delta_1 = 1$), which will result in an axial displacement along the length of the beam given by eq. (4.5):

$$u_1(x) = 1 - \frac{x}{L}. \tag{4.5) repeated}$$

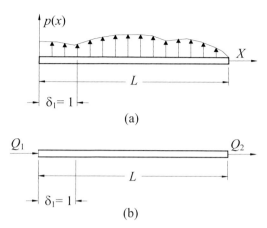

Fig. P4.2   (a) Beam element supporting a general axial force $p(x)$
(b) Beam element supporting the equivalent nodal forces Q_1 and Q_2

We require that the resulting virtual work of the forces Q_i be equal to the virtual work performed by the externally applied axial load $p(x)$. The work W_E^* of the equivalent forces Q_i is simply

$$W_E^* = Q_1 \delta_1 = Q_1 \qquad\qquad\text{(a)}$$

since Q_1 is the only equivalent force undergoing displacement and $\delta_1 = 1.0$.

The work performed on a differential element "dx" by the external force $p(x)\ dx$ is equal to $p(x)\ u_1(x)dx$ and the total work W_E^{**} is then

$$W_E^{**} = \int_0^L p(x)\ u_1\ (x)dx. \qquad\qquad\text{(b)}$$

Equating these two calculations of the virtual work results in the following expression for determining the equivalent nodal force Q_1:

$$Q_1 = \int_0^L p(x)\ u_1\ (x)dx.$$

In general, the expression for the equivalent nodal force Q_i is then given by

$$Q_i = \int_0^L p(x)\, u_i(x)dx \qquad\qquad\text{Q.E.D.}$$

where u_i $(i = 1, 2)$ is given by eq. (4.5) or by eq. (4.6).

Analytical Problem 4.3
Demonstrate that the mass influence coefficient m_{ij} (force at nodal coordinate i due to a unit acceleration at nodal coordinate j) for axial effect may be calculated by eq. (4.23), repeated here for convenience

$$m_{ij} = \int_0^L \overline{m}(x)u_i(x)u_j(x)dx \qquad (4.23) \text{ repeated}$$

in which $\overline{m}(x)$ is the mass per unit of length, and u_i and u_j displacement functions given by eqs. (4.5) and (4.6).

Solution:
The displacement $u(x,t)$ at any section x of a beam element due to dynamic nodal displacements, $\delta_1(t)$ and $\delta_2(t)$, is obtained by superposition. Hence,

$$u(x,t) = u_1(x)\delta_1(t) + u_2(x)\delta_2(t) \qquad (a)$$

in which $u_1(x)$ and $u_2(x)$ are given by eqs. (4.5) and (4.6).

Now consider the beam of Figure P4.3 undergoing a unit acceleration, $\ddot{\delta}_1(t) = 1$, which by eq. (a) results in an acceleration at x given by

$$\ddot{u}_1(x,t) = u_1(x)\ddot{\delta}_1(t)$$

or

$$\ddot{u}_1(x,t) = u_1(x) \qquad (b)$$

since $\ddot{\delta}_1(t) = 1$.

The external work due to a virtual displacement, $\delta_2 = 1.0$ is then given by

$$W_E = m_{21}\delta_2$$

or

$$W_E = m_{21} \qquad (c)$$

since $\ddot{\delta}_1(t) = 1.0$.

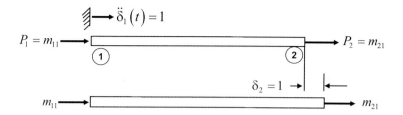

Fig. P4.3 Beam element with unit displacement at node 2 ($\delta_2 = 1$) undergoing a unit axial acceleration at node 1 [$\ddot{\delta}_1$ (t) = 1].

The internal force "f" on an element dx is equal to:

$$f = \overline{m}(x)\ddot{u}_1(x)dx$$

or in view of eq. (b)

$$f = \overline{m}(x)u_1(x)dx \tag{d}$$

and the internal work of the inertial force f during this virtual displacement ($\delta_2 = 1.0$) is

$$\delta W_I = f\, u_2(x)$$

or, from eq. (d)

$$\delta W_I = \overline{m}\,(x)\, u_1\,(x)\, u_2\,(x).$$

Hence, the total internal work is

$$W_I = \int_0^L \overline{m}(x)\, u_1(x)u_2(x)dx. \tag{d}$$

Finally, equating the external work, eq. (c) and the internal work, eq. (d) yields

$$m_{21} = \int_0^L \overline{m}(x)u_1(x)u_2(x)dx \tag{e}$$

or, in general,

$$m_{ij} = \int_0^L \overline{m}(x)u_i(x)u_j(x)dx. \qquad \text{Q.E.D.} \quad (4.23)$$

4.16 Practice Problems

Problem 4.4

For the plane frame shown in Figure P4.4, determine:

(a) Displacements at node 1
(b) Reaction at the supports

All Members:
$E = 30\ GP_a$
$A = 80\ cm^2$
$I = 50,000\ cm^4$

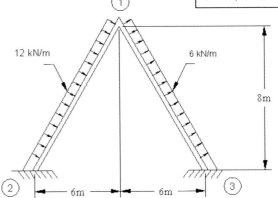

Fig. P4.4

Problem 4.5

Use SAP2000 to solve the plane frame shown in Figure P4.5, which has an inclined roller support.

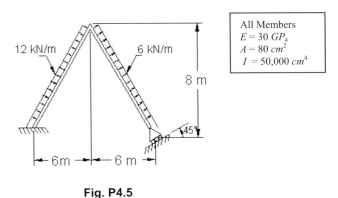

All Members
$E = 30\ GP_a$
$A = 80\ cm^2$
$I = 50,000\ cm^4$

Fig. P4.5

Problems 4.6, 4.7, and 4.8

Determine the joint displacements, element end forces, and support reactions for the frames shown in Figs.P4.6, P4.7, and P4.8.

Problem 4.6

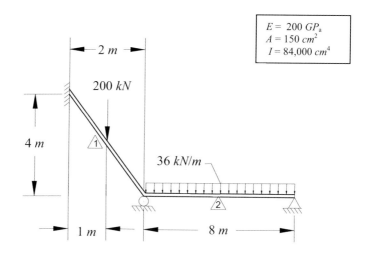

$E = 200\ GP_a$
$A = 150\ cm^2$
$I = 84{,}000\ cm^4$

Fig. P4.6

Problem 4.7

$E = 29{,}500\ ksi$
$A = 12\ in^2$
$I = 312\ in^4$

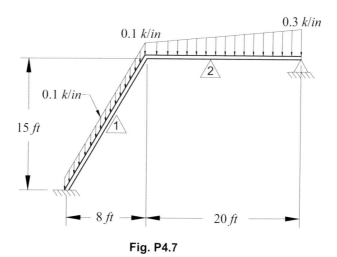

Fig. P4.7

Problem 4.8

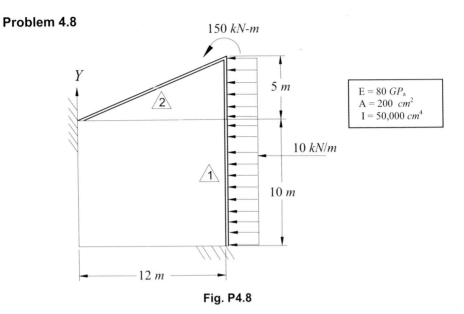

Fig. P4.8

E = 80 GP_a
A = 200 cm^2
I = 50,000 cm^4

Problem 4.9

For the concrete frame shown in Figure P4.9, determine:

(a) Joint displacements
(b) Element forces
(c) Reactions at the supports

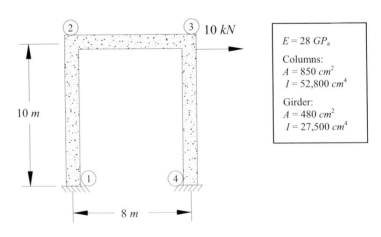

E = 28 GP_a

Columns:
A = 850 cm^2
I = 52,800 cm^4

Girder:
A = 480 cm^2
I = 27,500 cm^4

Fig. P4.9

Problem 4.10

For the frame shown in Figure P4.10 determine:

(a)   Joint displacements
(b)   Member end forces
(c)   Support reactions

The braced diagonals
are hinged at the ends.

Girder and Columns:
$A = 100\ in^2$
$I = 800\ in^4$

Braced Diagonals:
$A = 5.0\ in^2$
$I = 84\ in^4$

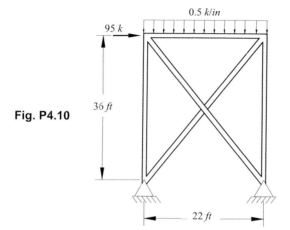

Fig. P4.10

Problem 4.11

For the plane frame shown in Figure P4.11, determine:

(a)   The displacements at the joints
(b)   Member end forces
(c)   Reactions at the supports

Lower story girders: $I = 280\ in^4;\ A = 28\ in^2$
Upper story girder: $I = 250\ in^4;\ A = 16\ in^2$
Lower story columns: $I = 188\ in^4;\ A = 9.8\ in^2$
Upper story columns: $I = 108\ in^4;\ A = 8.6\ in^2$

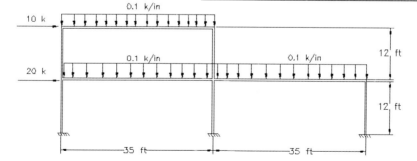

Fig. P4.11

Problem 4.12

For the plane frame shown in Figure P4.12 determine:

(a)   Joint displacements
(b)   Member end forces
(c)   Reactions at the supports

$E = 32\ GP_a$
$A = 250\ cm^2$
$I = 18,000\ cm^4$

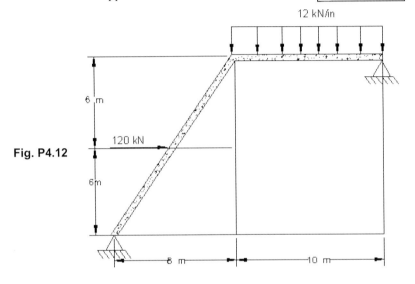

Fig. P4.12

Problem 4.13

For the plane frame shown in Figure P4.13 determine: (a) Joint Displacements, (b) Member end forces, and (c) Support reactions.

$E = 4500\ ksi$

Columns:
$A = 100\ in^2$
$I = 600\ in^4$

Girder:
$A = 120\ in^2$
$I = 1500\ in^4$

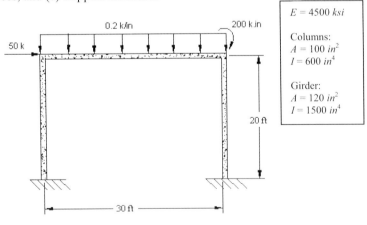

Fig. P4.13

Problem 4.14

For the plane frame shown in Figure P4.14 determine:

(a)   Joint displacements
(b)   Member end forces
(c)   Support reactions

All Members:
E = 29,500 ksi
A = 24 in²
I = 1200 in⁴

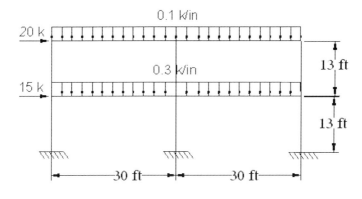

Fig. P4.14

Problem 4.15

For the plane frame shown in Figure P4.15 determine:

(a)   Joint displacements
(b)   Member end forces
(c)   Support reactions

E = 29,500 ksi
I = 800 in⁴
A = 22 in²

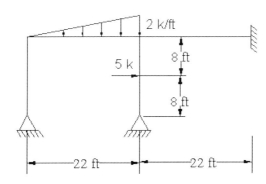

Fig. P4.15

Problem 4.16
Determine the natural frequencies and corresponding normal modes for the frame shown in Figure P4.15.

Problem 4.17
Determine the response of the frame shown in Figure P4.15 when it is acted upon by a force $F(t) = 1.0$ *kip* suddenly applied horizontally on the girder for 0.5 *sec*. Neglect damping in the system.

Problem 4.18
Determine the maximum response of the frame shown in Figure P4.18(a) when subjected to the triangular impulsive load shown in Figure 4.18(b) applied along the nodal coordinate 2. Neglect damping in the system.

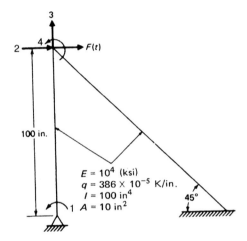

Fig. P4.18(a)

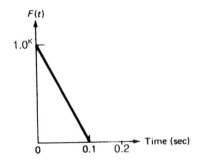

Fig. P4.18(b)

Problem 4.19
Determine the steady-state response of the frame shown in Figure P4.18(a) when subjected to harmonic force $F(t) = 10 \sin 30t$ kip along nodal coordinate 2. Neglect damping in the system.

Problem 4.20
Repeat Problem 4.19 assuming 10% of the critical damping.

Problem 4.21
The frame shown in Figure P4.21, determine the natural frequencies and corresponding normalized mode shapes.

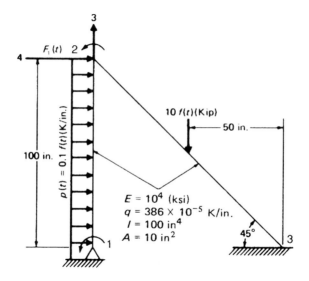

Fig. P4.21

Problem 4.22
Determine the response of the frame shown in Figure P4.21 when subjected to dynamic forces contained in the function $f(t) = F_1(t)$ shown in Figure 4.23(b).

Problem 4.23
Determine the response for the plane frame shown in Figure P4.23(a) when subjected to the force $F_3(t)$ depicted in Figure 4.23(b) acting along nodal coordinate 1. Assume 5% damping in all of the modes.

Problem 4.24

Determine the steady-state response of the frame in Figure P4.23(a) subjected to the harmonic force $F_1(t) = 10\cos 50t$ kip applied at nodal coordinate 1. Neglect damping in the system

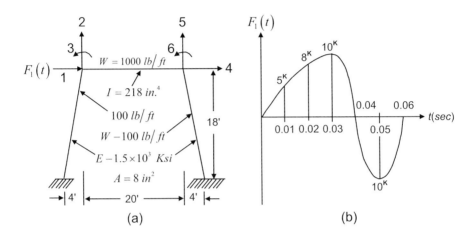

(a)   (b)

Fig. P4.23

Problem 4.25

Determine the response of the plane frame shown in Figure P4.23(a) when acted upon by the force $F_1(t)$ depicted in Figure P4.25 and applied in the horizontal direction at nodal coordinate 1. Assume 10% damping in all the modes.

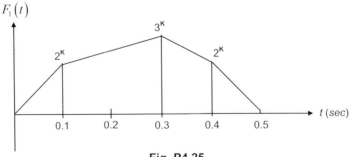

Fig. P4.25

5 Grid Frames

5.1   Introduction

In Chapter 4 consideration was given to the structural analysis of plane frames with loads applied in the plane of the frame. When the plane structural frame is subjected to loads applied normally to its plane, the structure is referred to as a *grid frame*. These structures also could be treated as three-dimensional frames (Chapter 6.) However, structures modeled as plane frames or as grid frames are treated as special cases because there is an immediate reduction in the number of nodal coordinates of the structure. When analyzing the plane frame under the action of loads in its plane, only the nodal displacements that have to be considered are the translations in the X and Y directions, and rotation about the Z axis: thus resulting in a total of three nodal coordinates at each node of the structure. Also, for a grid frame located on the plane *X-Y* and loaded normal to the plane of the structure, only three nodal coordinates are considered at each node: translation in the Z direction and rotations about the *X* and *Y* axes. Consequently, the analysis of structures modeled either as plane frames or as grid frames requires consideration of only three nodal coordinates at each node of the structure, while treating these structures as three dimensional frames (presented in Chapter 6) would require six nodal coordinates at each node, a considerable increase in the size of the problem to be analyzed.

5.2   Torsional Effects

The analysis of grid frames using the stiffness method, that is, for plane frames subjected to normal loads, requires the determination of the torsional stiffness coefficients of an element of the grid frame. The derivation of these coefficients is essentially identical to the derivation for the axial stiffness coefficients of a beam element. Similarity between these two derivations occurs because the differential equation for both problems has the same mathematical form. For the axial problem, the differential equation for the displacement function is given by eq. (4.4) as:

$$\frac{du}{dx} = \frac{P}{AE}.$$ (5.1)

Likewise, the differential equation for torsional angular displacement is

$$\frac{d\theta}{dx} = \frac{T}{JG}$$ (5.2)

where,

u	=	linear displacement	θ =	angular torsional displacement
P	=	axial force	T =	torsional moment
E	=	modulus of elasticity	G =	modulus of elasticity in shear
A	=	cross-section area	J =	torsional constant
				(Polar moment of inertia for circular sections)

As a consequence of the analogy between eqs. (5.1) and (5.2), we can express the displacement functions for torsional effects as corresponding functions for displacements for axial effects; hence by analogy to eqs. (4.5) and (4.6) and in reference to the nodal coordinates shown in Figure 5.1, we have

$$\theta_1(x) = \left(1 - \frac{x}{L}\right)$$ (5.3)

and

$$\theta_2(x) = \frac{x}{L}$$ (5.4)

in which the angular displacement function $\theta_1(x)$ corresponds to the linear displacement function $u_1(x)$ and the angular displacement function $\theta_2(x)$ corresponds to the linear displacement function $u_2(x)$. Also analogously to eq. (4.7), the stiffness coefficients for torsional effects may be calculated from

$$k_{ij} = \int_0^L JG\,\theta_i'(x)\,\theta_j'(x)\,dx$$ (5.5)

in which $\theta_i(x)$ and $\theta_j(x)$ are the derivatives with respect to x of the angular displacement functions $\theta_i(x)$ and $\theta_j(x)$ given by eqs. (5.3) or (5.4).

Fig. 5.1 Nodal torsional coordinates for a beam element

5.3   Stiffness Matrix for an Element of a Grid Frame

The application of eq. (5.5) for a uniform element of a grid frame yields the stiffness coefficients $k_{11} = k_{22} = -k_{12} =- k_{21} = JG/L$.

Thus, in matrix notation, the relationship between the torsional moments T_1 and T_2 and the angular displacements θ_1 and θ_2 at the two nodes of the element is

$$\begin{Bmatrix} T_1 \\ T_2 \end{Bmatrix} = \frac{JG}{L} \begin{bmatrix} 1 & -1 \\ -1 & 1 \end{bmatrix} \begin{Bmatrix} \theta_1 \\ \theta_2 \end{Bmatrix}. \tag{5.6}$$

Finally, the torsional stiffness matrix in eq. (5.6) is combined with the flexural stiffness matrix in eq. (1.11) to obtain the stiffness matrix for an element of a grid frame. In reference to the local coordinate system indicated in Figure 5.2(a), the stiffness equation for a uniform grid element is

$$\begin{Bmatrix} P_1 \\ P_2 \\ P_3 \\ P_4 \\ P_5 \\ P_6 \end{Bmatrix} = \frac{EI}{L^3} \begin{bmatrix} JGL^2/EI & 0 & 0 & -JGL^2/EI & 0 & 0 \\ 0 & 4L^2 & -6L & 0 & 2L^2 & 6L \\ 0 & -6L & 12 & 0 & -6L & -12 \\ -JGL^2/EI & 0 & 0 & JGL^2/EI & 0 & 0 \\ 0 & 2L^2 & -6L & 0 & 4L^2 & 6L \\ 0 & 6L & -12 & 0 & 6L & 12 \end{bmatrix} \begin{Bmatrix} \delta_1 \\ \delta_2 \\ \delta_3 \\ \delta_4 \\ \delta_5 \\ \delta_6 \end{Bmatrix} \tag{5.7}$$

or in condensed notation

$$\{P\} = [k]\{\delta\} \tag{5.8}$$

in which $\{P\}$ and $\{\delta\}$ are, respectively, the force and the displacement vectors at the nodes of the grid element and $[k]$ is the element stiffness matrix defined in eq. (5.7) in reference to the local system of coordinates.

5.4   Transformation of Coordinates

The stiffness matrix [eq. (5.7)] is expressed in reference to the local system of coordinates, as shown in Figure 5.2(a).

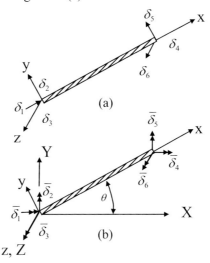

Fig. 5.2   Components of nodal displacements for a grid element
(a) Local coordinate system, (b) Global coordinate system

Therefore, before the assemblage of the stiffness matrix for the structure, it is necessary to transform the reference of the element stiffness matrix to the global system of coordinates as shown in Figure 5.2(b). Since the z axis in the local coordinate system coincides with the Z axis for the global system, it is only necessary to perform a transformation of coordinates in the x-y plane. The corresponding matrix for this transformation may be obtained by establishing the relationship between components in these two systems of coordinates of the moments at the nodes. In reference to Figure 5.3, these relations when written for node 1 are

$$P_1 = \overline{P}_1 \cos \theta + \overline{P}_2 \sin \theta$$
$$P_2 = -\overline{P}_1 \sin \theta + \overline{P}_2 \cos \theta \qquad\qquad (5.9)$$
$$P_3 = \overline{P}_3$$

and for node 2

$$P_4 = \overline{P}_4 \cos \theta + \overline{P}_5 \sin \theta$$
$$P_5 = -\overline{P}_4 \sin \theta + \overline{P}_5 \cos \theta \qquad\qquad (5.10)$$
$$P_6 = \overline{P}_6$$

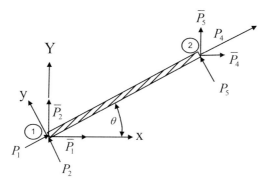

Fig. 5.3   Components in the local and global coordinates of the nodal moments in an element of a grid frame

It should be noticed that eqs. (5.9) and (5.10) are identical to eqs. (4.10) and (4.11) derived for the transformation of coordinates for the nodal forces of an element of a plane frame.

Equations (5.9) and (5.10) may be written in matrix notation as

$$
\begin{Bmatrix} P_1 \\ P_2 \\ P_3 \\ P_4 \\ P_5 \\ P_6 \end{Bmatrix}
=
\begin{bmatrix}
\cos\theta & \sin\theta & 0 & 0 & 0 & 0 \\
-\sin\theta & \cos\theta & 0 & 0 & 0 & 0 \\
0 & 0 & 1 & 0 & 0 & 0 \\
0 & 0 & 0 & \cos\theta & \sin\theta & 0 \\
0 & 0 & 0 & -\sin\theta & \cos\theta & 0 \\
0 & 0 & 0 & 0 & 0 & 1
\end{bmatrix}
\begin{Bmatrix} \overline{P}_1 \\ \overline{P}_2 \\ \overline{P}_3 \\ \overline{P}_4 \\ \overline{P}_5 \\ \overline{P}_6 \end{Bmatrix}
\tag{5.11}
$$

or in short notation

$$
\{P\} = [T]\{\overline{P}\} \qquad \text{and} \qquad \{\overline{P}\} = [T]^T\{P\} \tag{5.12}
$$

in which $\{P\}$ and $\{\overline{P}\}$ are the vectors of the nodal forces of an element of a grid frame, respectively, in local and global coordinates and $[T]$ the transformation matrix defined in eq. (5.11). The same transformation matrix $[T]$ also serves to transform the nodal components of the displacements from a global to a local system of coordinates. In condensed notation, this transformation is given by

$$
\{\delta\} = [T]\{\overline{\delta}\} \tag{5.13}
$$

where $\{\delta\}$ and $\{\overline{\delta}\}$ are, respectively, the components of nodal displacements in local and global coordinates. The substitution of eqs. (5.12) and (5.13) in the stiffness equation, eq. (5.8) results in

$$[T]\{\overline{P}\} = [k][T]\{\overline{\delta}\}$$

or since $[T]$ is an orthogonal matrix $([T]^{-1} = [T]^{T})$, it follows that

$$\{\overline{P}\} = [T]^{T} [k][T]\{\overline{\delta}\}$$

or

$$\{\overline{P}\} = [\overline{k}]\{\overline{\delta}\} \tag{5.14}$$

in which

$$[\overline{k}] = [T]^{T} [k] [T] \tag{5.15}$$

is the element stiffness matrix in reference to the global coordinate system.

5.5   Analysis of Grid Frames

The structural analysis of grid frames is mathematically identical to the analysis of beams or plane frames presented in Chapters 1 and 4, respectively. These analyses differ only in the selection of nodal coordinates and the expressions corresponding to the stiffness matrix for the elements in each structure. The following numerical example provides a detailed analysis of a simple grid frame.

Illustrative Example 5.1
For the grid frame shown in Figure 5.4, perform the structural analysis to determine the following:

   (a) Displacements at the joints between elements.
   (b) End forces on the elements.
   (c) Reactions of the supports.

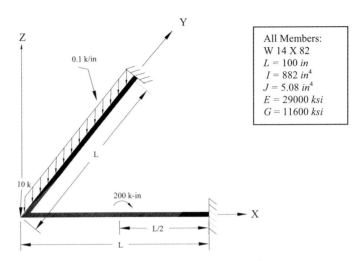

Fig. 5.4 Grid frame for Illustrative Example 5.1

Solution:

1.   Model the grid frame.

The grid frame of Illustrative Example 5.1 has been modeled with two elements, three nodes and nine system nodal coordinates as shown in Figure 5.5 The first three system nodal coordinates correspond to the free nodal coordinates and the last six to the fixed nodal coordinates as labeled in Fig. 5.5.

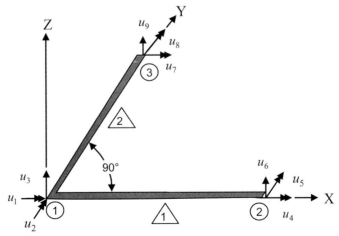

Fig. 5.5   Modeled grid frame of Illustrative Example 5.1 showing the nodal coordinates u_1 through u_9

2. Element stiffness matrices (local coordinates).

ELEMENT 1 or 2

Substituting numerical values into the stiffness matrix in eq. (5.7) yields

$$[k]_1 = [k]_2 = \begin{array}{cccccc} 1 & 2 & 3 & 4 & 5 & 6 \\ \begin{bmatrix} 589.3 & 0 & 0 & -589.3 & 0 & 0 \\ 0 & 1.023E6 & -1.535E4 & 0 & 5.116E5 & 1.535E4 \\ 0 & -1.535E4 & 306.9 & 0 & -1.535E4 & -3.069E2 \\ -589.3 & 0 & 0 & 589.3 & 0 & 0 \\ 0 & 5.116E5 & -1.535E4 & 0 & 1.023E6 & 1.535E4 \\ 0 & 1.535E4 & -3.069E2 & 0 & 1.535E4 & 3.069E2 \end{bmatrix} & \begin{array}{c} 1 \\ 2 \\ 3 \\ 4 \\ 5 \\ 6 \end{array} \end{array} \quad (a)$$

3. Element transformation matrices.

ELEMENT 1: Since $\theta = 0$

$$[T]_1 = [I]$$

ELEMENT 2

From eq. (5.11) with $\theta = 90°$:

$$[T]_2 = \begin{bmatrix} 0 & 1 & 0 & 0 & 0 & 0 \\ -1 & 0 & 0 & 0 & 0 & 0 \\ 0 & 0 & 1 & 0 & 0 & 0 \\ 0 & 0 & 0 & 0 & 1 & 0 \\ 0 & 0 & 0 & -1 & 0 & 0 \\ 0 & 0 & 0 & 0 & 0 & 1 \end{bmatrix} \quad (b)$$

4. Element stiffness matrices (global coordinates).

ELEMENT 1: Since $\theta = 0$:

$$[\bar{k}]_1 = [k]_1$$

ELEMENT 2: with $\theta = 90°$

The substitution of the stiffness matrix $[k]_2$ from eq. (a) and the transformation matrix $[T]_2$ from eq. (b) into the expression for the element stiffness matrix in global coordinates, eq. (5.15)

$$[\bar{k}]_2 = [T]_2^T [k]_2 [T]_2 \qquad (5.15) \text{ repeated}$$

yields:

$$[\bar{k}]_2 = \begin{array}{c c c c c c c} \quad 1 & 2 & 3 & 7 & 8 & 9 \\ \begin{bmatrix} 1.023E6 & 0 & 1.535E4 & 5.116E5 & 0 & -1.535E4 \\ 0 & 5.893E2 & 0 & 0 & -5.839E2 & 0 \\ 1.535E4 & 0 & 3.069E2 & 1.535E4 & 0 & -3.069E2 \\ 5.116E5 & 0 & 1.535E4 & 1.023E6 & 0 & -1.535E4 \\ 0 & -5.839E2 & 0 & 0 & 5.893E2 & 0 \\ -1.535E4 & 0 & -3.069E2 & -1.535E4 & 0 & 3.069E2 \end{bmatrix} & \begin{matrix} 1 \\ 2 \\ 3 \\ 7 \\ 8 \\ 9 \end{matrix} \end{array} \quad \text{(c)}$$

5. Assemblage of the reduced system stiffness matrix.

The transfer to the reduced system stiffness matrix of coefficients corresponding to the free coordinates, from element stiffness matrices, eqs. (a) and (c), to the locations indicated at the top and on the right of these matrices yields

$$[K]_R = \begin{bmatrix} 5.893E2 + 1.023E6 & 0+0 & 0+1.535E4 \\ 0+0 & 1.023E6 + 5.893E2 & -1.535E4 + 0 \\ 0+1.535E4 & -1.535E4 & 3.069E2 + 3.069E2 \end{bmatrix}$$

or

$$[K]_R = \begin{bmatrix} 1.024E6 & 0 & 1.535E4 \\ 0 & 1.024E6 & -1.535E4 \\ 1.535E4 & -1.535E4 & 6.139E2 \end{bmatrix} \qquad \text{(d)}$$

6. Equivalent nodal forces.

ELEMENT 1

From Appendix I, Case (b) (concentrated moment)

$$Q_1 = -\frac{6ML_1}{L^3}L_2 = \frac{6 x 200 x 50^2}{100^3} = 3.0 \; kip$$

$$Q_2 = \frac{ML_2}{L^2}(L_2 - 2L_1) = \frac{-200 x 50}{100^2}(50 - 100) = 50 \; kip{\cdot}in$$

$$Q_3 = \frac{6ML_1L_2}{L^3} = -3.0 \; kip$$

$$Q_4 = \frac{ML_1}{L^2}(L_1 - 2L_2) = 50 \; kip{\cdot}in$$

Arranging these values of the equivalent nodal forces in a vector, and in accordance with the global nodal coordinates shown in Figure 5.5 results in

$$\{\bar{Q}\}_1 = \{Q\}_1 = \begin{Bmatrix} 0 \\ 50 \\ 3.0 \\ 0 \\ 50 \\ -3.0 \end{Bmatrix} \begin{matrix} 1 \\ 2 \\ 3 \\ 4 \\ 5 \\ 6 \end{matrix} \qquad (e)$$

$\{\bar{Q}\}_1 = \{Q\}_1$ for element 1, because $[T]_1 = [I]$

ELEMENT 2

From Appendix I, Case (c) (uniform distributed force)

$$Q_1 = \frac{wL}{2} = -\frac{0.1x100}{2} = -5.0 \ kip$$

$$Q_2 = \frac{wL^2}{12} = -\frac{0.1x100^2}{12} = -83.33 \ kip$$

$$Q_3 = \frac{wL}{L} = -5.0$$

$$Q_4 = \frac{wL^2}{12} = +83.33$$

or using eq. (5.12) in vector form in accordance with the global nodal coordinate for element 2 indicated in Figure 5.5 gives

$$\{\bar{Q}\}_2 = \begin{Bmatrix} -83.33 \\ 0 \\ -5.0 \\ 83.33 \\ 0 \\ -5.0 \end{Bmatrix} \begin{matrix} 1 \\ 2 \\ 3 \\ 7 \\ 8 \\ 9 \end{matrix} \qquad (f)$$

7. Reduced system force vector.

The transfer to the system force vector of the coefficients corresponding to the free nodal coordinates 1, 2, and 3, [from eqs. (e) and (f)] yields:

$$\{F\}_R = \begin{Bmatrix} 0-83.33 \\ 50+0 \\ 3.0-5 \end{Bmatrix} + \begin{Bmatrix} 0 \\ 0 \\ -10 \end{Bmatrix} = \begin{Bmatrix} -83.33 \\ 50 \\ -12.0 \end{Bmatrix} \tag{g}$$

Equation (g) includes the force equal to –10 *kips* directly applied at nodal coordinate 3.

8. Reduced system stiffness equation.

Substitution of the reduced system stiffness matrix [eq. (d)] and of the reduced system force vector, [eq. (g)] into the stiffness equation $\{F\}_R=[K]_R\{u\}$ results in

$$\begin{bmatrix} -83.33 \\ 50 \\ -12.0 \end{bmatrix} = \begin{bmatrix} 1.024E6 & 0 & 1.535E4 \\ 0 & 1.024E6 & -1.535E4 \\ 1.535E4 & -1.535E4 & 6.139E2 \end{bmatrix} \begin{bmatrix} u_1 \\ u_2 \\ u_3 \end{bmatrix}. \tag{h}$$

9. Solution of the unknown displacements at the joints

Solution of eq. (h) yields:

$u_1 = 1.040E{-}3 \; rad$

$u_2 = -1.170E{-}3 \; rad$

$u_3 = -0.0748 \; in$ Ans.

10. Element nodal displacements (local coordinates).

The nodal displacements for elements 1 and 2 in local coordinates identified from Figure 5.5 results in

$$\{\delta\}_1 = \begin{Bmatrix} 1.040E{-}3 \\ -1.170E{-}3 \\ -0.0748 \\ 0 \\ 0 \\ 0 \end{Bmatrix} \quad \text{and} \quad \{\delta\}_2 = \begin{Bmatrix} -1.170E{-}3 \\ -1.040E{-}3 \\ -0.0748 \\ 0 \\ 0 \\ 0 \end{Bmatrix} \tag{i}$$

11. Element end forces.

Element end forces are calculated as in eq. (4.20) by

$$\{P\} = [k]\,\{\delta\} - \{Q\} \qquad\qquad (4.20)\ \text{repeated}$$

ELEMENT 1:

$$\begin{Bmatrix} P_1 \\ P_2 \\ P_3 \\ P_4 \\ P_5 \\ P_6 \end{Bmatrix}_1 = \begin{bmatrix} 589.3 & 0 & 0 & -589.3 & 0 & 0 \\ 0 & 1.023E6 & -1.535E4 & 0 & 5.116E5 & 1.535E4 \\ 0 & -1.535E4 & 306.9 & 0 & -1.535E4 & -3.069E2 \\ -589.3 & 0 & 0 & 589.2 & 0 & 0 \\ 0 & 5.116E5 & -1.535E4 & 0 & 1.023E6 & 1.535E4 \\ 0 & 1.535E4 & -3.069E2 & 0 & 1.535E4 & 3.069E2 \end{bmatrix} \begin{Bmatrix} 1.040E\!-\!3 \\ -1.170E\!-\!3 \\ -0.0748 \\ 0 \\ 0 \\ 0 \end{Bmatrix}$$

$$-\begin{Bmatrix} 0 \\ 50 \\ 3.0 \\ 0 \\ 50 \\ -3.0 \end{Bmatrix} = \begin{Bmatrix} -0.589 \\ 0.663 \\ 8.000 \\ -0.589 \\ -599 \\ 8.\,000 \end{Bmatrix} \qquad \text{Ans.}$$

ELEMENT 2

Analogously, substituting numerical values for element 2 into eq. (4.20), yields

$$\begin{Bmatrix} P_1 \\ P_2 \\ P_3 \\ P_4 \\ P_5 \\ P_6 \end{Bmatrix}_2 = \begin{Bmatrix} 0.663 \\ 0.589 \\ 2.000 \\ 0.663 \\ -699 \\ 12.000 \end{Bmatrix} \qquad \text{Ans.}$$

12. Reactions at supports

Support at joint 2:

$$R_4 = \overline{P}_{41} = -0.589\ kip \cdot in$$

$$R_5 = \overline{P}_{51} = 599\ kip \cdot in$$

$$R_6 = \overline{P}_{61} = 8.00\ kip \qquad\qquad \text{Ans.}$$

where $\{\overline{P}\}$ is calculated using eq. (5.12).

Support at joint 3:

$$R_7 = \overline{P}_{42} = -699 \; kip \cdot in$$

$$R_8 = \overline{P}_{52} = 0.663 \; kip \cdot in$$

$$R_9 = \overline{P}_{62} = 12.00 \; kip \qquad\qquad\qquad \text{Ans.}$$

where $\{\overline{P}\}$ is calculated using eq. (5.12).

Notation: P_{ij} = Force at nodal coordinate i of element j.

5.6   Analysis of Grid Frames Using SAP2000

Illustrative Example 5.2
Use SAP2000 to solve Illustrative Example 5.1.

Solution:
The following commands are implemented in SAP2000:

Begin: Open SAP2000.

Enter: Click OK to close the "Tip of the Day".

Hint: Maximize both windows for a full view.

Model: In the lower right-hand corner of the window, use the drop-down menu select "kip-in".

Select: From the main menu select:
FILE>NEW MODEL
Coordinate System Definition:
$x = 1$, $y = 0$ and $z = 1$ divisions.
Grid Spacing:
$x = 100$, $y = 1$ and $z = 100$.
Click OK.

Edit: Maximize the screen showing the 3-D view and click on the PAN icon, then drag the plot to the center of the screen.

Draw: From the main menu select:
DRAW>DRAW FRAME ELEMENT.
Click at the origin of the coordinates and drag the cursor along the y axis to the next grid line ($y = 100$), click twice and press Enter.

Click at the origin of the coordinates and drag the cursor along the x axis to the next grid line ($x = 100$), click twice and press Enter.

Label: For viewing convenience, label joints and elements:
From the main menu select:
VIEW>SET ELEMENTS.
Click on "Joint Labels" and "Frame Labels".
Click OK.

Boundaries: Click on Joints 2 and 3 and from the main menu select:
ASSIGN>JOINT>RESTRAINTS.
Select restraints in all directions.
Click OK.

Material: From the main menu select:
DEFINE>MATERIALS.
Click on STEEL.
Click Modify/Show Material.
    Set: Modulus of Elasticity = 29000.
    Click OK.

Define Section: From the main menu select:
DEFINE >FRAME SECTION.
Click: Import/Wide Flange.
Click OK.
    Select : "Sections.pro".
    Click OPEN.
        Select: W14 X 82.
        Click OK.
    Click: Modification Factors.
    Set: Shear area in 2 directions = 0.
    Click OK, OK, OK.

Assign Section: Click on beam elements 1 and 2.
From the main menu select:
ASSIGN>FRAMES>SECTIONS.
Select W14 X 82.
Click OK.

Define Loads: From the main menu select:
    DEFINE>STATIC LOAD CASES.
    Select: LIVE LOAD and set Self-Weight Multiplier = 0.
    Click: Change Load.
    Click OK.

Assign Load: Click on joint 1.
    From the main menu select:
    ASSIGN > JOINT > FORCE.
    Select: Direction Z = −10.
    Click OK.

Assign Loads: Click on Beam element 1.
    From the main menu select:
    ASSIGN>FRAME STATIC LOADS>POINT AND UNIFORM.
    Select: Moments, Direction: Global Y
    Enter: Distance = 0.5
          Load   = 200
    Click OK.

    Click on Beam element 2.
    From the main menu select:
    ASSIGN>FRAME STATIC LOADS>POINT AND UNIFORM.
    Select: Distributed Load.
    Click on Direction Global Z.
    Enter: Distance = 0   load = −0.1
             Distance = 1.0   load = −0.1
    Click OK.

Analyze: From the main menu select:
    ANALYZE>SET OPTIONS.
    Check Available degrees of freedom UZ, RX and RY (or simply click the
    plane grid diagram).
    Click OK.

    From the main menu select:
    ANALYZE>RUN.
    Enter File Name: "Example 5.2".
    Click SAVE.
    At the conclusion of the calculations, click OK.

Print Input Tables: FILE>PRINT INPUT TABLES

(Table 5.1 provides the edited Input Tables for Illustrative Example 5.2.)

Table 5.1 Edited Input Tables for Illustrative Example 5.2 (Units: kips, inches)

JOINT DATA

JOINT	GLOBAL-X	GLOBAL-Y	GLOBAL-Z	RESTRAINTS
1	0.00000	0.00000	0.00000	0 0 0 0 0 0
2	100.00000	0.00000	0.00000	1 1 1 1 1 1
3	0.00000	100.00000	0.00000	1 1 1 1 1 1

FRAME ELEMENT DATA

FRAME	JNT-1	JNT-2	SECTION	RELEASES	SEGMENTS	LENGTH
1	1	2	W14X82	000000	4	100.000
2	1	3	W14X82	000000	4	100.000

JOINT FORCES Load Case LOAD1

JOINT	GLOBAL-X	GLOBAL-Y	GLOBAL-Z	GLOBAL-XX	GLOBAL-YY	GLOBAL-ZZ
1	0.000	0.000	−10.000	0.000	0.000	0.000

FRAME SPAN DISTRIBUTED LOADS Load Case LOAD1

FRAME	TYPE	DIRECTION	DISTANCE-A	VALUE-A	DISTANCE-B	VALUE-B
2	FORCE	GLOBAL-Z	0.0000	−0.1000	1.0000	−0.1000

FRAME SPAN POINT LOADS Load Case LOAD1

FRAME	TYPE	DIRECTION	DISTANCE	VALUE
1	MOMENT	GLOBAL-Y	0.5000	200.0000

Print Output Tables: FILE>PRINT OUTPUT TABLES

(Table 5.2 provides the edited Output Tables for Illustrative Example 5.2.)

Table 5.2 Edited Output Tables for Illustrative Example 5.2 (Units: kips-inches)

JOINT DISPLACEMENTS

JOINT	LOAD	UX	UY	UZ	RX	RY	RZ
1	LOAD1	0.0000	0.0000	−0.0748	1.040E-03	−1.170E-03	0.0000
2	LOAD1	0.0000	0.0000	0.0000	0.0000	0.0000	0.0000
3	LOAD1	0.0000	0.0000	0.0000	0.0000	0.0000	0.0000

JOINT REACTIONS

JOINT	LOAD	F1	F2	F3	M1	M2	M3
2	LOAD1	0.0000	0.0000	8.0006	−0.5894	599.3922	0.0000
3	LOAD1	0.0000	0.0000	11.9994	−699.3553	0.6632	0.0000

FRAME ELEMENT FORCES

FRAME	LOAD	LOC	P	V2	V3	T	M2	M3
1	LOAD1							
		0.00	0.00	8.00	0.00	−5.894E-01	0.00	6.632E-01
		25.00	0.00	8.00	0.00	−5.894E-01	0.00	−199.35
		50.00	0.00	8.00	0.00	−5.894E-01	0.00	−199.36
		75.00	0.00	8.00	0.00	−5.894E-01	0.00	−399.38
		100.00	0.00	8.00	0.00	−5.894E-01	0.00	−599.39

Table 5.2 (continued)

2	LOAD1						
	0.00	0.00	2.00	0.00	6.632E-01	0.00	5.894E-01
	25.00	0.00	4.50	0.00	6.632E-01	0.00	-80.65
	50.00	0.00	7.00	0.00	6.632E-01	0.00	-224.38
	75.00	0.00	9.50	0.00	6.632E-01	0.00	-430.62
	100.00	0.00	12.00	0.00	6.632E-01	0.00	-699.36

Plot Deformed Shape: From the main menu select:
DISPLAY>SHOW DEFORMED SHAPE.
FILE>PRINT GRAPHICS.

(The deformed shape shown on the screen is reproduced in Figure 5.6.)

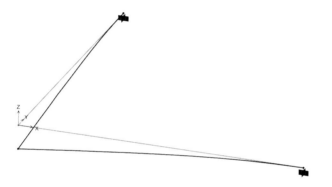

Fig. 5.6 Deformed shape for the grid frame in Illustrative Example 5.2

Illustrative Example 5.3

Consider the grid frame shown in Figure 5.7, which has fixed supports at Joints 3 and 7 and simple support at Joints 3 and 9. Use SAP2000 to analyze this grid frame supporting the loads shown in the figure.

Solution:

The following commands are implemented in SAP2000:

Begin: Open SAP2000.

Enter: Click "OK" to close the "Tip of the Day".

Hint: Maximize both windows for a full view.

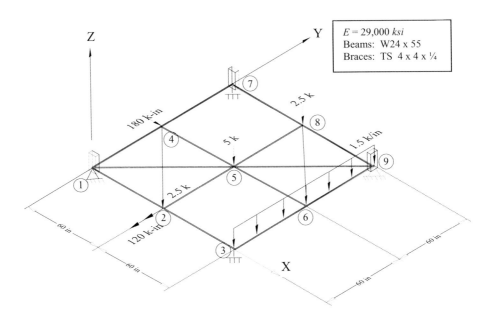

Fig. 5.7 Grid frame for Illustrative Example 5.3

Units: In the lower right-hand corner of the window, use the drop-down menu to select "kip-in".

Model: From the main menu select:
FILE>NEW MODEL.
Set divisions: x = 2, y = 2, z = 0.
Change grid spacing: x = 60, y = 60, z = 60.
Click OK.

Edit: Maximize the 3-D screen and click on the PAN icon and drag the plot to the center of the window.

View: From the main menu select:
VIEW>SET 3-D VIEW.
Set view angles: Plan = 290°; Elevation = 16°; and Aperture = 0°.
Click OK.

Draw: From the main menu select:
DRAW>QUICK DRAW ELEMENT.
Draw the Beam element by clicking successively on every grid line parallel to the axis X and Y.
From the main menu select:

DRAW >FRAME ELEMENT.
Click successively on two joints to draw the beam elements along the diagonals as shown in Figure 5.7.

Move Axes: From the main menu select:
SELECT>SELECT ALL.
From the main menu select:
EDIT>MOVE.
Set x = 60
   y = 60
Click OK.

Cancel Grid: From the main menu select:
VIEW.
Uncheck: Show grid.

Material: From the main menu select:
DEFINE>MATERIALS>STEEL.
Click on Modify/Show Material.
   Set: Modulus of Elasticity = 29500
        Weight per unit length = 0
        Click OK, OK.

Define Sections: From the main menu select:
DEFINE>FRAME SECTIONS>IMPORT/WIDE FLANGE.
Click OK.
   Click on Sections.pro.
   Click Open.
      Click on W24 X 55.
      Click OK, OK, OK.

From the main menu select:
DEFINE>FRAME SECTIONS>IMPORT BOX/TUBE.
(Use drop-down list)
Click on Sections.pro.
Click OPEN.
   Select TS4 X 4 X 1/4.
   Click OK.

Assign Sections: Mark (click on) all beams parallel to the axes X and Y.
From the main menu select:
ASSIGN>FRAMES>SECTIONS.
Click on W24 X 55.
Click OK.

Mark all the braces (diagonal members).
From the main menu select:
ASSIGN>FRAMES>SECTIONS.
Click on ST4 X 4 X 1/4.
Click OK.

Labels: From the main menu select:
VIEW>SET ELEMENTS.
Check Joint Labels and Frame Labels.
Click OK.

Boundaries: Mark (click) Joints 3 and 7.
From the main menu select:
ASSIGN>JOINT RESTRAINTS.
Check restraints in all local directions.
Click OK.

Mark (click) Joints 1 and 9.
From the main menu select:
ASSIGN>JOINT RESTRAINTS.
Check restraint Translations 1, 2, and 3.
Click OK.

Define Loads: From the main menu select:
DEFINE>STATIC LOAD CASES.
Change Load Type "DEAD" to "LIVE".
Set Self-Weight Multiplier = 0.
Click on Change Load.
Click OK.

Assign Loads: Mark Joints 2 and 8 and from the main menu select:
ASSIGN>JOINT STATIC LOADS>FORCES.
Set: Force Global $Z = -2.5$.
Click OK.

Mark Joint 5 and from the main menu select:
ASSIGN>JOINT STATIC LOADS>FORCES.
Set Force Global $Z = -5.0$.
Click OK.

Mark Joint 2 and from the main menu select:
ASSIGN>JOINT STATIC LOADS>FORCES.
Set Moment Global $YY = -120$.
Click OK.

Mark Joint 4 and from the main menu select:

ASSIGN>JOINT STATIC LOADS>FORCES.
    Set Moment Global X = 180.
Click OK.

Mark beam elements 10 and 13 and from the main menu select:
ASSIGN>FRAME STATIC LOADS>POINT AND UNIFORM FORCE.
    Direction: Global Z.
    Uniform Load = −1.5.
Click OK.

Set Options: From the main menu select:
    ANALYZE>SET OPTIONS.
    Available DOFs: Check UZ, RX and RY.
    Click OK.

Analyze: From the main menu select:
    ANALYZE>RUN.
    File Name: Example 5.3.
    Click Save.
    At the conclusion of calculation, click OK.

Plots: From the main menu select:
    DISPLAY>UNDEFORMED SHAPE.
    FILE>PRINT GRAPHICS.

(The plot of the Grid Frame of Illustrative Example 5.2 is reproduced as Figure 5.8.)

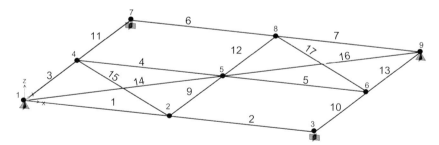

Fig. 5.8 Undeformed plot of grid frame for Illustrative Example 5.3

Print Tables: From the main menu select:
    FILE>PRINT INPUT TABLES.
    Check: Print to File.
    Click OK.

    From the main menu select:
    FILE>PRINT OUTPUT TABLES.
    Check: Displacements, Reactions and Frame Forces.

Check: Print to File and Append.
Click OK.

(The edited Input and Output Tables for Illustrative Example 5.3 are shown in Tables 5.3 and 5.4, respectively.)

Table 5.3 Edited Input Tables for Illustrative Example 5.3 (Units: kips, inches)

JOINT DATA

JOINT	GLOBAL-X	GLOBAL-Y	GLOBAL-Z	RESTRAINTS	ANG-A	ANG-B	ANG-C
1	0.00000	0.00000	0.00000	1 1 1 0 0 0	0.000	0.000	0.000
2	60.00000	0.00000	0.00000	0 0 0 0 0 0	0.000	0.000	0.000
3	120.00000	0.00000	0.00000	1 1 1 1 1 1	0.000	0.000	0.000
4	0.00000	60.00000	0.00000	0 0 0 0 0 0	0.000	0.000	0.000
5	60.00000	60.00000	0.00000	0 0 0 0 0 0	0.000	0.000	0.000
6	120.00000	60.00000	0.00000	0 0 0 0 0 0	0.000	0.000	0.000
7	0.00000	120.00000	0.00000	1 1 1 1 1 1	0.000	0.000	0.000
8	60.00000	120.00000	0.00000	0 0 0 0 0 0	0.000	0.000	0.000
9	120.00000	120.00000	0.00000	1 1 1 0 0 0	0.000	0.000	0.000

FRAME ELEMENT DATA

FRM	JNT-1	JNT-2	SECTN	RELEASES	SGMNT	R1	R2	FACTOR	LENGTH
1	1	2	W24X55	000000	4	0.000	0.000	1.000	60.000
2	2	3	W24X55	000000	4	0.000	0.000	1.000	60.000
3	1	4	W24X55	000000	4	0.000	0.000	1.000	60.000
4	4	5	W24X55	000000	4	0.000	0.000	1.000	60.000
5	5	6	W24X55	000000	4	0.000	0.000	1.000	60.000
6	7	8	W24X55	000000	4	0.000	0.000	1.000	60.000
7	8	9	W24X55	000000	4	0.000	0.000	1.000	60.000
9	2	5	W24X55	000000	4	0.000	0.000	1.000	60.000
10	3	6	W24X55	000000	4	0.000	0.000	1.000	60.000
11	4	7	W24X55	000000	4	0.000	0.000	1.000	60.000
12	5	8	W24X55	000000	4	0.000	0.000	1.000	60.000
13	6	9	W24X55	000000	4	0.000	0.000	1.000	60.000
14	1	5	TS4X4X1/4	000000	4	0.000	0.000	1.000	84.853
15	2	4	TS4X4X1/4	000000	4	0.000	0.000	1.000	84.853
16	5	9	TS4X4X1/4	000000	4	0.000	0.000	1.000	84.853
17	6	8	TS4X4X1/4	000000	4	0.000	0.000	1.000	84.853

JOINT FORCES Load Case LOAD1

JOINT	GLOBAL-X	GLOBAL-Y	GLOBAL-Z	GLOBAL-XX	GLOBAL-YY	GLOBAL-ZZ
2	0.000	0.000	-2.500	0.000	-120.000	0.000
8	0.000	0.000	-2.500	0.000	0.000	0.000
5	0.000	0.000	-5.000	0.000	0.000	0.000
4	0.000	0.000	0.000	180.000	0.000	0.000

Table 5.3 (continued)

FRAME SPAN DISTRIBUTED LOADS Load Case LOAD1

FRAME	TYPE	DIRECTION	DISTANCE-A	VALUE-A	DISTANCE-B	VALUE-B
10	FORCE	GLOBAL-Z	0.0000	−1.5000	1.0000	−1.5000
13	FORCE	GLOBAL-Z	0.0000	−1.5000	1.0000	−1.5000

Table 5.4 Output Tables for Illustrative Example 5.3 (Units: kips-inches)

JOINT DISPLACEMENTS

JOINT	LOAD	UX	UY	UZ	RX	RY	RZ
1	LOAD1	0.0000	0.0000	0.0000	4.417E-05	1.356E-04	0.0000
2	LOAD1	0.0000	0.0000	−6.724E-03	−2.990E-04	−4.767E-05	0.0000
3	LOAD1	0.0000	0.0000	0.0000	0.0000	0.0000	0.0000
4	LOAD1	0.0000	0.0000	2.756E-03	5.503E-05	3.974E-04	0.0000
5	LOAD1	0.0000	0.0000	−0.0225	1.897E-06	5.850E-04	0.0000
6	LOAD1	0.0000	0.0000	−0.0675	−1.106E-04	7.755E-04	0.0000
7	LOAD1	0.0000	0.0000	0.0000	0.0000	0.0000	0.0000
8	LOAD1	0.0000	0.0000	−6.560E-03	3.056E-04	1.071E-05	0.0000
9	LOAD1	0.0000	0.0000	0.0000	1.587E-03	−1.236E-04	0.0000

JOINT REACTIONS

JOINT	LOAD	F1	F2	F3	M1	M2	M3
1	LOAD1	0.0000	0.0000	4.2328	0.0000	0.0000	0.0000
3	LOAD1	0.0000	0.0000	113.1427	2412.0447	187.1865	0.0000
7	LOAD1	0.0000	0.0000	1.7965	93.0201	−190.7070	0.0000
9	LOAD1	0.0000	0.0000	70.8280	0.0000	0.0000	0.0000

FRAME ELEMENT FORCES

FRAME	LOAD	LOC	P	V2	V3	T	M2	M3
1	LOAD1							
		0.00	0.00	−4.01	0.00	−7.657E-02	0.00	1.45
		15.00	0.00	−4.01	0.00	−7.657E-02	0.00	61.54
		30.00	0.00	−4.01	0.00	−7.657E-02	0.00	121.64
		45.00	0.00	−4.01	0.00	−7.657E-02	0.00	181.74
		60.00	0.00	−4.01	0.00	−7.657E-02	0.00	241.83
2	LOAD1							
		0.00	0.00	5.19	0.00	6.672E-02	0.00	124.08
		15.00	0.00	5.19	0.00	6.672E-02	0.00	46.22
		30.00	0.00	5.19	0.00	6.672E-02	0.00	−31.64
		45.00	0.00	5.19	0.00	6.672E-02	0.00	−109.50
		60.00	0.00	5.19	0.00	6.672E-02	0.00	−187.36
3	LOAD1							
		0.00	0.00	−2.158E-01	0.00	5.842E-02	0.00	7.294E-01
		15.00	0.00	−2.158E-01	0.00	5.842E-02	0.00	3.97
		30.00	0.00	−2.158E-01	0.00	5.842E-02	0.00	7.20
		45.00	0.00	−2.158E-01	0.00	5.842E-02	0.00	10.44
		60.00	0.00	−2.158E-01	0.00	5.842E-02	0.00	3.68

Bending Moment: From the main menu select:
>DISPLAY>SHOW ELEMENT FORCES/STRESSES>FRAMES.
>Click on Moment 3-3 and Fill Diagram.
>Click OK.
>FILE>PRINT GRAPHICS.

(Figure 5.9 reproduces the bending moment diagram for the grid frame of Illustrative Example 5.3.)

Note: By clicking on a beam element and then selecting the desired type of internal force (bending moment, shear force, etc.) the window will show the beam element with the plot of the force specified. Also, as the cursor is moved along the beam element, the window will display the value of the plotted force.

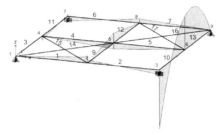

Fig. 5.9 Bending moment diagram for the grid frame of Illustrative Example 5.3

5.7   Dynamic Analysis of Grid Frames

The stiffness matrix analysis of structures subjected to dynamic load is based on the determination of the stiffness and mass matrices for an element in reference to the local coordinate system followed by the transformation of these matrices to the global system of coordinates. The development of the stiffness matrix for an element of a grid frame as well as the transformation matrix were presented in Sections 5.1 through 5.4; thus it remains only to develop the mass matrix. We begin by developing the coefficients of the mass matrix due to torsion as presented in the next section.

5.8   Lumped Mass Matrix for an Element of a Grid Frame

The lumped mass allocation to the nodal coordinates of a grid member is obtained from static considerations analogously to the lumped mass allocations for a loaded element of a beam as given in Figure 2.1. For a uniform member having a uniformly distributed mass along its length, the nodal rotational mass is simply one-half of the total rotational mass of the element, $I_{\overline{m}}L$, in which $I_{\overline{m}}$ is the polar mass moment of inertia per unit length along the beam element and L its length; thus this nodal allocation is equal to $I_{\overline{m}}L/2$. The rotational mass moment of inertia may conveniently be expressed as the product of the mass $\overline{m}$ per unit length times the radius of gyration squared, k^2. The radius of gyration may, in turn, be determined as the ratio I_0/A, where I_0 is the polar moment of inertia of the cross-sectional area and A the cross-sectional area. Therefore, the mass polar moment of inertia per unit length $I_{\overline{m}}$ may be calculated as

$$I_{\overline{m}} = \overline{m}\,\frac{I_0}{A}. \tag{5.16}$$

However, as stated in Chapter 2 in the formulation of the lumped mass matrix, the inertial effect due to angular displacements is neglected. Thus the mass matrix relating forces and accelerations at nodal coordinates of an element of a grid frame is then given by the following equation, in which only the translation inertial mass is considered:

$$\begin{Bmatrix} P_1 \\ P_2 \\ P_3 \\ P_4 \\ P_5 \\ P_6 \end{Bmatrix} = \frac{\overline{m}L}{2} \begin{bmatrix} 0 & & & & & \\ & 0 & & & & \\ & & 1 & & & \\ & & & 0 & & \\ & & & & 0 & \\ & & & & & 1 \end{bmatrix} \begin{Bmatrix} \ddot{\delta}_1 \\ \ddot{\delta}_2 \\ \ddot{\delta}_3 \\ \ddot{\delta}_4 \\ \ddot{\delta}_5 \\ \ddot{\delta}_6 \end{Bmatrix} \tag{5.17}$$

or briefly

$$\{P\} = \lceil m \rfloor_L \{\ddot{\delta}\} \tag{5.18}$$

in which $\lceil m \rfloor_L$ is the diagonal lumped mass matrix for an element of a grid frame.

5.9   Consistent Mass Matrix for an Element of a Grid Frame

As it was shown in Section 5.2, there exists a mathematical analogy between the axial and torsional effects on a beam element. Therefore analogous to eq. (4.26), the consistent mass for torsional effects establishing the relationship between the torsional moments and the angular acceleration at the nodal coordinates (Figure 5.1) of a uniform beam element is given by

$$\left\{\begin{matrix} T_1 \\ T_2 \end{matrix}\right\} = \frac{I_{\bar{m}}L}{6}\begin{bmatrix} 2 & 1 \\ 1 & 2 \end{bmatrix}\left\{\begin{matrix} \ddot{\theta}_1 \\ \ddot{\theta}_2 \end{matrix}\right\} \tag{5.19}$$

where $I_{\bar{m}}$ is the mass polar moment of inertia per unit length, which may be calculated by eq. (5.16).

The combination of the consistent mass matrix for flexural effects, eq. (2.16) with the mass matrix for torsional effects eq. (5.19) and after using eq. (5.16) results in the consistent mass matrix relating the acceleration and the forces at the nodal coordinates for a member of a grid frame, namely

$$\left\{\begin{matrix} P_1 \\ P_2 \\ P_3 \\ P_4 \\ P_5 \\ P_6 \end{matrix}\right\} = \frac{\bar{m}L}{420}\begin{bmatrix} 140\,I_0/A & & & & Symmetric & \\ 0 & 4L^2 & & & & \\ 0 & 22L & 156 & & & \\ 70\,I_0/A & 0 & 0 & 140\,I_0/A & & \\ 0 & -3L^2 & -13L & 0 & 4L^2 & \\ 0 & 13L & 54 & 0 & -22L & 156 \end{bmatrix}\left\{\begin{matrix} \delta_1 \\ \delta_2 \\ \delta_3 \\ \delta_4 \\ \delta_5 \\ \delta_6 \end{matrix}\right\} \tag{5.20}$$

$$\{P\} = [m]_C\{\ddot{\delta}\} \tag{5.21}$$

in which $[m]_C$ is the consistent mass matrix for a uniform member of a grid frame.

5.10   Coordinate Transformation

The lumped mass matrix in eq. (5.17) and the consistent mass matrix in eq. (5.20) are in reference to the local system of coordinates. In order to assemble the system mass matrix, it is necessary to introduce a transformation of coordinates to refer the element mass matrices to the global system of coordinates. The transformation matrix for the mass matrix is the same as the transformation matrix developed in Section 5.4. for the transformation of the stiffness matrix [eq. (5.11)], namely

$$[T] = \begin{bmatrix} \cos\theta & \sin\theta & 0 & 0 & 0 & 0 \\ -\sin\theta & \cos\theta & 0 & 0 & 0 & 0 \\ 0 & 0 & 1 & 0 & 0 & 0 \\ 0 & 0 & 0 & \cos\theta & \sin\theta & 0 \\ 0 & 0 & 0 & -\sin\theta & \cos\theta & 0 \\ 0 & 0 & 0 & 0 & 0 & 1 \end{bmatrix} \tag{5.22}$$

Hence, the relationship between the element nodal forces $\{P\}$ in reference to local coordinates and $\{\overline{P}\}$, the nodal forces referred to the global system of coordinates is expressed as

$$[P]=[T]\{\overline{P}\} \tag{5.23}$$

where the transformation matrix $[T]$ is given by the matrix in eq. (5.22).

Also the nodal acceleration components, $\{\ddot{\delta}\}$ in reference to local system of coordinates is related to the acceleration components, $\{\overline{\ddot{\delta}}\}$ referred to global coordinates through the same transformation matrix, namely

$$\{\ddot{\delta}\} = [T]\{\overline{\ddot{\delta}}\} \tag{5.24}$$

The substitution into eq. (5.18) or into eq. (5.21) of $\{P\}$ and $\{\ddot{\delta}\}$ respectively, from eqs. (5.23) and (5.24) results in

$$[T]\{\overline{P}\} = [m]\ [T]\{\overline{\ddot{\delta}}\}$$

in which $[m]$ is the element mass matrix determined as the lumped or consistent mass matrix in reference to the local system of coordinates.

Since $[T]$ is an orthogonal matrix ($[T]^{-1} = [T]^{T}$), it follows that

$$\{\overline{P}\} = [T]^{T}\ [m]\ [T]\{\overline{\ddot{\delta}}\}$$

or

$$\{\overline{P}\} = [\overline{m}]\ \{\overline{\ddot{\delta}}\} \tag{5.25}$$

where the mass matrix $[\overline{m}]$ now in reference to the global system of coordinates is given by

$$[\overline{m}] = [T]^{T}\ [m]\ [T] \tag{5.26}$$

Illustrative Example 5.4

Figure 5.10 shows a grid frame on the horizontal plane consisting of two prismatic beam elements subjected to a force of 5,000 *lb* suddenly applied upward at the joint of the two beam elements.

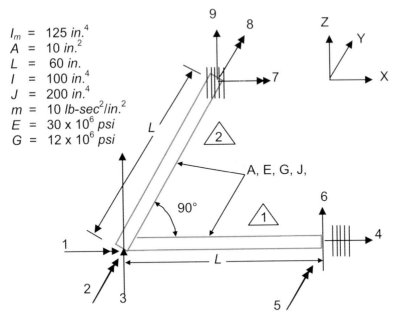

I_m = 125 in.^{4}
A = 10 in.^{2}
L = 60 in.
I = 100 in.^{4}
J = 200 in.^{4}
m = 10 lb-sec^2/in.^{2}
E = 30 x 10^6 psi
G = 12 x 10^6 psi

Fig. 5.10 Grid frame of Illustrative Example 5.4

Determine the lowest natural frequency and the response to the applied force. Model the structure with 2 elements and use the lumped mass method. (Neglect damping.)

Solution:
The stiffness matrix for elements 1 or 2 of the grid frame calculated in reference to the local system of coordinates by eq. (5.7) results in

$$[k]_1 = [k]_2 = 10^6 \begin{matrix} & 1 & 2 & 3 & 4 & 5 & 6 & \\ & 40 & 0 & 0 & -40 & 0 & 0 & 1 \\ & 0 & 200 & -5 & 0 & 100 & 5 & 2 \\ & 0 & -5 & 0.167 & 0 & -5 & -0.167 & 3 \\ & -40 & 0 & 0 & 40 & 0 & 0 & 4 \\ & 0 & 100 & -5 & 0 & 200 & 5 & 5 \\ & 0 & 5 & -0.167 & 0 & 5 & 0.167 & 6 \end{matrix} \qquad (a)$$

The transformation matrix for element 1 with $\theta = 0°$ is simply the unit matrix $[T_1] = [I]$. Hence, by eq. (5.15) the stiffness matrix for element 1 in reference to the global system of coordinates is

$$[\bar{k}]_1 = [T]_1^T [k]_1 [T]_1 = [k]_1 .$$

By eq. (5.23), the transformation matrix for element 2 with $\theta = 90°$ is

$$[T]_2 = \begin{bmatrix} 0 & 1 & 0 & 0 & 0 & 0 \\ -1 & 0 & 0 & 0 & 0 & 0 \\ 0 & 0 & 1 & 0 & 0 & 0 \\ 0 & 0 & 0 & 0 & 1 & 0 \\ 0 & 0 & 0 & -1 & 0 & 0 \\ 0 & 0 & 0 & 0 & 0 & 1 \end{bmatrix} \tag{b}$$

Then substituting $[k]_2$ and $[T]_2$, respectively, from eqs. (a) and (b) into

$$[\bar{k}]_2 = [T]_2^T [k]_2 [T]_2$$

results in

$$[\bar{k}]_2 = 10^6 \begin{array}{cccccc} 1 & 2 & 3 & 4 & 5 & 6 \\ \begin{bmatrix} 200 & 0 & 5 & 100 & 0 & 5 \\ 0 & 40 & 0 & 0 & -40 & 0 \\ 5 & 0 & 0.167 & -5 & 0 & 0.167 \\ 100 & 0 & 5 & 200 & 0 & -5 \\ 0 & -40 & 0 & 0 & 40 & 0 \\ -5 & 0 & -0.167 & -5 & 0 & 0.167 \end{bmatrix} \begin{array}{c} 1 \\ 2 \\ 3 \\ 4 \\ 5 \\ 6 \end{array} \end{array} \tag{c}$$

Finally, the reduced system stiffness matrix $[K]_R$ (considering only the free coordinates 1, 2, and 3) assembled from $[\bar{k}]_1$ and $[\bar{k}]_2$ [eqs. (a) and c)] yields

$$[K]_R = 10^6 \begin{bmatrix} 240 & 0 & 5 \\ 0 & 240 & -5 \\ 5 & -5 & 0.333 \end{bmatrix} \tag{d}$$

The mass matrix in local coordinates for element 1 or 2 is given by the matrix in eq. (5.18), which after substituting numerical values result in

$$
[m]_1 = [m]_2 = 300
\begin{array}{c}
\begin{array}{cccccc} 1 & 2 & 3 & 4 & 5 & 6 \end{array} \\
\begin{bmatrix}
0 & 0 & 0 & 0 & 0 & 0 \\
0 & 0 & 0 & 0 & 0 & 0 \\
0 & 0 & 1 & 0 & 0 & 0 \\
0 & 0 & 0 & 0 & 0 & 0 \\
0 & 0 & 0 & 0 & 0 & 0 \\
0 & 0 & 0 & 0 & 0 & 1
\end{bmatrix}
\begin{array}{c} 1 \\ 2 \\ 3 \\ 4 \\ 5 \\ 6 \end{array}
\end{array}
\qquad \text{(e)}
$$

The mass matrix for element 1 in reference to the global coordinates is

$$[\overline{m}]_1 = [m]_1$$

since

$$[T]_1 = [I]$$

The lumped mass matrix for element 2 in global coordinates calculated from eq. (5.27) results in

$$
[\overline{m}]_2 = 300
\begin{array}{c}
\begin{array}{cccccc} 1 & 2 & 3 & 7 & 8 & 9 \end{array} \\
\begin{bmatrix}
0 & 0 & 0 & 0 & 0 & 0 \\
0 & 0 & 0 & 0 & 0 & 0 \\
0 & 0 & 1 & 0 & 0 & 0 \\
0 & 0 & 0 & 0 & 0 & 0 \\
0 & 0 & 0 & 0 & 0 & 0 \\
0 & 0 & 0 & 0 & 0 & 1
\end{bmatrix}
\begin{array}{c} 1 \\ 2 \\ 3 \\ 7 \\ 8 \\ 9 \end{array}
\end{array}
\qquad \text{(f)}
$$

The reduced system mass matrix is then assemble from eqs. (e) and (f) to obtain

$$
[M]_R = 300
\begin{bmatrix}
0 & 0 & 0 \\
0 & 0 & 0 \\
0 & 0 & 2
\end{bmatrix}
\qquad \text{(g)}
$$

Because the presence of zeros in the main diagonal of the mass matrix $[M]_R$ is necessary to condense the first two nodal coordinates, the application of the condensation procedure presented in Chapter 1 results in the following condensed stiffness, mass, and transformation matrices:

$$[\overline{K}] = 1.2843\,E5 \qquad \text{(h)}$$

$$[\overline{M}] = 600 \tag{i}$$

$$[T] = \begin{bmatrix} -0.000851 \\ 0.000820 \\ 0.040825 \end{bmatrix} \tag{j}$$

The natural frequencies and mode shapes are obtained from the solution of the following system of homogeneous linear equations [see eq. (2.11)]:

$$\{ [K]_R - \omega^2 [M]_R \} \{a\} = \{0\} \tag{k}$$

For a non-trivial solution, it is required that the determinant of the coefficient of the unknown vector {a} be equal to zero (see eq. (2.12), that is,

$$\left| [K] - \omega^2 [M] \right| = \{0\} \tag{2.12 repeated}$$

After substituting into eq. (2.12) $[\overline{K}]$ from eq. (h) and $[\overline{M}]$ from eq. (i), we obtain from the solution of eq. (2.12) the following value for ω^2 (square of the natural frequency):

$$\omega_1^2 = 214.05 \; (rad/sec)^2 \tag{1}$$

Then

$$\omega_1 = 14.63 \; rad/sec \qquad \text{and} \qquad f_1 = \omega_1/2\pi = 2.33 \; cps \qquad \text{Ans.}$$

The corresponding normalized eigenvector is equal to the transformation matrix [T] in eq. (j):

$$[\Phi] = \begin{bmatrix} -0.000851 \\ 0.000820 \\ 0.040825 \end{bmatrix} \tag{m}$$

The modal equation is given in general by eqs. (2.18) and (2.19) as

$$\ddot{z}_j + \omega_j^2 z_j = P_j \qquad (j = 1) \tag{n}$$

where

$$P_j = \sum_{i=1}^{3} \phi_{ij} F_i \tag{o}$$

In eq. (o), F_i ($i = 1, 2, 3$) are the external forces applied at the nodal coordinates. In this example, these forces are: $F_1 = F_2 = 0$ and $F_3 = 5000 \; lb$.

Then, the substitution of numerical values into eq. (n) results in the following differential equation:

$$\ddot{z}_1 + 214.05 \, z_1 = 0.040825 \times 5000 = 204.13. \qquad\qquad (p)$$

The solution of eq. (p) with zero initial conditions [see eq. (b) of Illustrative Example 2.3] is

$$z_1 = \frac{204.13}{214.05}(1 - \cos 14.63t). \qquad\qquad (q)$$

The displacements at the nodal coordinates are calculated from eq. (2.16)

$$\{u\} = [\Phi] \, \{z\}_1. \qquad\qquad (2.16) \text{ repeated}$$

The maximum modal displacements may be estimated by substituting the cosine functions for ± 1. Thus substituting $[\Phi]$ and z_1, respectively, from eqs. (m) and (q) into eq. (2.16) results in

$$u_{1,max} = 0.00162 \ rad \qquad u_{2,max} = 0.00156 \ rad \qquad u_{3,max} = 0.0778 \ in \qquad \text{(Ans.)}$$

5.11   Dynamic Analysis of Grid Frames Using SAP2000

Illustrative Example 5.5
Use SAP200 to analyze the grid frame of Illustrative Example 5.4.

The following commands are implemented in SAP2000:

Begin: Open SAP2000

Hint: Maximize both windows for a full view.

Select: Use the drop-down menu located in the lower right-hand corner of the window to select "lb-in".

Select: From the main menu select:
FILE>NEW MODEL.
Number of Grid Spaces:
        X direction = 2
        Y direction = 2
        Z direction = 0
Grid Spaces
        X direction = 60
        Y direction = 60
        Z direction = 1
Click OK.

Edit: Click on 3D View icon on the main menu.
             Maximize the 3D View window and use the PAN icon on the toolbar to
             drag the figure to the center of the window.

Draw: From the main menu select:
             DRAW>DRAW FRAME ELEMENT.
             Click on (0,0,0) and drag the cursor to (60,0,0), located at the
             intersections of the grid lines. (One frame section completed.)
             Click on (0,0,0) and drag the cursor to (0,60,0). (Second frame element
             completed.)

Label: From the main menu select:
             VIEW>SET ELEMENTS.
             Click on Joint Labels and Frame Labels.
             Click OK.
             Use the PAN icon on the toolbar to center the figure in the window.

Material: From the main menu select:
             DEFINE>MATERIALS.
             Under "Materials", select STEEL.
             Click Modify/Show Material.
                 Material Name = STEEL
                 Mass per Unit Volume = 1
                 Weight per Unit Volume = 0
                 Modulus of Elasticity = 30E6
                 Poisson's Ratio = 0.3
                 Accept remaining default values.
             Click OK, OK.

Frame Sections: From the main menu select:
             DEFINE>FRAME SECTIONS.
             Click on Add I/Wide Flange and scroll down to Add General.
             Enter: Cross Sectional Area = 10
                     Torsional Constant = 200
                     Moment of Inertia about 3 axis = 100
                     Moment of Inertia about 2 axis = 100
                     Shear Area in directions 2 and 3 = 0
                     Accept remaining default values.
                     Material select STEEL.
                     Click OK, OK, OK.

Assign: Select all the elements of the frame by clicking on them.
             From the main menu select:
             ASSIGN>FRAME>SECTIONS.

           Select FSEC2.
           Click OK.

Boundary: Select joints 2 and 3 by clicking on them and enter:
           From the main menu select:
           ASSIGN>JOINT>RESTRAINTS.
           Select restraints in all directions.
           Click OK.

Load: From the main menu select:
           DEFINE>STATIC LOAD CASES.
           Enter: Load = LOAD1
                    Type = LIVE
                    Self Weight Multiplier = 0
                    Click Add New Load.
                    Click OK.

Functions: From the main menu select:
           DEFINE>TIME HISTORY FUNCTIONS.
           Click on Add New Function and accept default name FUNC1.
           Enter: Time = 0
                    Value = 5000 Then click ADD.
           Enter: Time = 0.5
                    Value = 5000 Then click ADD.
           Click OK, OK.

Load Case: From the main menu select:
           DEFINE>TIME HISTORY CASES.
           Enter: Add New History
                    Analysis Type = Linear
                    Number of Output Time Steps = 51
                    Time Steps = 0.01
                    Click "Envelopes"
                    Load Assignments:
                        Load = LOAD1.
                        Function = FUNC1. Accept all other default values.
                        Click ADD.
                        Click OK, OK.

Assign Loads: Click on Joint 1 and from the main menu select:
           ASSIGN>JOINT STATIC LOADS>FORCES.
           For LOAD1: Enter Load, Force Global Z = 1
           Click OK.

Analyze: From the main menu select:
ANALYZE>SET OPTIONS.
Check only UZ, RX, RY in Available DOFs (Degrees of Freedom).
Check Dynamic Analysis and then click on Set Dynamic Parameters.
Enter: Number of Modes = 3.
Click OK, OK.

From the main menu select:
ANALYZE>RUN.
Enter filename: EXD 5.5.
Click SAVE.
At the conclusion of the calculations, after checking for errors, click OK.
Select 3D view.

Print Input Tables: From the main menu select:
FILE>PRINT INPUT TABLES.
Click "Print to File" and accept all other default values.
Click OK.
Use Notepad or other word editor to retrieve and edit the input table file
EXD 5.5.txt.
(The edited Input Tables of Illustrative Example 5.5 are reproduced in Table 5.5.)

Table 5.5   Edited Input Tables for Illustrative Example 5.5

```
STATIC LOAD CASES
    STATIC     CASE   SELF WT
    CASE       TYPE   FACTOR
    LOAD1      LIVE   0.0000

TIME HISTORY CASES
    HISTORY   HISTORY      NUMBER OF      TIME STEP
    CASE      TYPE         TIME STEPS     INCREMENT
    HIST1     LINEAR       50             0.01000

JOINT DATA
    JOINT  GLOBAL-X      GLOBAL-Y      GLOBAL-Z   RESTRAINTS
    1      0.00000       0.00000       0.00000     0 0 0 0 0 0
    2      60.00000      0.00000       0.00000     1 1 1 1 1 1
    3      0.00000       60.00000      0.00000     1 1 1 1 1 1

FRAME ELEMENT DATA
FRAME JNT1 JNT2 SCT    ANGLE    RLSS   SGMNT R1      R2     FACTOR LENGTH
1      1    2   FSEC2  0.000    000000   4   0.000   0.000  1.000  60.000
2      1    3   FSEC2  0.000    000000   4   0.000   0.000  1.000  60.000

JOINT FORCES  Load Case LOAD1
    JOINT   GLOBAL-X   GLOBAL-Y   GLOBAL-Z  GLOBAL-XX  GLOBAL-YY  GLOBAL-ZZ
    1       0.000      0.000      1.000     0.000      0.000      0.000
```

Plot Displacement Function: From the main menu select:
    DISPLAY>SHOW TIME HISTORY TRACES.
    Click Define Functions.
    Click to "Add Joint Disps/Forces".
        Joint ID = 1
        Vector Type = Displacement
        Component = UZ
        Mode Number = Include All
        Click OK.
    Under List of Functions click once on "Joint1".
    Click ADD to move "Joint1" to Plot Functions column.
    Axis Labels
        Horizontal = Time (*sec*)
        Vertical = Displacement Component Z at Joint 1 (*in*)
    Click "Display" to view the Displacement function.
    Select FILE>PRINT GRAPHICS to obtain a printed version of this plot.

(Figure 5.11 reproduces the plot for the displacement component Z at Joint 1 of the plane grid structure of Illustrative Example 5.5.)

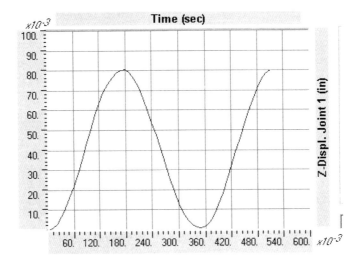

Fig. 5.11   Displacement in the Z direction at node 1 for the grid frame of Illustrative Example 5.5

    Note: Displacement values may be viewed on the screen by moving the cursor along the curves. Table 5.6 gives the maximum displacement in the Z direction at node 1 for the grid frame of Illustrative Example 5.5.

Table 5.6 Maximum Displacements at node 1 in Z direction for the grid frame of Illustrative Example 5.5

Direction	Maximum Displacement
Z	0.0808 *in*
RX	1.696E-3 *rad*
RY	1.696E-3 *rad*

In the Time History form select:
FILE>PRINT TABLES TO FILE.

At the prompt, save the file as " displacement EXD 5.5.txt".

Alternatively, a table of the displacement function in the Z direction at joint 1 may be obtained by selecting:
FILE>PRINT TABLES TO FILE.

At the prompt, Save the file as "Displacement EXD5.5.txt".

Then using Notepad or similar word processor, the displacement function may be accessed and edited. An abbreviated table of the displacement component function in the Z direction for joint 1 is reproduced in Table 5.7. The values in this table may be plotted using Excel or similar software.

Table 5.7 Displacement Function at Joint 1 in the Z direction for the grid frame of Illustrative Example 5.4

Displacement at Joint 1	
Time (*sec*)	Z-direction (*in*)
0.00	0.000
0.01	0.001
0.02	0.003
0.03	0.007
0.04	0.011
0.05	0.016
----------------------	--------------------
----------------------	--------------------
0.49	0.040
0.50	0.043

Natural Periods and Frequencies:
Using Notepad or other word editor, open the following file:
C:/SAP2000e/EXD 5.5.OUT.
The Natural Period and Frequency table may be found in this file.

(Table 5.8 provides the edited Natural Period and Frequency values of Illustrative Example 5.5.)

Table 5.8   Natural Period and Frequency values for the plane grid structure of Illustrative Example 5.5

NATURAL PERIODS AND FREQUENCIES

MODE	PERIOD (SEC)	FREQUENCY (CYC/SEC)	FREQUENCY (RAD/SEC)	EIGENVALUE (RAD/SEC)^2
1	0.438	2.284	14.35	205.92

Plot Bending Moment: From the main menu select:
    DISPLAY>SHOW TIME HISTORY TRACES.
    Click "Define Functions".
    Click to "Add Frame Element Forces".
        Element ID = 4
        Station = 1
        Component = Moment 3-3
        Accept all other default settings.
        Click OK, OK.
    Click once on "Frame3" and then ADD to move "Frame3" to the Plot Function column.
    Highlight "Joint1" and click Remove to move "Joint1" back to List of Functions column.
    Axis Labels
        Horizontal = Time (*sec*)
        Vertical = Bending Moment at joint 1 of Element 2 (*lb-in*)
    Click Display to view the plot of the Bending moment and its maximum and minimum values.
        Max = 2.256*E*04
        Min = −6.191*E*04

  (Figure 5.12 provides a plot of the bending moment time function for frame element 3 of the grid frame of Illustrative Example 5.5.)

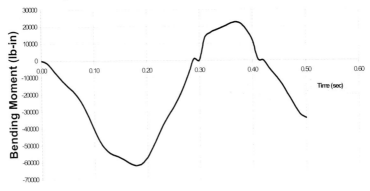

Fig 5.12   Bending moment time function for element 3, at section 1, of the plane grid of Illustrative Example 5.5

Note: Bending moment values may be viewed on screen by moving the cursor along the curve.

Bending Moment Function:
A table for the bending moment function (or other internal force) for a member may be obtained by selecting:
FILE > PRINT TABLES TO FILE.
At the prompt, enter the filename "bending moment joint 1 element 3 EXD 5.5.txt".

Then using Notepad or a similar word processor, the file "bending moment EXD 5.4.txt" may be edited to obtain a table of the Bending Moment Time History Function. (An abbreviated table of the Bending Moment History Function for the beam element 3 of the plane grid of Illustrative Example 5.5 is shown in Table 5.9.)

Table 5.9 Bending moment time function for joint 1, element 3, of the plane grid structure of Illustrative Example 5.5

Bending Moment Frame 3	
Time (*sec*)	Value (*lb-in*)
0.00	0.0
0.01	−1195.0
0.02	−4318.0
0.03	−8263.0
0.04	−11971.0
0.05	−15056.0
--------------------	--------------------
--------------------	--------------------
0.49	−32194.0
0.50	−34139.0

Illustrative Example 5.6
Solve Illustrative Example 5.5 after modeling the grid frame of that example into 4, 8, and 16 frame elements using the command "Divide Frames".

Open the file EXD 5.5 in SAP2000.

Enter the following commands:

Unlock file EXD 5.5: Click on the padlock icon in the toolbar.
Click Yes to confirm.

Divide Frames: Select all the frame elements by clicking on each frame or alternatively by clicking on any point and then dragging the pointer to wrap the grid frame.
From the main menu select:
EDIT > DIVIDE FRAMES.
Accept the default "Divide frame in 2".
Click OK.

ANALYZE: From the main menu select:
ANALYZE > RUN.
At the conclusion of the calculations, check for errors and then click OK.

Natural Periods and Frequencies:
Using Notepad or other editor, open the following file:
C:/SAP2000e/EXD 5.5.OUT.
The Natural Period and Frequency table may be found in this file.

(Table 5.10 provides the edited Natural Period and Frequency values of Illustrative Example 5.6.)

Table 5.10   Natural Period and Frequency values for the plane grid structure of Illustrative Example 5.6 modeled with 4 beam elements

NATURAL PERIODS AND FREQUENCIES

MODE	PERIOD (SEC)	FREQUENCY (CYC/SEC)	FREQUENCY (RAD/SEC)	EIGENVALUE (RAD/SEC)^2
1	0.346	2.890	18.158	329.72
2	0.085	11.757	73.871	5456.99
3	0.076	13.144	82.586	6820.47

Displacements: From the main menu select:
DISPLAY>SHOW TIME HISTORY TRACES.
Click Define Functions.
Click to "Add Joint Disps/Forces".
    Joint ID = 1
    Vector Type = Displacement
    Component = UZ
    Mode Number = Include All
    (UZ displacement function is now stored in "Joint1".)
    Click OK.

Click Define Functions.
Select "Add Joint Disps/Forces".
    Joint ID = 1

Vector Type = Displacement
Component = RX
Mode Number = Include All
(RX displacement function is now stored in "Joint1-1".)
Click OK.

Click on Define Functions.
Select "Add Joint Disps/Forces".
Joint ID = 1
Vector Type = Displacement
Component = RY
Mode Number = Include All
(RY displacement function is now stored in "Joint1-2".)
Click OK.

Under List of Functions, click once on "Joint1".
Click ADD to move "Joint1" to Plot Functions column.
Axis Labels
Horizontal = Time (*sec*)
Vertical = Displacement Component Z at Joint 1 (*in*)
Click Display to view the plot Displacement function.
To obtain a printed version of this plot, select:
FILE>PRINT GRAPHICS.

Max. and Min. Displacement values are shown on the left side of the
window.
(Maximum absolute value of the displacement UZ = 0.0803 *in.*)
Click OK.

Click on Joint1 and on Remove.
Under List of Functions, click once on "Joint1-1".
Click ADD to move "Joint1-1" to the Plot Functions column.
Axis Labels
Horizontal = Time (*sec*)
Vertical = Displacement Component RX at Joint 1 (*rad*)

Click Display to view the plot Displacement function.
To obtain a printed version of this plot, enter:
FILE>PRINT GRAPHICS.
Max. and Min. Displacement values are shown on the left side of the
window.
[Maximum absolute value of the displacement RX = 1.671E-3 (*rad*).]
Click OK.

Click on Joint1-1 and on Remove.

Under List of Functions, click once on "Joint1-2".
Click ADD to move "Joint1-2" to Plot Functions column.
Axis Labels
    Horizontal = Time (*sec*)
    Vertical = Displacement Component RY at Joint 1 (*rad*)

Click Display to view the Displacement function.
To obtain a printed version of this plot select:
FILE>PRINT GRAPHICS.
Max. and Min. Displacement values are shown on the left side of the
window.
[Maximum absolute value of the displacement RY = 1.671*E*-3 (*rad*).]
Click OK and DONE.

To analyze the grid frame of this example with 8 and then with 16 frame elements,
repeat the preceding commands.

Table 5.11 provides the natural frequencies obtained in the analysis of the grid
frame of Illustrative Example 5.6 modeled with 2, 4, 8 and 16 frame elements.

Table 5.12 provides the maximum values for the displacements of Joint 1 of the grid
frame of Illustrative Example 5.6 modeled with 2, 4, 8 and 16 frame elements.

Table 5.11 Natural frequencies for the grid frame of Illustrative Example 5.6
modeled with 2, 4, 8 and 16 frame elements

Mode #	Number of Frame Elements in the Model (*cps*)			
	2 frames	4 frames	8 frames	16 frames
1	2.284	2.890	3.076	3.126
2		11.757	12.345	12.363
3		13.144	16.479	17.529
4			38.034	38.788
5			41.684	46.546
6			70.284	80.122

Table 5.12 Maximum displacements for the grid frame of Illustrative
Example 5.6 modeled with 2, 4, 8 and 16 frame elements

Displacement	Maximum Displacements			
	2 frames	4 frames	8 frames	16 frames
UZ (*in*)	0.0778	0,0803	0.0804	0.0805
RX (*rad*)	1.696*E*-3	1.671*E*-3	1.654*E*-3	1.649*E*-3
RY (*rad*)	1.696*E*-3	1.671*E*-3	1.654*E*-3	1.649*E*-3

5.12 Problems in Structural Analysis

Problems 5.1 through 5.6
For the grid frames shown in Figs. 5.1 through 5.6 determine:

(a) The joint displacements
(b) Member end forces
(c) Support reactions

Neglect self-weight of the members and shear deformation.

Problem 5.1:

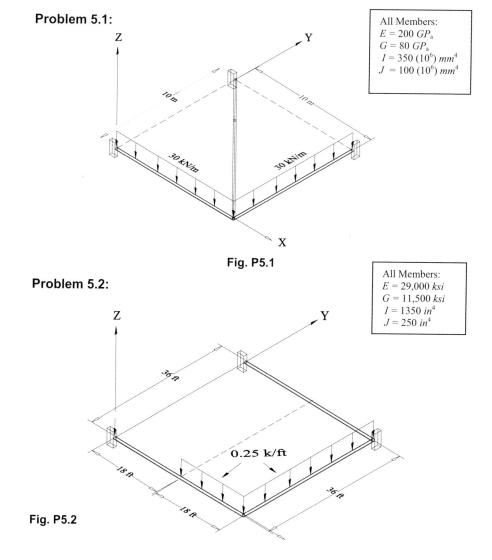

All Members:
$E = 200 \ GP_a$
$G = 80 \ GP_a$
$I = 350 \ (10^6) \ mm^4$
$J = 100 \ (10^6) \ mm^4$

Fig. P5.1

Problem 5.2:

All Members:
$E = 29,000 \ ksi$
$G = 11,500 \ ksi$
$I = 1350 \ in^4$
$J = 250 \ in^4$

Fig. P5.2

Problem 5.3

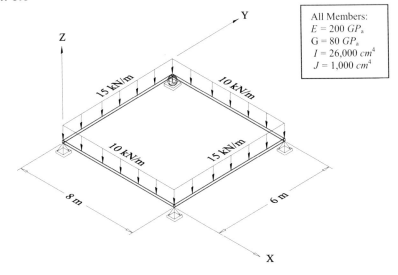

All Members:
$E = 200\ GP_a$
$G = 80\ GP_a$
$I = 26{,}000\ cm^4$
$J = 1{,}000\ cm^4$

Fig. P5.3

Problem 5.4:

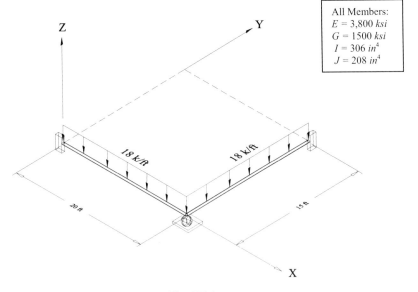

All Members:
$E = 3{,}800\ ksi$
$G = 1500\ ksi$
$I = 306\ in^4$
$J = 208\ in^4$

Fig. P5.4

Problem 5.5:

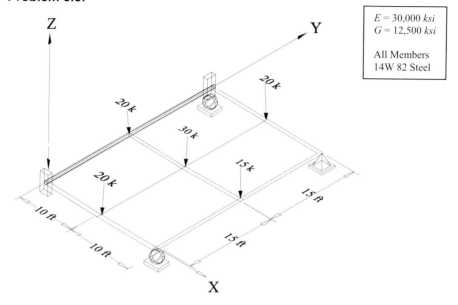

Fig. P5.5

Problem 5.6:

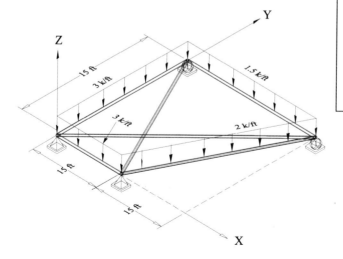

Fig. P5.6

5.13 Problems in Structural Dynamics

Problem 5.7
For the grid frame shown in Figure P5.7, determine the system stiffness and mass matrices. Base the analysis on the three nodal coordinates indicated in the figure. Use the consistent mass method.

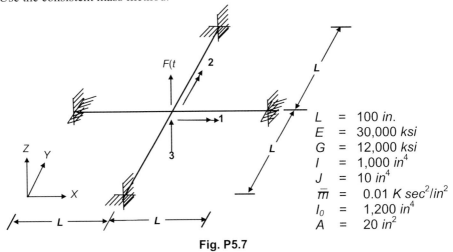

$$
\begin{aligned}
L &= 100 \ in.\\
E &= 30{,}000 \ ksi\\
G &= 12{,}000 \ ksi\\
I &= 1{,}000 \ in^4\\
J &= 10 \ in^4\\
\bar{m} &= 0.01 \ K \ sec^2/in^2\\
I_0 &= 1{,}200 \ in^4\\
A &= 20 \ in^2
\end{aligned}
$$

Fig. P5.7

Problem 5.8
Use static condensation to eliminate the rotational degrees of freedom and determine the transformation matrix and the reduced stiffness and mass matrices in Problem 5.7.

Problem 5.9
Determine the natural frequency for the reduced system in Problem 5.8.

Problem 5.10
Determine the natural frequencies and corresponding mode shapes for the grid frame analyzed in Problem 5.7.

Problem 5.11
Determine the response of the grid frame shown in Figure P5.7 when acted upon by a force $F(t) = 10 \ kip$ suddenly applied for one second at the nodal coordinate 3 as shown in the figure. Use results of Problem 5.8 to obtain the equation of motion for the condensed system. Assume 10% modal damping.

Problem 5.12

Use results from Problem 5.10 to solve Problem 5.11 on the basis of the three nodal coordinates as indicated in Figure P5.7.

Problem 5.13

Determine the steady-state response of the grid frame shown in Figure P5.7 when subjected to harmonic force $F(t) = 10 \sin 50t$ *kip* applied along nodal coordinate 3. Neglect damping in the system.

Problem 5.14

Repeat Problem 5.7 assuming that the damping is proportional to the stiffness of the system, $[C] = a_0 [K]$, where $a_0 = 0.3$.

Problem 5.15

Determine the equivalent nodal forces for a member of a grid frame loaded with a dynamic force, $P(t) = P_0 f(t)$, uniformly distributed along its length.

Problem 5.16

Determine the equivalent nodal forces for a member of a grid frame supporting a concentrated dynamic force $F(t)$ as shown in Figure P5.8.

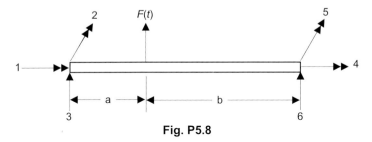

Fig. P5.8

Problem 5.17

Use SAP2000 to determine the natural frequencies and corresponding mode shapes for the grid frame shown in Figure P5.7

Problem 5.18

Determine the response of the grid frame shown in Figure P5.7 when acted upon in the direction of the nodal coordinate 3 by the force depicted in Figure P5.9. Neglect damping in the system.

Problem 5.19
Repeat Problem 5.18 assuming 15% damping in all the modes.

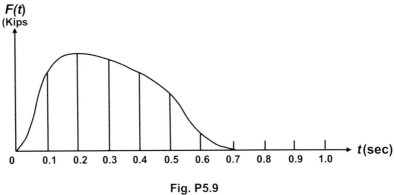

Fig. P5.9

6 Space Frames

6.1 Introduction

The stiffness method for analysis presented in Chapter 4 for plane frames and in Chapter 5 for grid frames can readily be extended to the analysis of space or three dimensional frames. The plane frame or the grid frame has three nodal coordinates at each joint; however, the *space frame* has a total of six nodal coordinates at each joint: three translatory components along the X, Y, and Z axes and three rotational components about these axes. Consequently, a beam element of a space frame has for its two joints a total of 12 nodal coordinates; hence, the resulting element stiffness matrix will be of dimension 12×12.

The analysis of space frames results in a comparatively longer computer analysis, requiring more input data as well as substantially more computational time. Except for size, the analysis of space frames by the stiffness method is identical to the analysis of plane frames or grid frames.

6.2 Element Stiffness Matrix

Figure 6.1 shows the beam element of a space frame with its 12 nodal coordinates numbered consecutively. The convention adopted is to first label the three translatory displacements of the first joint, followed by the three rotational displacements of the same joint, then continuing with the three translatory displacements of the second joint and finally the three rotational displacements of the second joint. The double arrows used in Figure 6.1 serve to indicate rotational nodal coordinates; hence, these are distinguished from translational nodal coordinates for which single arrows are used.

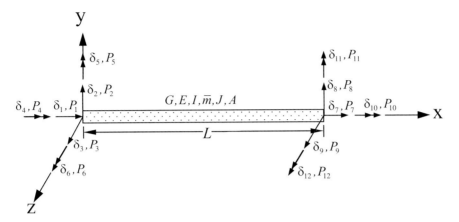

Fig. 6.1 Beam element of a space frame showing forces and displacements at the nodal coordinates.

The stiffness matrix for a three-dimensional uniform beam segment is readily written by the combination of the axial stiffness matrix from eq. (4.3), the torsional stiffness matrix from eq. (6.6) and the flexural stiffness matrix from eq. (1.11). The flexural stiffness matrix is used twice in forming the stiffness matrix of an element of a space frame to account for the flexural effects in the two principal planes of the cross-section. Proceeding to combine these matrices in an appropriate manner, we obtain in eq. (6.1) the stiffness equation for a uniform beam element of a space frame, namely

$$
\begin{Bmatrix} P_1 \\ P_2 \\ P_3 \\ P_4 \\ P_5 \\ P_6 \\ P_7 \\ P_8 \\ P_9 \\ P_{10} \\ P_{11} \\ P_{12} \end{Bmatrix}
=
\begin{bmatrix}
\dfrac{EA}{L} & & & & & & & & & & & \\[4pt]
0 & \dfrac{12EI_z}{L^3} & & & & \text{symmetric} & & & & & & \\[4pt]
0 & 0 & \dfrac{12EI_y}{L^3} & & & & & & & & & \\[4pt]
0 & 0 & 0 & \dfrac{GJ}{L} & & & & & & & & \\[4pt]
0 & 0 & \dfrac{-6EI_y}{L^2} & 0 & \dfrac{4EI_y}{L} & & & & & & & \\[4pt]
0 & \dfrac{6EI_z}{L^2} & 0 & 0 & 0 & \dfrac{4EI_z}{L} & & & & & & \\[4pt]
\dfrac{-EA}{L} & 0 & 0 & 0 & 0 & 0 & \dfrac{EA}{L} & & & & & \\[4pt]
0 & \dfrac{-12EI_z}{L^3} & 0 & 0 & 0 & \dfrac{-6EI_z}{L^2} & 0 & \dfrac{12EI_z}{L^3} & & & & \\[4pt]
0 & 0 & \dfrac{-12EI_y}{L^3} & 0 & \dfrac{6EI_y}{L^2} & 0 & 0 & 0 & \dfrac{12EI_y}{L^3} & & & \\[4pt]
0 & 0 & 0 & \dfrac{-GJ}{L} & 0 & 0 & 0 & 0 & 0 & \dfrac{GJ}{L} & & \\[4pt]
0 & 0 & \dfrac{-6EI_y}{L^2} & 0 & \dfrac{2EI_y}{L} & 0 & 0 & 0 & \dfrac{6EI_y}{L^2} & 0 & \dfrac{4EI_y}{L} & \\[4pt]
0 & \dfrac{6EI_z}{L^2} & 0 & 0 & 0 & \dfrac{2EI_z}{L} & 0 & \dfrac{-6EI_z}{L^2} & 0 & 0 & 0 & \dfrac{4EI_z}{L}
\end{bmatrix}
\begin{Bmatrix} \delta_1 \\ \delta_2 \\ \delta_3 \\ \delta_4 \\ \delta_5 \\ \delta_6 \\ \delta_7 \\ \delta_8 \\ \delta_9 \\ \delta_{10} \\ \delta_{11} \\ \delta_{12} \end{Bmatrix}
$$

(6.1)

or in condensed notation

$$\{P\} = [k]\{\delta\}.$$
(6.2)

6.3   Transformation of Coordinates

The stiffness matrix given in eq. (6.1) refers to local coordinate axes fixed on the beam element. Inasmuch as the coefficients of these matrices (corresponding to the same nodal coordinates of the structure) should be added to obtain the system stiffness matrix, it is necessary to first transform these matrices to the same reference system, the global system of coordinates. Figure 6.2 shows these two reference systems, the (x, y, z) axes representing the local system of coordinates and the (X, Y, Z) axes representing the global system of coordinates.

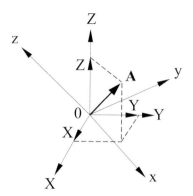

Fig. 6.2 Components X, Y, and Z in global coordinates of a general vector **A**

Also shown in this figure is a general vector **A** with its components X, Y, and Z along the global coordinate axes. The vector **A** may represent any force or displacement at the nodal coordinates of one of the joints of the structure. To obtain the components of vector **A** along one of the local axes x, y or z, it is necessary to add the projections along that axis of the components X, Y, Z. For example, the component x of vector **A** along the x coordinate is given by

$$x = X \cos xX + Y \cos xY + Z \cos xZ$$
(6.3a)

in which $\cos xY$ is the cosine of the angle between axes x and Y and corresponding definitions for other cosines. Similarly, the y and z components of the vector **A** are

$$y = X \cos yX + Y \cos yY + Z \cos yZ$$
(6.3b)

$$z = X \cos zX + Y \cos zY + Z \cos zZ.$$
(6.3c)

These equations are conveniently written in matrix notation as

$$\begin{Bmatrix} x \\ y \\ z \end{Bmatrix} = \begin{bmatrix} \cos xX & \cos xY & \cos xZ \\ \cos yX & \cos yY & \cos yZ \\ \cos zX & \cos zY & \cos zZ \end{bmatrix} \begin{Bmatrix} X \\ Y \\ Z \end{Bmatrix} \qquad (6.4)$$

or in short notation

$$\{A\} = [T]_1 \{\overline{A}\} \qquad (6.5)$$

in which $\{A\}$ and $\{\overline{A}\}$ are, respectively, the components in the local and global systems of the general vector $\mathbf{A}$ and $[T]_1$ the transformation matrix given by

$$[T]_1 = \begin{bmatrix} \cos xX & \cos xY & \cos xZ \\ \cos yX & \cos yY & \cos yZ \\ \cos zX & \cos zY & \cos zZ \end{bmatrix}. \qquad (6.6)$$

The cosines required in eq. (6.6) to determine the transformation matrix $[T]_1$ are usually calculated from the global coordinates of the two points at the nodes of the beam element in addition to one more piece of information. This additional information could be either the *angle of roll* or the global coordinates of a third point located in the local plane *x-y* of the beam element. The plane *x-y* is defined by the local axis *x*, which is an axis along the centroid of the beam element, and the axis *y*, which is located in a plane perpendicular to x and directed along the minor principal axis of the cross-sectional area of the beam element. The local axes *x* and *y*, together with a third axis *z* forming a right-hand system of coordinates, constitute the so called element or local coordinate system. The positive direction of axis *x* is dictated by the order in which two nodes of the beam element are specified.

The *angle of roll* is the angle by which the local axis *y* has been rotated from the "standard" orientation. This standard orientation exists when the local plane, formed by the local axes *x'* and *y'*, is vertical, that is, parallel to the global axis *Z* as shown in Figure 6.3.

The angle of roll, ϕ is positive for the case in which the local axis *y* appears rotated around the local axis *x* counter-clockwise from the standard orientation of the plane defined by the axes *x* and *y* when observed from the positive end of axis *x*. The direction cosines for the local axis *x*, conveniently designated $c_1 = \cos xX$, $c_2 = \cos xY$, and $c_3 = \cos xZ$, are readily calculated from the coordinates (X_1, Y_1, Z_1) and (X_2, Y_2, Z_2) of the two nodes of the beam element as

$$c_1 = \frac{X_2 - X_1}{L} \qquad c_2 = \frac{Y_2 - Y_1}{L} \qquad c_3 = \frac{Z_2 - Z_1}{L} \qquad (6.7)$$

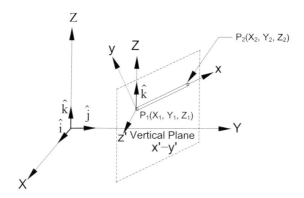

Fig. 6.3 Global system of coordinates (X, Y, Z) and local coordinate system (x', y', z)
with the beam element in the "standard" orientation (plane x'-y' vertical).

in which L is the length of the beam element given by

$$L = \sqrt{(X_2 - X_1)^2 + (Y_2 - Y_1)^2 + (Z_2 - Z_1)^2} . \qquad (6.8)$$

As already stated, the transformation matrix $[T]_1$ from the global coordinate system
(X, Y, Z) to the local system (x, y, z) may be expressed in terms of the angle of roll
ϕ and the direction cosines c_1, c_2, and c_3 of the local axis x (see Analytical Problem
6.1 in Section 6.10). The final expression for the transformation matrix $[T]_1$ is

$$[T]_1 = \begin{bmatrix} c_1 & c_2 & c_3 \\ -\dfrac{c_1 c_3}{d}\cos\phi - \dfrac{c_2}{d}\sin\phi & -\dfrac{c_1}{d}\sin\phi - \dfrac{c_2 c_3}{d}\cos\phi & d\cos\phi \\ \dfrac{c_1 c_3}{d}\sin\phi + \dfrac{c_2}{d}\cos\phi & -\dfrac{c_1}{d}\cos\phi + \dfrac{c_2 c_3}{d}\sin\phi & -d\sin\phi \end{bmatrix} \qquad (6.9)$$

in which

$$d = \sqrt{c_1^2 + c_2^2} . \qquad (6.10)$$

It should be noted that the transformation matrix $[T]_1$ is not defined if the local axis
x is parallel to the global axis Z. In this case, eqs. (6.7) and (6.10) result in $c_1 = 0$,
$c_2 = 0$, and $d = 0$. If the centroidal axis of the beam element is vertical, that is, the
local axis x and the global axis Z are parallel, the angle of roll is then defined as the

angle that the local axis y has been rotated about the axis x from the "standard" direction defined for a vertical beam element. The "standard" direction in this case exists when the local axis y is parallel to the global axis X as shown in Figure 6.4.

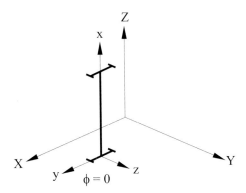

Fig. 6.4 Standard orientation of the cross-section for a vertical beam element

For a vertical beam element, the angle of roll ϕ is positive if the local axis y appears rotated from the "standard orientation" counter-clockwise when observed from the positive end of the local axis x.

The final expression for the transformation matrix $[T]_1$ (see Analytical Problem 6.2 of Section 6.10) for the case of a vertical beam element is given by

$$[T]_1 = \begin{bmatrix} 0 & 0 & \lambda \\ \cos\phi & \lambda\sin\phi & 0 \\ -\sin\phi & \lambda\cos\phi & 0 \end{bmatrix} \qquad (6.11)$$

where ϕ is the angle of roll between the global axis X and the local axis y for the case in which the centroidal axis of the beam has a vertical orientation. The coefficient $\lambda = +1$ when the local axis x and the global axis Z have the same sense and $\lambda = -1$ for the opposite sense of these axes. Figures 6.5(a) through 6.5(d) illustrate the measurement of the angle of roll for several cases.

It has been shown that the knowledge of the coordinates of the two ends of the beam element and of the angle of roll suffices to calculate the direction cosines of the transformation matrix $[T]_1$ in eq. (6.6) given in the general case by eq. (6.9) and by eq. (6.11) for the special case in which the local axis x is parallel to the global axis Z.

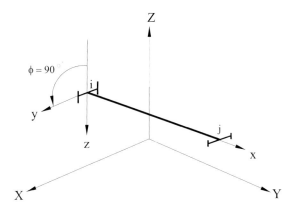

Fig. 6.5 (a) Local axis *x* is parallel to +Y Axis
Local axis *y* is rotated 90° from Z-x Plane

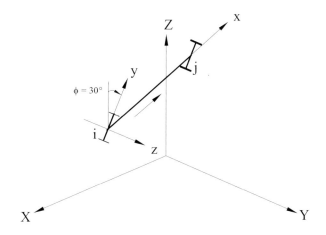

Fig. 6.5(b) Local x Axis is not parallel to X, Y or Z Axes
Local y Axis is rotated 30° from Z-x Plane

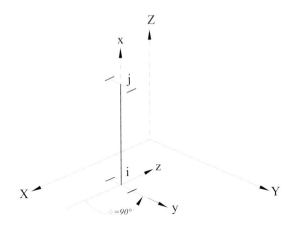

Fig. 6.5(c)   Local axis *x* is parallel to +Z Axis
                    Local Axis *y* is rotated 90° from being parallel to X

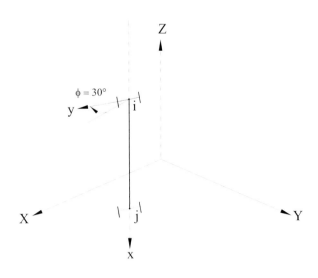

Fig. 6.5(d)   Local Axis *x* is parallel to –Z Axis
                  Local Axis *y* is rotated 30° from y Parallel to X

Alternatively, as already stated, the direction cosines in eq. (6.6) may be calculated from the coordinates of the two points at the two nodes of a beam element and the coordinates of a third point located in the x-y local plane, in which y is the minor principal axis of the cross-section and x is the axis along the beam element. Analytical Problem 6.3 of Section 6.10 includes the development of the relationship to calculate the transformation matrix $[T]_1$ from the coordinates of these three points.

For the beam element of a space frame, the transformations of the nodal displacement vectors involve the transformation of linear and angular displacement vectors at each node of the segment. Therefore, a beam element of a space frame requires the transformation of a total of four displacement vectors for the two nodes. The transformation of the 12 nodal displacements $\{\overline{\delta}\}$ in global coordinates to the displacements $\{\delta\}$ in the local coordinates may be written in abbreviated form as

$$\{\delta\} = [T]\{\overline{\delta}\} \qquad (6.12)$$

in which

$$[T] = \begin{bmatrix} [T]_1 & & & \\ & [T]_1 & & \\ & & [T]_1 & \\ & & & [T]_1 \end{bmatrix} \qquad (6.13)$$

Analogously, the transformation of the nodal forces $\{\overline{P}\}$ in global coordinates to the nodal forces $\{P\}$ in local coordinates is given by

$$\{P\} = [T]\{\overline{P}\}. \qquad (6.14)$$

Finally, to obtain the stiffness matrix $[\overline{k}]$ in reference to the global system of coordinates, we simply substitute into eq. (6.2) $\{\delta\}$ from eq. (6.12) and $\{P\}$ from eq. (6.14) to obtain

$$[T]\{\overline{P}\} = [k][T]\{\overline{\delta}\}$$

or since $[T]$ is an orthogonal matrix $[\{T\}^{-1} = \{T\}^T]$,

$$\{\overline{P}\} = [T]^T[k][T]\{\overline{\delta}\}. \qquad (6.15)$$

Equation eq.(6.15), may be written as

$$\{\overline{P}\} = [\overline{K}]\{\overline{\delta}\} \qquad (6.16)$$

in which the element stiffness matrix $[\bar{k}]$ in reference to the global system of coordinates is given by

$$[\bar{k}] = [T]^T [k][T].$$ (6.17)

The reduced system stiffness matrix $[K]_R$ (only free coordinates are considered) is assembled by transferring the coefficients of the element stiffness matrices to the appropriate locations in the system stiffness matrix, as presented in the preceding chapters for beams, plane frames and grid frames. Thus, the relationship between the unknown nodal displacements $\{u\}$ and the forces $\{F\}_R$ at the free nodal coordinates is established through the reduced system stiffness matrix:

$$\{F\}_R = [K]_R \{u\}$$ (6.18)

in which the reduced force vector $\{F\}_R$ contains the equivalent nodal forces for forces applied on the beam elements of the structure and the forces applied directly to the free nodal coordinates.

6.4   Analysis of Space Frames Using SAP2000

Illustrative Example 6.1
Use SAP2000 to analyze the space frame shown in Figure 6.6. The corresponding input data is given in Table 6.1.

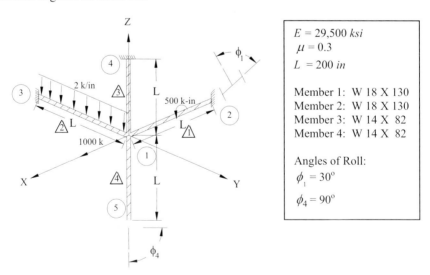

$E = 29,500\ ksi$
$\mu = 0.3$
$L = 200\ in$

Member 1: W 18 X 130
Member 2: W 18 X 130
Member 3: W 14 X 82
Member 4: W 14 X 82

Angles of Roll:
$\phi_1 = 30^\circ$
$\phi_4 = 90^\circ$

Fig. 6.6 Space frame for Illustrative Example 6.1

Table 6.1   Edited input data for Illustrative Example 6.1 (Units: kips, inches)

JOINT DATA

JOINT	GLOBAL-X	GLOBAL-Y	GLOBAL-Z	RESTRAINTS
1	0.00000	0.00000	0.00000	0 0 0 0 0 0
2	−200.00000	0.00000	0.00000	1 1 1 1 1 1
3	0.00000	−200.00000	0.00000	1 1 1 1 1 1
4	0.00000	0.00000	200.00000	1 1 1 1 1 1
5	0.00000	0.00000	−200.00000	1 1 1 1 1 1

FRAME ELEMENT DATA

FRAME	JNT-1	JNT-2	SECTION	ANGLE	RELEASES	SEGMENTS	LENGTH
1	1	2	W18X130	30.000	000000	4	200.000
2	1	3	W18X130	0.000	000000	4	200.000
3	1	4	W14X82	0.000	000000	2	200.000
4	1	5	W14X82	90.000	000000	2	200.000

JOINT FORCES   Load Case LOAD1

JOINT	GLOBAL-X	GLOBAL-Y	GLOBAL-Z	GLOBAL-XX	GLOBAL-YY	GLOBAL-ZZ
1	0.000	1000.000	0.000	0.000	0.000	0.000

FRAME SPAN DISTRIBUTED LOADS

FRAME	TYPE	DIRECTION	ISTANCE-A	VALUE-A	DISTANCE-B	VALUE-B
2	DISTRIBUTED	GLOBAL-Z	0.0000	−2.0000	1.0000	−2.0000

FRAME SPAN POINT LOADS   Load Case LOAD1

FRAME	TYPE	DIRECTION	DISTANCE	VALUE
1	MOMENT	GLOBAL-Y	0.5000	500.0000

Solution:

Begin: Open SAP2000

Enter: "OK" to close the Tip of the Day.

Hint: Click the Windows Maximize button for full views of both screens.

Units: In the lower right-hand corner of the screen, use the drop-down menu to select "kip-in".

Model: From the main menu select:
FILE>NEW MODEL
Number of Grid Spaces
x direction = 4
y direction = 4
z direction = 4

           Grid Spaces
               x direction = 200
               y direction = 200
               z direction = 200
       Click OK.

Select: Default screen to 3-D screen view
       Then enter from the main menu select:
       VIEW>SET 3D VIEW.
       Rotate plane view to approximately 30° to have a frontal view of X axis.

Enter: From the main menu select
       VIEW>2-D VIEW.
       Set   Plane Y-Z to $X = 0$
                Plane X-Z to $Y = 0$
                Plane X-Y to $Z = 0$
       Click OK.

Select: Click the 2-D view.
       Set   Plane Y-Z to $X = 0$
                Plane X-Z to $Y = 0$
                Plane X-Y to $Z = 0$
       Click OK.

Add Grid: From the main menu select:
       DRAW>EDIT GRID.
       Click on Z
                Enter –200, Add grid.
       Click OK.

Draw: Set the X-Z plane view at $Y = 0$.
       From the main menu select:
       DRAW>DRAW FRAME ELEMENT.
       Start with the cursor at the origin and drag to the negative of the X axis
       until the next grid line. Then double click to stop drawing elements.

Draw: Set the Y-Z plane view at $X = 0$.
       From the main menu select:
       DRAW>DRAW FRAME ELEMENT.
       Draw the element from the origin along the Y axis in the negative direction
       until the next grid line, then double click. Next, start at the origin in the
       positive direction of the Z axis until the next grid line and double click.
       Finally, start at the origin and drag the cursor in the negative direction of
       the Z axis until the next grid line and double click.

Labels: Click the 3-D view
From the main menu select:
VIEW>SET ELEMENTS.
Click on Joint Labels and Frame Labels.
Click OK.

Restraints: Select Joints 2, 3, 4, and 5 with a click of the mouse.
From the main menu select:
ASSIGN>JOINTS>RESTRAINTS.
Select: "Restraints all".
Click OK.

Materials: From the main menu select:
DEFINE>MATERIALS.
Select STEEL.
Click Modify/Show Material.
Set $E = 29500$ and Poisson's Ratio $= 0.3$.
Click OK.

Sections: From the main menu select:
DEFINE>FRAME SECTIONS.
Click IMPORT I/WIDE FLANGE.
Click OK.
Select Sections.prop.
Click OPEN.
Scroll down and select W18 X 130.
Hold down the Ctrl key and select W14 X 82.
Click OK.

Click W18 X 130 and Modification Factors.
Set: Shear area in Z direction $= 0$
Shear area in Y direction $= 0$.
Click OK.

Click W14 X 82 and Modification Factors.
Set: Shear area Z direction $= 0$
Shear area Y direction $= 0$.
Click OK, OK.

Assign: Click elements 1 and 2 and from the main menu select:
ASSIGN>FRAME SECTIONS.
Select W18 X 130.
Click OK.

Click elements 3 and 4 and from the main menu select:
ASSIGN>FRAME SECTIONS.
Select W14 X 82.
Click OK.

Loads: From the main menu select:
DEFINE>STATIC LOAD CASES.
Change DEAD to LIVE LOAD.
Set Self-Weight Multiplier to 0.
Click Change Load.
Click OK.

Assign: Select Joint 1 and from the main menu select:
ASSIGN>JOINT STATIC LOADS>FORCES.
Enter Force Global $X = 1000$.
Click OK.

Assign: Element 2 and from the main menu select:
ASSIGN>FRAME STATIC LOADS>POINT AND UNIFORM.
Check: Force
Select Global Z.
Enter Uniform load $= -2.0$.
Click OK.

Assign: Select element 1 and from the main menu select:
ASSIGN>FRAME STATIC LOADS>POINT AND UNIFORM.
Select: Moments
Select: Direction Global Y
          Distance 0.5 Load $= 500$
Click OK.
Warning: Make certain that Uniform Load value is set to zero.

Rolling angle: Select element 1 and from the main menu select:
ASSIGN>FRAME LOCAL AXES.
Enter Angle in degree $= 30$.
Click OK.

Select element 4 and from the main menu select:
ASSIGN>FRAME>LOCAL AXES, AND ENTER.
Enter Angle in degree $= 90$.
Click OK.

Options: From the main menu select:
ANALYZE>SET OPTIONS.
Select all six degrees of freedom.

Click OK
From the main menu select:
ANALYZE>RUN.
Enter File name "Example 6.1".
Click SAVE.
When the analysis is completed, click OK.
Use the "Rubber Band Zoom" icon and the "Pan" icon to enlarge and center the deformed frame in the view.

Results: From the main menu select:
FILE>PRINT INPUT TABLES.
Click OK.
Check "Print to File".
Click OK.
(Edited Input Tables are shown as Table 6.1.)

From the main menu select:
FILE>PRINT OUTPUT TABLES.
Check "Print to File" and "append".
Click OK.
(Edited Output tables are shown in Table 6.2.)

Table 6.2 Edited output tables for Illustrative Example 6.1 (Units: kips, inches)

JOINT DISPLACEMENTS

JOINT	LOAD	UX	UY	UZ	RX	RY	RZ
1	LOAD1	0.1865	0.0316	−0.0280	3.095E−03	−9.256E−03	−0.0275
2	LOAD1	0.0000	0.0000	0.0000	0.0000	0.0000	0.0000
3	LOAD1	0.0000	0.0000	0.0000	0.0000	0.0000	0.0000
4	LOAD1	0.0000	0.0000	0.0000	0.0000	0.0000	0.0000
5	LOAD1	0.0000	0.0000	0.0000	0.0000	0.0000	0.0000

JOINT REACTIONS

JOINT	LOAD	F1	F2	F3	M1	M2	M3
2	LOAD1	−1050.6785	−311.5120	−35.5336	−2.5811	2334.0000	−16600.2305
3	LOAD1	31.5414	−178.1268	236.7357	9217.1768	7.7191	−2026.2944
4	LOAD1	26.3624	1.8055	99.3989	112.9852	−1432.0647	7.9267
5	LOAD1	−7.2252	−12.1667	99.3989	814.0115	−520.4639	7.9267

FRAME ELEMENT FORCES

FRAME	LOAD	LOC	P	V2	V3	T	M2	M3
1	LOAD1							
		0.00	1050.68	63.47	181.00	2.58	6108.39	−1984.38
		50.00	1050.68	63.47	181.00	2.58	−2941.72	−5157.93
		100.00	1050.68	−186.53	−252.01	2.58	−11991.83	−8331.48
		150.00	1050.68	−186.53	−252.01	2.58	608.70	994.97
		200.00	1050.68	−186.53	−252.01	2.58	13209.20	10321.42

Table 6.2 (continued)

2 LOAD1

0.00	178.13	−163.26	−31.54	−7.72	−4281.98	−1870.03
50.00	178.13	−63.26	−31.54	−7.72	−2704.91	3793.18
100.00	178.1	36.74	−31.54	−7.72	−1127.84	4456.40
150.00	178.13	136.74	−31.54	−7.72	449.23	119.61
200.00	178.13	236.74	−31.54	−7.72	2026.20	−9217.18

3 LOAD1

0.00	99.40	26.36	1.81	7.93	248.12	3840.41
100.00	99.40	26.36	1.81	7.93	67.57	1204.17
200.00	99.40	26.36	1.81	7.93	−112.99	−1432.06

4 LOAD1

0.00	−99.40	12.17	7.23	−7.93	924.59	1619.33
100.00	−99.40	12.17	7.23	−7.93	202.06	402.66
200.00	−99.40	12.17	7.23	−7.93	−520.46	−814.01

Plot Deformed Frame: From the main menu select:
   DISPLAY>SHOW DEFORMED SHAPE, then select
   FILE>PRINT GRAPHICS.
   (The deformed shape is reproduced in Figure 6.7.)

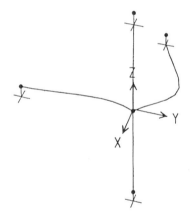

Fig. 6.7 Deformed frame of Illustrative Example 6.1

Plot the Bending Moment: From the main menu select:
   DISPLAY>SHOW FORCES/STRESSES>FRAMES.
   Select Moment 3-3.
   Click OK.
   Select:
   FILE>PRINT GRAPHICS.
   (The bending moment diagram is depicted in Figure 6.8.)

Fig. 6.8 Bending moment diagram for the space frame of Illustrative Example 6.1

Illustrative Example 6.2

For the space frame shown in Figure 6.9 determine:

(a) Joint displacements
(b) End force members
(c) Support reactions

All the beams of this space frame support uniformly distributed vertical loads of magnitude $w = -0.25$ *kip/in.* In addition, the frame is subjected to horizontal loads of magnitude 50 *kips* parallel to the *Y* axis applied to joints 5, 6, 11, and 12 as shown in Figure 6.9.

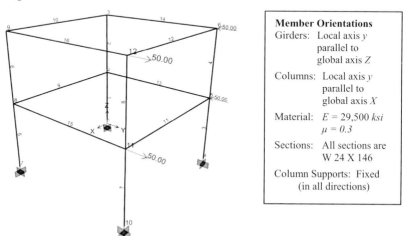

Member Orientations

Girders: Local axis y
          parallel to
          global axis Z

Columns: Local axis y
          parallel to
          global axis X

Material: $E = 29,500$ *ksi*
           $\mu = 0.3$

Sections: All sections are
           W 24 X 146

Column Supports: Fixed
           (in all directions)

Fig. 6.9 Space frame for Illustrative Example 6.2

Solution:

The implementation of the pertinent data for this Illustrative example and the execution of SAP2000 provide the following:

1. Edited tables for input data and output results are reproduced as Tables 6.3 and 6.4, respectively.

2. The deformed shape and the bending moment diagrams for the frame of Illustrative Example 6.2 are reproduced in Figs. 6.10 and 6.11, respectively.

Table 6.3 Edited input tables for Illustrative Example 6.2 (Units: kips-inches)

JOINT DATA

JOINT	GLOBAL-X	GLOBAL-Y	GLOBAL-Z	RESTRAINTS
1	0.00000	0.00000	0.00000	1 1 1 1 1 1
2	0.00000	0.00000	144.00000	0 0 0 0 0 0
3	0.00000	0.00000	288.00000	0 0 0 0 0 0
4	0.00000	288.00000	0.00000	1 1 1 1 1 1
5	0.00000	288.00000	144.00000	0 0 0 0 0 0
6	0.00000	288.00000	288.00000	0 0 0 0 0 0
7	288.00000	0.00000	0.00000	1 1 1 1 1 1
8	288.00000	0.00000	144.00000	0 0 0 0 0 0
9	288.00000	0.00000	288.00000	0 0 0 0 0 0
10	288.00000	288.00000	0.00000	1 1 1 1 1 1
11	288.00000	288.00000	144.00000	0 0 0 0 0 0
12	288.00000	288.00000	288.00000	0 0 0 0 0 0

FRAME ELEMENT DATA

FRAME	JNT-1	JNT-2	SECTION	ANGLE	RELEASES	SEGMENTS	LENGTH
1	1	2	W24X146	0.000	000000	2	144.000
2	2	3	W24X146	0.000	000000	2	144.000
3	4	5	W24X146	0.000	000000	2	144.000
4	5	6	W24X146	0.000	000000	2	144.000
5	7	8	W24X146	0.000	000000	2	144.000
6	8	9	W24X146	0.000	000000	2	144.000
7	10	11	W24X146	0.000	000000	2	144.000
8	11	12	W24X146	0.000	000000	2	144.000
9	2	8	W24X146	0.000	000000	4	288.000
10	3	9	W24X146	0.000	000000	4	288.000
11	5	11	W24X146	0.000	000000	4	288.000
12	6	12	W24X146	0.000	000000	4	288.000
13	2	5	W24X146	0.000	000000	4	288.000
14	3	6	W24X146	0.000	000000	4	288.000
15	8	11	W24X146	0.000	000000	4	288.000
16	9	12	W24X146	0.000	000000	4	288.000

JOINT FORCES Load Case LOAD1

JOINT	GLOBAL-X	GLOBAL-Y	GLOBAL-Z	GLOBAL-XX	GLOBAL-YY	GLOBAL-ZZ
11	0.000	50.000	0.000	0.000	0.000	0.000
12	0.000	50.000	0.000	0.000	0.000	0.000
5	0.000	−50.000	0.000	0.000	0.000	0.000
6	0.000	−50.000	0.000	0.000	0.000	0.000

Table 6.3 (continued)

FRAME SPAN DISTRIBUTED LOADS Load Case LOAD1

FRAME	TYPE	DIRECTION	DIST-A	VALUE-A	DIST-B	VALUE-B
9	FORCE	GLOBAL-Z	0.0000	−0.2500	1.0000	−0.2500
10	FORCE	GLOBAL-Z	0.0000	−0.2500	1.0000	−0.2500
11	FORCE	GLOBAL-Z	0.0000	−0.2500	1.0000	−0.2500
12	FORCE	GLOBAL-Z	0.0000	−0.2500	1.0000	−0.2500
13	FORCE	GLOBAL-Z	0.0000	−0.2500	1.0000	−0.2500
14	FORCE	GLOBAL-Z	0.0000	−0.2500	1.0000	−0.2500
15	FORCE	GLOBAL-Z	0.0000	−0.2500	1.0000	−0.2500
16	FORCE	GLOBAL-Z	0.0000	−0.2500	1.0000	−0.2500

Table 6.4 Edited output tables for the space frame Illustrative Example 6.2 (Units: kips-inches)

JOINT DISPLACEMENTS

JOINT	LOAD	UX	UY	UZ	RX	RY	RZ
1	LOAD1	0.0000	0.0000	0.0000	0.0000	0.0000	0.0000
2	LOAD1	0.0591	−0.9291	−0.0190	$5.011E{-}04$	$5.723E{-}04$	$3.426E{-}03$
3	LOAD1	0.1266	−1.4566	−0.0277	$-7.196E{-}04$	$6.596E{-}04$	$5.484E{-}03$
4	LOAD1	0.0000	0.0000	0.0000	0.0000	0.0000	0.0000
5	LOAD1	−0.0614	−0.9339	−0.0137	$2.473E{-}03$	$-1.735E{-}04$	$3.455E{-}03$
6	LOAD1	−0.1230	−1.4639	−0.0213	$1.803E{-}03$	$1.731E{-}04$	$5.514E{-}03$
7	LOAD1	0.0000	0.0000	0.0000	0.0000	0.0000	0.0000
8	LOAD1	0.0614	0.9282	−0.0137	$-2.461E{-}03$	$1.735E{-}04$	$3.426E{-}03$

JOINT	LOAD	UX	UY	UZ	RX	RY	RZ
9	LOAD1	0.1230	1.4582	−0.0213	$-1.804E{-}03$	$-1.731E{-}04$	$5.484E{-}03$
10	LOAD1	0.0000	0.0000	0.0000	0.0000	0.0000	0.0000
11	LOAD1	−0.0591	0.9348	−0.0190	$-5.130E{-}04$	$-5.723E{-}04$	$3.455E{-}03$
12	LOAD1	−0.1266	1.4622	−0.0277	$7.210E{-}04$	$-6.596E{-}04$	$5.514E{-}03$

JOINT REACTIONS

JOINT	LOAD	F1	F2	F3	M1	M2	M3
1	LOAD1	−6.8109	40.3840	167.5066	−2947.7888	−1027.3541	−3.6172
4	LOAD1	18.5768	34.1778	120.4934	−2658.8940	1500.2927	−3.6481
7	LOAD1	−18.5768	−33.9613	120.4934	2642.3523	−1500.2927	−3.6172
10	LOAD1	6.8109	−40.6005	167.5066	2964.3303	1027.3541	−3.6481

FRAME ELEMENT FORCES

FRAME	LOAD	LOC	P	V2	V3	T	M2	M3
1	LOAD1							
		0.00	−167.51	6.81	−40.38	3.62	−2947.79	1027.35
		72.00	−167.51	6.81	−40.38	3.62	−40.14	536.97
		144.00	−167.51	6.81	−40.38	3.62	2867.51	46.59
2	LOAD1							
		0.00	−76.49	−8.06	−24.56	2.17	−1670.59	−498.77
		72.00	−76.49	−8.06	−24.56	2.17	97.78	81.91
		144.00	−76.49	−8.06	−24.56	2.17	1866.16	662.58
3	LOAD1							
		0.00	−120.49	−18.58	−34.18	3.65	−2658.89	−1500.29
		72.00	−120.49	−18.58	−34.18	3.65	−198.09	−162.76
		144.00	−120.49	−18.58	−34.18	3.65	2262.71	1174.77

Table 6.4 (Continued)

FRAME	LOAD	LOC	P	V2	V3	T	M2	M3
4	LOAD1							
		0.00	−67.51	−23.43	−10.05	2.17	−669.57	−1361.67
		72.00	−67.51	−23.43	−10.05	2.17	53.68	325.21
		144.00	−67.51	−23.43	−10.05	2.17	776.92	2012.09
5	LOAD1							
		0.00	−120.49	18.58	33.96	3.62	2642.35	1500.29
		72.00	−120.49	18.58	33.96	3.62	197.14	162.76
		144.00	−120.49	18.58	33.96	3.62	−2248.08	−1174.77
6	LOAD1							
		0.00	−67.51	23.43	10.08	2.17	673.00	1361.67
		72.00	−67.51	23.43	10.08	2.17	−52.61	−325.21
		144.00	−67.51	23.43	10.08	2.17	−778.22	−2012.09
7	LOAD1							
		0.00	−167.51	−6.81	40.60	3.65	2964.33	−1027.35
		72.00	−167.51	−6.81	40.60	3.65	41.09	−536.97
		144.00	−167.51	−6.81	40.60	3.65	−2882.14	−46.59
8	LOAD1							
		0.00	−76.49	8.06	24.53	2.17	1667.16	498.77
		72.00	−76.49	8.06	24.53	2.17	−98.85	−81.91
		144.00	−76.49	8.06	24.53	2.17	−1864.86	−662.58
9	LOAD1							
		0.00	9.86	−29.09	−5.01	−1.56	−721.98	−545.75
		72.00	9.86	−11.09	−5.01	−1.56	−360.99	900.68
		144.00	9.86	6.91	−5.01	−1.56	0.00	1051.10
		216.00	9.86	24.91	−5.01	−1.56	360.99	−94.47
		288.00	9.86	42.91	−5.01	−1.56	721.98	−2536.04
10	LOAD1							
		0.00	−15.75	−31.32	−7.69	−5.726E−01	−1107.20	−662.84
		72.00	−15.75	−13.32	−7.69	−5.726E−01	−553.60	943.91
		144.00	−15.75	4.68	−7.69	−5.726E−01	0.00	1254.66
		216.00	−15.75	22.68	−7.69	−5.726E−01	553.60	269.41
		288.00	−15.75	40.68	−7.69	−5.726E−01	1107.20	−2011.84
11	LOAD1							
		0.00	9.86	−42.91	−5.03	−1.58	−724.35	−2536.04
		72.00	9.86	−24.91	−5.03	−1.58	−362.18	−94.47
		144.00	9.86	−6.91	−5.03	−1.58	0.00	1051.10
		216.00	9.86	11.09	−5.03	−1.58	362.18	900.68
		288.00	9.86	29.09	−5.03	−1.58	724.35	−545.75
12	LOAD1							
		0.00	−15.75	−40.68	−7.71	−5.711E−01	−1109.56	−2011.84
		72.00	−15.75	−22.68	−7.71	−5.711E−01	−554.78	269.41
		144.00	−15.75	−4.68	−7.71	−5.711E−01	0.00	1254.66
		216.00	−15.75	13.32	−7.71	−5.711E−01	554.78	943.91
		288.00	−15.75	31.32	−7.71	−5.711E−01	1109.56	−662.84
13	LOAD1							
		0.00	−20.84	−61.93	5.01	−3.937E-01	20.54	−4536.54
		72.00	−20.84	−43.93	5.01	−3.937E-01	359.68	−725.73
		144.00	−20.84	−25.93	5.01	−3.937E-01	−1.17	1789.08
		216.00	−20.84	−7.93	5.01	−3.937E-01	−362.03	3007.89
		288.00	−20.84	10.07	5.01	−3.937E-01	−722.88	2930.71
14	LOAD1							
		0.00	−32.25	−45.17	7.68	−2.568E-01	1105.02	−1865.58
		72.00	−32.25	−27.17	7.68	−2.568E-01	551.92	738.90

Table 6.4 (Continued)

FRAME	LOAD	LOC	P	V2	V3	T	M2	M3
		144.00	-32.25	-9.17	7.68	-2.568E-01	-1.18	2047.38
		216.00	-32.25	8.83	7.68	-2.568E-01	-554.29	2059.87
		288.00	-32.25	26.83	7.68	-2.568E-01	1107.39	776.35
15	LOAD1							
		0.00	28.90	-10.07	5.01	-3.937E-01	720.54	2919.52
		72.00	28.90	7.93	5.01	-3.937E-01	359.68	2996.71
		144.00	28.90	25.93	5.01	-3.937E-01	-1.17	1777.90
		216.00	28.90	43.93	5.01	-3.937E-01	-362.03	-736.92
		288.00	28.90	61.93	5.01	-3.937E-01	-722.88	-4547.73
16	LOAD1							
		0.00	17.77	-26.83	7.68	-2.568E-01	1105.02	777.65
		72.00	17.77	-8.83	7.68	-2.568E-01	551.92	2061.17
		144.00	17.77	9.17	7.68	-2.568E-01	-1.18	2048.68
		216.00	17.77	27.17	7.68	-2.568E-01	-554.29	740.20
		288.00	17.77	45.17	7.68	-2.568E-01	-1107.39	1864.28

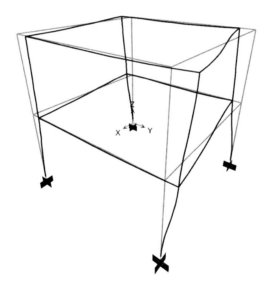

Fig. 6.10 Deformed shape for the space frame of Illustrative Example 6.2

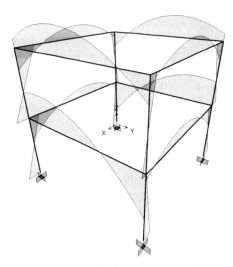

Fig. 6.11 Bending moment diagram for the space frame in Illustrative Example 6.2

6.5   Dynamic Analysis of Space Frames

The analysis of space frames subjected to dynamic loads is analogous to the analysis of these frames under static loading. The difference is that the analysis under dynamic loads requires the inclusion of inertial forces and that the load is time dependent. The inertial forces are due to the action of the acceleration over the mass of the structure. Analogous to the development of the stiffness matrix to obtain the elastic forces, the development of the mass matrix serves to account for the inertial forces. Also, the system mass matrix is obtained by transferring the coefficients of the mass matrix of the elements in the same procedure as it is done to assemble the system stiffness matrix. Therefore, the element mass matrix in reference to local coordinates fixed on the beam element is prepared first and then transformed to the global coordinates system for the structure.

6.6   Element Mass Matrix

The lumped mass matrix for the uniform beam element of a three-dimensional frame is simply a diagonal matrix in which the coefficients corresponding to translatory displacements are equal to one-half of the total mass of the beam element, while the coefficients corresponding to flexural and torsional rotations are assumed to be zero. The diagonal lumped mass matrix for the uniform beam of distributed mass $\overline{m}$ per unit of length may be written conveniently as

$$[M_L] = \frac{\overline{m}L}{2}\begin{bmatrix} 1 & 1 & 1 & 0 & 0 & 0 & 1 & 1 & 1 & 0 & 0 & 0 \end{bmatrix} \tag{6.19}$$

in which L is the length of the beam element.

The consistent mass matrix for a uniform beam segment of a three-dimensional frame is readily obtained combining the consistent mass matrices, eq. (4.3) for axial effects, eq. (6.6) for torsional effects, and eq. (2.6) for flexural effects. The appropriate combination of these matrices results in the consistent mass matrix for the uniform beam element of a three-dimensional frame, namely,

$$
\begin{bmatrix} P_1 \\ P_2 \\ P_3 \\ P_4 \\ P_5 \\ P_6 \\ P_7 \\ P_8 \\ P_9 \\ P_{10} \\ P_{11} \\ P_{12} \end{bmatrix} = \frac{\overline{m}L}{420}
\begin{bmatrix}
140 & & & & & & & & & & & \\
0 & 156 & & & & & & & & & & \\
0 & 0 & 156 & & & & \text{Symmetric} & & & & & \\
0 & 0 & 0 & \frac{140I_o}{A} & & & & & & & & \\
0 & 0 & -22L & 0 & 4L^2 & & & & & & & \\
0 & 22L & 0 & 0 & 0 & 4L^2 & & & & & & \\
70 & 0 & 0 & 0 & 0 & 0 & 140 & & & & & \\
0 & 54 & 0 & 0 & 0 & 13L & 0 & 156 & & & & \\
0 & 0 & 54 & 0 & -13L & 0 & 0 & 0 & 156 & & & \\
0 & 0 & 0 & \frac{70I_o}{A} & 0 & 0 & 0 & 0 & 0 & \frac{140I_o}{A} & & \\
0 & 0 & 13L & 0 & -3L^2 & 0 & 0 & 0 & 22L & 0 & 4L^2 & \\
0 & -13L & 0 & 0 & 0 & -3L^2 & 0 & -22L & 0 & 0 & 0 & 4L^2
\end{bmatrix}
\begin{bmatrix} \delta_1 \\ \delta_2 \\ \delta_3 \\ \delta_4 \\ \delta_5 \\ \delta_6 \\ \delta_7 \\ \delta_8 \\ \delta_9 \\ \delta_{10} \\ \delta_{11} \\ \delta_{12} \end{bmatrix} \tag{6.20}
$$

6.7   Element Damping Matrix

The damping matrix for a uniform beam segment of a three-dimensional frame could be obtained in a manner entirely similar to those of the stiffness, eq. (6.1), and mass, eq. (6.20), matrices. Nevertheless, in practice, damping is generally expressed in terms of the damping ratio for each mode of vibration. Therefore, if the response is sought using the modal superposition method, damping ratios are directly introduced in the modal equations [see eq. (2.20)]. When the damping matrix is required explicitly, it may be determined from given values of damping ratios by the methods presented in Chapter 20 of Paz and Leigh 2004.

6.8   Differential Equation of Motion

The differential equations of motion that are obtained by establishing the dynamic equilibrium among the inertial, damping, and elastic forces with the external forces may be expressed in matrix notation as

$$[M]\ \{\ddot{u}\} + [C]\ \{\dot{u}\} + [K]\ \{u\} = \{F(t)\} \qquad\qquad (6.21)$$

in which $[M]$, $[C]$, and $[K]$ are, respectively, the system mass, damping, and stiffness matrices; $\{\ddot{u}\}, \{\dot{u}\}, \{u\}$ are, respectively, the acceleration, velocity, and displacement vectors at the system nodal coordinates; and $\{F(t)\}$ is the force vector, which includes the forces applied directly to the nodes of the structure and the equivalent nodal forces for the forces applied on the beam elements of the structure.

6.9   Dynamic Analysis of Space Frames Using SAP2000

The integration of the differential equations of motion, eq. (6.21), may be accomplished by the modal superposition method as presented in the previous chapters to obtain the response of structures modeled as beams, plane frames, or grid frames. As it has been presented in the preceding chapters, the analysis using modal superposition method requires the solution of an eigenproblem to uncouple the differential equations resulting in the modal equations of motion.

Illustrative Example 6.3
For the three-dimensional frame shown in Figure (6.12a), determine:

(a) The first 6 natural frequencies and corresponding modal shapes.

(b) The dynamic response when the frame is acted upon by the impulsive force depicted in Figure 6.12(b) applied at Joint 2 of the frame parallel to the Y-direction, as shown in the figure.

The frame has concentrated masses of $m = 0.001$ (*kip sec^2/in*) at each of the four top joints.

Solution:
The following commands are implemented in SAP2000:

Begin: Open SAP2000

           Hint: Maximize both windows for a full view.

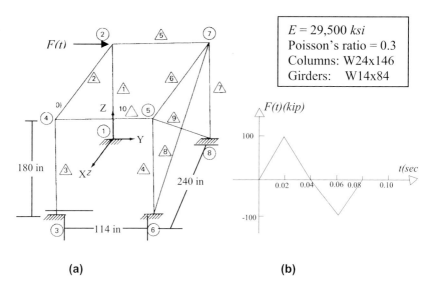

Fig 6.12 Space frame for Illustrative Example 6.3

Select: In the lower right-hand corner of the window, use the drop-down menu to select "kip-in".

Model: From the main menu select:
FILE>NEW MODEL FROM TEMPLATE.
Click on the space frame.
Number of grid spaces
    Number of stories = 1
    Number of Bays along X = 1
    Number of Bays along Y = 1
Grid Distance:
    Story Height = 180
    Bay width along X = 240
    Bay width along Y = 114.
    Click OK.

Edit: Maximize the 3-D screen and use the PAN icon on the tool bar to drag the figure to the center of the window.
From the main menu select:
VIEW>3-D VIEW.
Rotate the plan view to approximately 30 degrees for a frontal view of the X axis.

Label: From the main menu select:

VIEW>SET ELEMENTS.
Click Joint Labels and Frame Labels.
Click OK.

Axes   Translate the origin of the global coordinates to joint 1:
From the main menu select:
SELECT > SELECT ALL.
This command will mark all the members of the frame.
From the main menu select:
Enter: EDIT>MOVE.
  X = 120
  Y = 57
Click OK.

Material: From the main menu select:
DEFINE>MATERIALS.
Select STEEL.
Click Modify/Show Material.
  Set: Mass per unit of volume = 0
    Weight per unit of volume = 0
    Modules of elasticity = 29500
    Poisson ratio = 0.3.
    Click OK, OK.

Restraints: Select Joints 1, 3, 6, and 8 with a click of the mouse.
From the main menu select:
ASSIGN>JOINT>RESTRAINTS.
Select: "Restraint all."
Click OK.

Sections: From the main menu select:
DEFINE>FRAME SECTIONS.
Click Import/Wide Flange.
 Select section properties and click OPEN.
  Scroll to select W24x146.
  Hold the control key and scroll to select W14x82.
  Click OK,OK,OK.

Assign: Select all columns by clicking on them.
From the main menu select:
ASSIGN>FRAME>SECTIONS.
Select section W24x146.
Click OK.

Select all the girders by clicking on them.

From the main menu select:
ASSIGN>FRAME>SECTION.
Select section W14x82.
Click OK.

Mass: Select Joints 2, 4, 5, and 7 and from the main menu select:
ASSIGN>JOINT>MASSES.
   Enter Masses:
      Direction 1 = 0.001
      Direction 2 = 0.001
      Direction 3 = 0.001
      Click OK.

Load: From the main menu select:
DEFINE>STATIC LOAD CASES.
Change Type of load to LIVE.
Set the Self-Weight Multiplier to 0.
Click Change Load.
Click OK.

Load Function: From the main menu select:
DEFINE>TIME HISTORY FUNCTIONS.
Click "Add New Function".
Click "Add" to register the first point (0,0).
   Enter: Time= 0.02, Value = 1.0; click ADD.
   Enter: Time= 0.04, Value = 0.0; click ADD.
   Enter: Time= 0.06, Value = −1.0; click ADD.
   Enter: Time= 0.08, Value = 0.0; click ADD.
   Enter: Time= 0.10, Value = 0.0; click ADD.
   Click OK, OK.

Load Case: From the main menu select:
DEFINE>TIME HISTORY CASES.
Click Add New History.
Enter:
   Number of output time steps = 50
   Output time step = 0.002
   Check Envelopes
   Load Assignments
      Load = LOAD1
      FUNCTION = FUNC1
      Accept all of the default values.
      Click ADD. Click OK, OK.

Assign Load: Click on Joint 2, and from the main menu select:
ASSIGN>JOINT STATIC LOADS>FORCES.
Enter Force Global Y = 1.0. Click OK.

Options: From the main menu select:
ANALYZE>SET OPTIONS.
Accept check on all available DOFs.
Click on Dynamic Analyze and set Dynamic Parameters.
Enter Numbers of Modes = 6. Click OK, OK.

Analyze From the main menu select
ANALYZE>RUN.
Enter Filename EXD 6.3.
Click SAVE.
After calculations end and after checking for errors, click OK.

Print Input Tables From the main menu select:
FILES>PRINT INPUT TABLES.
Check "Print to File". Click OK.
Use NOTEPAD or another editor to retrieve, edit and print
the file C:\SAP2000\EXD 6.3.txt.

(The edited input tables for Illustrative Example 6.3 is given in Table 6.5.)

Table 6.5 Edited input tables for Illustrative Example 6.3 (Units: kips, Inches)

TIME HISTORY CASES

HISTORY CASE	HISTORY TYPE	NUMBER OF TIME STEPS	TIME STEP INCREMENT
HIST1	LINEAR	50	0.00200

JOINT DATA

JOINT	GLOBAL-X	GLOBAL-Y	GLOBAL-Z	RESTRAINTS
1	0.00000	0.00000	0.00000	1 1 1 1 1 1
2	0.00000	0.00000	180.00000	0 0 0 0 0 0
3	0.00000	114.00000	0.00000	1 1 1 1 1 1
4	0.00000	114.00000	180.00000	0 0 0 0 0 0
5	240.00000	0.00000	0.00000	1 1 1 1 1 1
6	240.00000	0.00000	180.00000	0 0 0 0 0 0
7	240.00000	114.00000	0.00000	1 1 1 1 1 1
8	240.00000	114.00000	180.00000	0 0 0 0 0 0

JOINT MASS DATA

JOINT	M-U1	M-U2	M-U3	M-R1	M-R2	M-R3
2	1.000E-03	1.000E-03	1.000E-03	0.000	0.000	0.000
4	1.000E-03	1.000E-03	1.000E-03	0.000	0.000	0.000
6	1.000E-03	1.000E-03	1.000E-03	0.000	0.000	0.000
8	1.000E-03	1.000E-03	1.000E-03	0.000	0.000	0.000

Table 6.5 (continued)

FRAME ELEMENT DATA

FRAME	JNT-1	JNT-2	SECTION	ANGLE	RELEASES	SEGMENTS	LENGTH
1	1	2	W24X146	0.000	000000	2	180.000
2	3	4	W24X146	0.000	000000	2	180.000
3	5	6	W24X146	0.000	000000	2	180.000
4	7	8	W24X146	0.000	000000	2	180.000
5	2	6	W14X82	0.000	000000	4	240.000
6	4	8	W14X82	0.000	000000	4	240.000
7	2	4	W14X82	0.000	000000	4	114.000
8	6	8	W14X82	0.000	000000	4	114.000

JOINT FORCES Load Case LOAD1

JOINT	GLOBAL-X	GLOBAL-Y	GLOBAL-Z	GLOBAL-XX	GLOBAL-YY	GLOBAL-ZZ
2	0.000	1.000	0.000	0.000	0.000	0.000

Plot Displacement Function: From the main menu select
DISPLAY>SHOW TIME HISTORY TRACES.
Click Define Functions.
Click to "Add Joint Disps/Forces".
    Joint ID = 2
    Vector Type = Displacement
    Component = UY
    Mode Number = Include All
    Click OK.
Under List of Functions, click once on Joint8.
Click ADD to move Joint2 to Plot Functions column.
Axis Labels
        Horizontal = Time (sec)
        Vertical = Y-Displacements Joint 8 (in)
Click "Display" to view Displacement functions.

Print displayed plot:
Select FILE>PRINT GRAPHICS.
(Figure 6.13 reproduces the plot of the Displacement Function in the Y direction for Joint 2 of Illustrative Example 6.3.)

Alternatively, a table of the displacement function may be obtained by selecting:
FILE>PRINT TABLES TO FILE.
The values in this table may be plotted using Excel or similar software.

(Table 6.6 provides a portion of the Displacement Function in the Y-direction for Joints 2, 4 and 5 and 7 of Illustrative Example 6.3.)

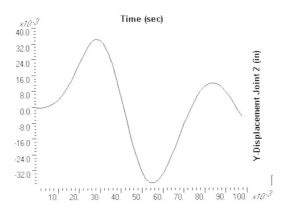

Fig. 6.13   Y-Displacement function for Joint 2 of the space frame in Illustrative Example 6.3

Table 6.6   Displacements in the Y-direction for Joints 2, 3 and 4 of the frame of Illustrative Example 6.3 (Units: inches)

Time (sec)	Joint 2	Joint 3	Joint 4
0	0	0	0
0.01	0.12446	0.25299	0.37763
0.02	0.48273	0.99197	1.4467
0.03	1.05737	2.16921	3.05747
0.04	1.8315	3.69292	5.02009
0.05	2.76758	5.41934	7.141
0.48	−2.37705	−4.51742	−5.79848
0.49	−1.40215	−2.7831	−3.69537

Natural Periods and Frequencies:

Using Notepad or other word editor, open the following file:
C:/SAP2000e/EXD 6.5.OUT.
The Natural Period and Frequency Table may be found in this file.

(Table 6.7 provides the edited values for Natural Period and Frequency of the space frame of Illustrative Example 6.3.)

Table 6.7   Natural Periods and Natural Frequencies for Illustrative Example 6.3

NORMAL PERIODS AND FREQUENCIES			
MODE	PERIOD (TIME)	FREQUENCY (CYC/TIME)	FREQUENCY (RAD/TIME)	EIGENVALUE $(RAD/TIME)^2$
1	0.044572	22.435627	140.967200	19871.752
2	0.040749	24.540317	154.191357	23774.975
3	0.020409	48.997895	307.862851	94779.535
4	0.018242	54.818423	344.434308	118634.992

Table 6.7 (continued)

MODE	PERIOD (TIME)	FREQUENCY (CYC/TIME)	FREQUENCY (RAD/TIME)	EIGENVALUE (RAD/TIME)^2
5	0.002565	389.905775	2449.850	6.0018E+06
6	0.002563	390.226543	2451.866	6.0116E+06

Mode Shapes: From the main menu select:
        DISPLAY>SHOW MODE SHAPE.
        Modal Shape 1: Click the "Wire Shadow" and "Cubic Curve" options.
        Scale Factor = 1. Click OK.

(Figure 6.14 reproduces the 1st mode shape for the space frame of Illustrative Example 6.3.)

(To view Higher Modes, click on the + icon on the toolbar located in the left-lower corner of the screen.)

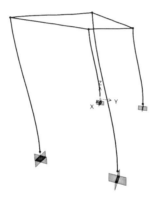

Fig. 6.14 First mode shape for the space frame of Illustrative Example 6.3

Numerical values for a mode shape may be viewed on screen by right clicking on a node in the mode shape.

6.10   Analytical Problems

Analytical Problem 6.1
Show that the transformation matrix $[T]_1$ in eq. (6.6) between global axes (X, Y, Z) and local axes (x, y, z) may be expressed as given by eq. (6.9) in terms of the coordinates of the two points at the two nodes of the beam element and of the angle of roll (ϕ).

Solution:
The direction cosines of the local axis x along the beam element are given by eq. (6.7) as

$$c_1 = \frac{X_2 - X_1}{L} \qquad c_2 = \frac{Y_2 - Y_1}{L} \qquad c_3 = \frac{Z_2 - Z_1}{L} \qquad \text{(a)}$$

in which (X_1, Y_1, Z_1) and (X_2, Y_2, Z_2) are the coordinates of the two points at the ends of the beam element and L is the beam element length given by

$$L = \sqrt{(X_2 - X_1)^2 + (Y_2 - Y_1)^2 + (Z_2 - Z_1)^2}. \qquad \text{(b)}$$

We first consider the special case in which the local axes y' and x' form a vertical plane parallel to the global axis Z (Figure P6.1).

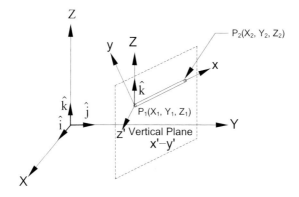

Fig. P6.1   Global system of coordinates (X,Y,Z) and local system of coordinates $\left(x', y', z'\right)$ with $x'y'$ plane vertical

For the particular case in which the plane defined by the local axes x',y' is vertical, the direction cosines for the local axes, y' and z', can also be expressed in terms of the coordinates of the two points at the two nodes of the beam element. Since in this particular case the local axis z' is perpendicular to the vertical plane defined by the local axis x', and an axis Z' parallel to the global axis Z as shown in Figure P6.1, a vector, z, along the local axis z' may be found as the cross product between the unit vector $\hat{x} = c_1\hat{i} + c_2\hat{j} + c_3\hat{k}$ along the x axis and the unit vector $\hat{k}$ along the Z' axis, namely

$$\mathbf{z} = \hat{x} \times \hat{k} = \begin{bmatrix} \hat{i} & \hat{j} & \hat{k} \\ c_1 & c_2 & c_3 \\ 0 & 0 & 1 \end{bmatrix} = c_2\hat{i} - c_1\hat{j} \tag{c}$$

in which

$$\hat{x} = c_1\hat{i} + c_2\hat{j} + c_3\hat{k} \tag{d}$$

is a unit vector along the local axis x, and $\hat{i}, \hat{j}, \hat{k}$, are respectively the unit vectors along the axes X, Y and Z of the global system.

A unit vector $\hat{z}$ along the local axis z' is then calculated from eq. (c) as

$$\hat{z} = \frac{\mathbf{z}}{d} = \frac{c_2\hat{i}}{d} - \frac{c_1\hat{j}}{d} \tag{e}$$

where

$$d = \sqrt{c_1^2 + c_2^2}. \tag{f}$$

Finally, a unit vector $\hat{y}$ along the local axis y' is then given by the cross product of the unit vectors $\hat{z}$ and $\hat{x}$, namely

$$\hat{y} = \hat{z} \times \hat{x} = \begin{bmatrix} i & j & k \\ \dfrac{c_2}{d} & -\dfrac{c_1}{d} & 0 \\ c_1 & c_2 & c_3 \end{bmatrix} \tag{g}$$

or

$$\hat{y} = -\frac{c_1 c_3}{d}\hat{i} - \frac{c_2 c_3}{d}\hat{j} + d\hat{k}. \tag{h}$$

Therefore, using the unit vectors ($\hat{x}, \hat{y}, \hat{z}$) from eqs. (d), (e), and (h), in this case, the transformation in eq. (6.9) is given by

$$
\begin{Bmatrix} x' \\ y' \\ z' \end{Bmatrix} = \begin{bmatrix} c_1 & c_2 & c_3 \\ -\dfrac{c_1 c_3}{d} & -\dfrac{c_2 c_3}{d} & d \\ \dfrac{c_2}{d} & -\dfrac{c_1}{d} & 0 \end{bmatrix} \begin{Bmatrix} X \\ Y \\ Z \end{Bmatrix} \tag{i}
$$

where d is given by eq. (f).

Equation (i) is the transformation matrix between local coordinates (x', y', z') and global coordinates (X, Y, Z) for the particular case in which the local plane $x'-y'$ is vertical. If this plane is not vertical, it is necessary to rotate the local plane $x'-y'$ in an angle ϕ (angle of roll) until the axis y' reaches the actual direction of the local axis y. Denoting by (x', y', z') the auxiliary coordinate system in which the plane $x'-y'$ is vertical and (x, y, z) the element local coordinate system, the transformation matrix, due to rotation (around the x axis) between these two local systems, is given by

$$
\begin{Bmatrix} x \\ y \\ z \end{Bmatrix} = \begin{bmatrix} 1 & 0 & 0 \\ 0 & \cos\phi & \sin\phi \\ 0 & -\sin\phi & \cos\phi \end{bmatrix} \begin{Bmatrix} x' \\ y' \\ z' \end{Bmatrix} \tag{j}
$$

in which ϕ is the angle of roll from the axis y' to the axis y. This angle is positive for a counter-clockwise rotation around axis x observing the rotation from the second end joint of the beam element to the first end joint. The transformation of the global system (X, Y, Z) to the local system (x, y, z) is then obtained by substituting eq. (i) into eq. (j), namely

$$
\begin{Bmatrix} x \\ y \\ z \end{Bmatrix} = \begin{bmatrix} 1 & 0 & 0 \\ 0 & \cos\phi & \sin\phi \\ 0 & -\sin\phi & \cos\phi \end{bmatrix} \begin{bmatrix} c_1 & c_2 & c_3 \\ -\dfrac{c_1 c_3}{d} & -\dfrac{c_2 c_3}{d} & d \\ \dfrac{c_2}{d} & -\dfrac{c_1}{d} & 0 \end{bmatrix} \begin{bmatrix} X \\ Y \\ Z \end{bmatrix}. \tag{k}
$$

The final expression for the transformation matrix $[T]_1$ is then given by the product of the two matrices in eq. (k):

$$[T]_1 = \begin{bmatrix} c_1 & c_2 & c_3 \\ -\dfrac{c_1 c_3}{d}\cos\phi-\dfrac{c_2}{d}\sin\phi & -\dfrac{c_1}{d}\sin\phi-\dfrac{c_2 c_3}{d}\cos\phi & d\cos\phi \\ \dfrac{c_1 c_3}{d}\sin\phi+\dfrac{c_2}{d}\cos\phi & -\dfrac{c_1}{d}\cos\phi+\dfrac{c_2 c_3}{d}\sin\phi & -d\sin\phi \end{bmatrix} \qquad (1)$$

in which

$$d = \sqrt{c_1^2 + c_2^2}.$$

It should be noted that the transformation matrix $[T]_1$ is not defined if the local axis x is parallel to the global axis Z. In this case, eq. (a) results in $c_1 = 0$, $c_2 = 0$, and $d = 0$. The transformation matrix between the global system of coordinates (X,Y,Z) and the local system (x, y, z) is then obtained as a special case as it is developed in Problem 6.2.

Analytical Problem 6.2
Develop the transformation matrix in eq. (6.11) between the global coordinate system (X,Y,Z) and the local (x, y, z) for the special case in which the local axis x is parallel to the global axis Z.

Solution:
If the centroidal axis of the beam element is vertical, that is, the local axis x and the global axis Z are parallel, the angle of roll is then defined as the angle that the local axis y has been rotated about the axis x from the "standard" direction defined for a vertical beam element. The "standard" direction in this case exists when the local axis y is parallel to the global axis X ($\phi = 0$ in Figure P6.2).

Let us designate by (x', y', z') an auxiliary system of coordinates in which the local axis y' is parallel to the global axis X, as shown in Figure P6.2 for $\phi = 0$. The transformation of coordinates between this auxiliary system (x', y', z') and the global system (X,Y,Z) obtained from Figure P6.2 is given by

$$\begin{Bmatrix} x' \\ y' \\ z' \end{Bmatrix} = \begin{bmatrix} 0 & 0 & \lambda \\ 1 & 0 & 0 \\ 0 & \lambda & 0 \end{bmatrix} \begin{Bmatrix} X \\ Y \\ Z \end{Bmatrix} \qquad (a)$$

in which $\lambda = 1$ when the local x has the sense of the global axis Z; otherwise, $\lambda = -1$.

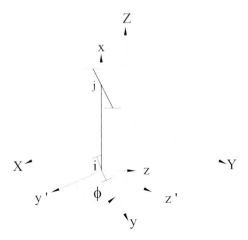

Fig. P6.2 Vertical beam element showing the local axes (*x, y, z*) and the auxiliary coordinate system (x', y', z') in which the local axis y' is parallel to the global axis *X*

The simple transformation of coordinates from (x', y', z') system to (*x, y, z*) is

$$\begin{Bmatrix} x \\ y \\ z \end{Bmatrix} = \begin{bmatrix} 1 & 0 & 0 \\ 0 & \cos\phi & \sin\phi \\ 0 & -\sin\phi & \cos\phi \end{bmatrix} \begin{Bmatrix} x' \\ y' \\ z' \end{Bmatrix} \qquad \text{(b)}$$

where the angle ϕ is the angle of roll around the local axis *x* measured from local axis *y* to the auxiliary axis *y'*.

As already stated, the angle of roll is positive for a counter-clockwise rotation observing the *x* axis from the second end toward the first end of the beam element.

The substitution of eq. (a) into eq. (b) yields

$$\begin{Bmatrix} x \\ y \\ z \end{Bmatrix} = \begin{bmatrix} 1 & 0 & 0 \\ 0 & \cos\phi & \sin\phi \\ 0 & -\sin\phi & \cos\phi \end{bmatrix} \begin{bmatrix} 0 & 0 & \lambda \\ 1 & 0 & 0 \\ 0 & \lambda & 0 \end{bmatrix} \begin{Bmatrix} X \\ Y \\ Z \end{Bmatrix}. \qquad \text{(c)}$$

The product of two matrices in eq. (c) is the transformation matrix between the global coordinate system (*X,Y,Z*) and the local system (*x, y, z*) for the particular case in which the local axis *x* is parallel to the global axis *Z*. Thus from eq. (c) we obtain the transformation matrix $[T]_1$ as

$$[T]_1 = \begin{bmatrix} 0 & 0 & \lambda \\ \cos\phi & \lambda\sin\phi & 0 \\ -\sin\phi & \lambda\cos\phi & 0 \end{bmatrix} \qquad (d)$$

in which $\lambda = 1$ when the local axis x has the same sense as the global axis Z; otherwise $\lambda = -1$, and ϕ is the angle of roll.

Analytical Problem 6.3

Demonstrate that direction cosines in eq. (6.6) for the transformation of global to local coordinates may be determined from the global coordinates of three points: two points defining the ends of the beam element along the local axis x and any third point located in local plane x-y in which y is the minor principal axis of the cross-sectional area of the member

$$[T_1] = \begin{bmatrix} \cos xX & \cos xY & \cos xZ \\ \cos yX & \cos yY & \cos yZ \\ \cos zX & \cos zY & \cos zZ \end{bmatrix}. \qquad (6.6)\ \text{repeated}$$

Solution:

Designate the coordinates of the points at the ends of a beam element as $P_i\ (X_i,\ Y_i,\ Z_i)$ and $P_j\ (X_j,\ Y_j,\ Z_j)$ and of a third point $P\ (X_P,\ Y_P,\ Z_P)$ located on the plane x-y of the cross-sectional area of the member. The direction cosines of the local axis x along the beam element are given by eq. (6.7) as

$$\cos xX = \frac{X_j - X_i}{L} \qquad \cos xY = \frac{Y_j - Y_i}{L} \qquad \cos xZ = \frac{Z_j - Z_i}{L} \qquad (6.7)\ \text{repeated}$$

where

$$L = \sqrt{(X_j - X_i)^2 + (Y_j - Y_i)^2 + (Z_j - Z_i)^2}. \qquad (6.8)\ \text{repeated}$$

The direction cosines of the z axis can be calculated from the condition that any vector $\mathbf{Z}$ along the z axis must be perpendicular to the plane formed by any two vectors in the local x-y plane. These two vectors could simply be the vector $\mathbf{X}$ from point i to point j along the x axis and the vector $\mathbf{P}$ from point i at the first joint of the beam element to point P. The orthogonality condition is expressed by the cross product between vectors $\mathbf{X}$ and $\mathbf{P}$ as

$$\mathbf{Z} = \mathbf{X} \times \mathbf{P} \qquad (a)$$

or substituting the components of these vectors as

$$Z_x \hat{i} + Z_y \hat{j} + Z_z \hat{k} = \begin{vmatrix} \hat{i} & \hat{j} & \hat{k} \\ X_j - X_i & Y_j - Y_i & Z_j - Z_i \\ X_P - X_i & Y_P - Y_i & Z_P - Z_i \end{vmatrix} \tag{b}$$

where $\hat{i}, \hat{j},$ and $\hat{z}$ are the unit vectors along the global coordinate axes $X,$ Y and $Z,$ respectively. Consequently, the direction cosines of axis z are given by

$$\cos zX = \frac{Z_x}{|Z|} \qquad \cos zY = \frac{Z_y}{|Z|} \qquad \cos zZ = \frac{Z_z}{|Z|} \tag{c}$$

where

$$\begin{aligned} z_x &= (Y_j - Y_i)(Z_P - Z_i) - (Z_j - Z_i)(Y_P - Y_i) \\ z_y &= (Z_j - Z_i)(X_P - X_i) - (X_j - X_i)(Z_P - Z_i) \\ z_z &= (X_j - X_i)(Y_P - Y_i) - (Y_j - Y_i)(X_P - X_i) \end{aligned} \tag{d}$$

and

$$|Z| = \sqrt{Z_x^2 + Z_y^2 + Z_z^2} \,. \tag{e}$$

Analogously, the direction cosines of the local axis y are calculated from the condition of orthogonality between a vector $\mathbf{Y}$ along the y axis and the unit vectors $\mathbf{X_1}$ and $\mathbf{Z_1}$ along the x and z axes, respectively. Hence,

$$\mathbf{Y} = \mathbf{X_1} \times \mathbf{Z_1}$$

or in expanded notation

$$Y_x \hat{i} + Y_y \hat{j} + Y_z \hat{k} = \begin{vmatrix} \hat{i} & \hat{j} & \hat{k} \\ \cos xX & \cos xY & \cos xZ \\ \cos zX & \cos zY & \cos zZ \end{vmatrix}. \tag{f}$$

Therefore,

$$\cos yX = \frac{Y_x}{|Y|} \qquad \cos yY = \frac{Y_y}{|Y|} \qquad \cos yZ \frac{Y_z}{|Y|} \tag{g}$$

where

$$Y_x = \cos xY \cos zZ - \cos xZ \cos zY$$
$$Y_y = \cos xZ \cos zX - \cos xX \cos zZ \qquad (h)$$
$$Y_z = \cos xX \cos zY - \cos xY \cos zX$$

and

$$|Y| = \sqrt{Y_x^2 + Y_y^2 + Y_z^2} \ . \qquad (i)$$

We have, therefore, shown that knowledge of points at the two ends of an element of a point P on the local plane x-y suffices to calculate the direction cosines of the transformation matrix $[T_1]$ in eq. (6.6). These direction cosines are given by eq. (6.7), eq. (g) and eq. (c), respectively, for rows 1, 2 and 3 of matrix $[T_1]$. The choice of point P is generally governed by the geometry of the structure and the orientation of the principal directions of the cross-section of the member. Quite often point P is selected as a known point in the structure that is placed on the local axis y, although as it has been shown, the point P could be any point in the plane formed by the local x-y axes.

6.11 Practice Problems

Problem 6.4
For the space frame in Figure P6.4 determine:

(a) Joint displacements
(b) End force members
(c) Support reactions

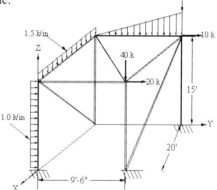

E = 29,500 ksi

Columns are W 24 X 146
Local axis y parallel to global axis X

Girders are W 14 X 82
Local axis y parallel to global axis Z

Diagonals are
Local y axis parallel to Z

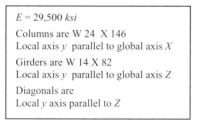

Fig. P6.4

Problem 6.5

For the space frame in Figure P6.5 determine:

(a) Joint displacements
(b) End force members
(c) Support reactions

All Members:
$E = 29,500\ ksi$
W 14 X 82

Girders have local axis y parallel to Z
Column has local axis y parallel to Y

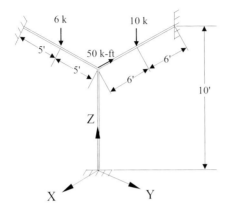

Fig. P6.5

Problem 6.6

For the space frame in Figure P6.6 determine:

(a) Joint displacements
(b) End-force members
(c) Support Reactions

Note: Girders have local axes y parallel to the global axis Z.
        Columns have local axes y parallel to the global axis X.

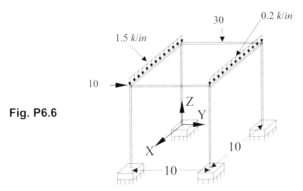

Fig. P6.6

All Members:
$E = 29,500\ ksi$
$G = 11650\ ksi$
$A = 42.5\ in^2$
$I_y = 324\ in^4$
$I_z = 3850\ in^4$
$J = 22.5\ in^4$

Problem 6.7

For the three-dimensional frame of Illustrative Example 6.5 reproduced for convenience in Figure P6.7, determine the dynamic response when the frame is acted upon by the harmonic force $F(t)= 2000 \sin 5.3t$ *lb* applied at joint 2 of the frame along the Y direction, as shown in the figure.

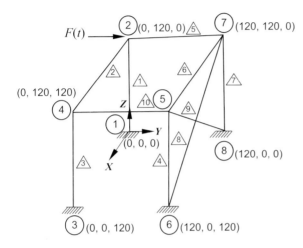

Problem Data (for all members):
$E = 30 \times 106\ psi$
$G = 12 \times 106\ psi$
$\bar{m} = 0.2\ lb\ sec^2 /in/in$
$m = 10\ lb\ sec^2 /in$
Columns: W24x146
Girders: W14x84

Fig. P6.7

Problem 6.8

Determine the dynamic response of the three-dimensional frame of Problem 6.7 subjected to the harmonic excitation of acceleration $a(t) = 0.3 \sin 5.3t$ applied at support 3 in the Y direction.

7 Plane Trusses

7.1 Introduction

Trusses are defined as structures assembled with longitudinal members assumed to be connected at their ends by frictionless pins. Furthermore, it is assumed that loads are applied only at connecting joints between members of the truss; thus the self-weight, when considered in the analysis, is simply allocated to the joints at the end of the member. Under these assumptions, the elements of the truss are two-force members and the problem is reduced to the determination of the axial forces (tension or compression) in the members of the truss.

7.2 Element Stiffness Matrix in Local Coordinates

A member of a plane truss has a total of four nodal coordinates with two nodal coordinates at each joint as shown in Figure 7.1. For small deflections, it may be assumed that the force-displacement relationship for the nodal coordinates along the axis of the member (coordinates 1 and 3 in Figure 7.1) is independent of the transverse displacements along nodal coordinates 2 and 4. This assumption is equivalent to stating that a displacement along nodal coordinates 1 or 3 does not produce forces along nodal coordinates 2 or 4 and visa versa.

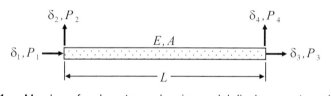

Fig. 7.1   Member of a plane truss showing nodal displacements and forces

The stiffness coefficients corresponding to axial nodal coordinates may be obtained using eq. (4.7) of Chapter 4. The application of this equation to a uniform element results in the following coefficients, using the notation in Figure 7.1:

$$k_{11} = k_{33} = \frac{AE}{L} \qquad k_{13} = k_{31} = -\frac{AE}{L} \qquad (7.1)$$

in which E is the modulus of elasticity, A the cross-sectional area, and L the length of the element.

The stiffness coefficients for pin-ended elements, corresponding to the nodal coordinates 2 and 4, are all equal to zero, since a force is not required to produce small displacements at these coordinates. Therefore, arranging the coefficients given by eq. (7.1), we obtain the stiffness equation for a uniform member of a truss as

$$\begin{Bmatrix} P_1 \\ P_2 \\ P_3 \\ P_4 \end{Bmatrix} = \frac{AE}{L} \begin{bmatrix} 1 & 0 & -1 & 0 \\ 0 & 0 & 0 & 0 \\ -1 & 0 & 1 & 0 \\ 0 & 0 & 0 & 0 \end{bmatrix} \begin{Bmatrix} \delta_1 \\ \delta_2 \\ \delta_3 \\ \delta_4 \end{Bmatrix} \qquad (7.2)$$

or in condensed notation

$$\{P\} = [k]\{\delta\} \qquad (7.3)$$

in which $[k]$ is the stiffness matrix for an element of a plane truss, $\{P\}$ and $\{\delta\}$ are, respectively, the force and the displacement vectors at the nodal coordinates.

7.3   Transformation of Coordinates

The stiffness matrix in eq. (7.2) was obtained in reference to the nodal coordinates associated with the *local* or *element system of coordinates*. As discussed in the preceding chapters on framed structures, it is necessary to transform these element matrices to a common system of reference, the *global coordinate system*. The transformation of displacements and forces at the nodal coordinates is accomplished, as was demonstrated in Chapter 4, by performing a rotation of coordinates. Deleting the angular coordinates in eq. (4.12) and relabeling the remaining coordinates result in the following transformation for the nodal forces:

$$\begin{Bmatrix} P_1 \\ P_2 \\ P_3 \\ P_4 \end{Bmatrix} = \begin{bmatrix} \cos\theta & \sin\theta & 0 & 0 \\ -\sin\theta & \cos\theta & 0 & 0 \\ 0 & 0 & \cos\theta & \sin\theta \\ 0 & 0 & -\sin\theta & \cos\theta \end{bmatrix} \begin{Bmatrix} \overline{P}_1 \\ \overline{P}_2 \\ \overline{P}_3 \\ \overline{P}_4 \end{Bmatrix} \tag{7.4}$$

where θ is the angle between the global axis X and the local axis x as shown in Figure 7.2. Equation 7.4 may be written in condensed notation as

$$\{P\} = [T]\{\overline{P}\} \tag{7.5}$$

in which $\{P\}$ and $\{\overline{P}\}$ are the nodal forces, respectively, in reference to local and global coordinates and $[T]$ the transformation matrix defined in eq. (7.4).

The same transformation matrix $[T]$ also serves to transform the nodal displacement vector $\{\overline{\delta}\}$ in the global coordinate system to the nodal displacement vector $\{\delta\}$ in local coordinates:

$$\{\delta\} = [T]\{\overline{\delta}\} \tag{7.6}$$

The substitution of eqs. (7.5) and (7.6) into the stiffness equation (7.3) results in

$$[T]\{\overline{P}\} = [k][T]\{\overline{\delta}\}.$$

Since $[T]$ is an orthogonal matrix ($[T]^{-1} = [T]^T$), it follows that

$$\{\overline{P}\} = [T]^T [k][T]\{\overline{\delta}\}$$

or

$$\{\overline{P}\} = [\overline{k}]\{\overline{\delta}\} \tag{7.7}$$

in which

$$[\overline{k}] = [T]^T [k] [T] \tag{7.8}$$

is the element stiffness matrix in the global coordinate system.

7.4   Element Stiffness Matrix in Global Coordinates

The substitution into eq. (7.8) of the stiffness matrix $[k]$ from eq. (7.2) and of the transformation matrix $[T]$ and its transpose from eq. (7.4) results in the element stiffness matrix, $[\overline{k}]$, in reference to the global system of coordinates:

$$[\bar{k}] = \frac{EA}{L} \begin{bmatrix} c^2 & cs & -c^2 & -cs \\ cs & s^2 & -cs & -s^2 \\ -c^2 & -cs & c^2 & cs \\ -cs & -s^2 & cs & s^2 \end{bmatrix}. \qquad (7.9)$$

In eq. (7.9), c and s designate $\cos\theta$ and $\sin\theta$, respectively.

7.5   Assemble the System Stiffness Matrix

The system stiffness matrix for a plane truss is assembled by appropriately transferring the coefficients of the element stiffness matrices using exactly the same procedure described in the preceding chapters for beams or for frames. Illustrative Example 7.1 that follows illustrates the application of the stiffness method for the analysis of a plane truss.

7.6   End Forces for an Element of a Truss

The end forces for an element or member of a truss may be determined as for the case frame elements presented in the previous chapters. These end forces in global coordinates will then be calculated as

$$\{\bar{P}\} = [\bar{k}]\{\bar{\delta}\} \qquad (7.10)$$

in which $\{\bar{P}\}$ and $\{\bar{\delta}\}$ are, respectively, the element nodal force vector and element nodal displacement vector and $[\bar{k}]$ is the element stiffness matrix. The vectors $\{\bar{P}\}$ and $\{\bar{\delta}\}$ as well as the matrix $[\bar{k}]$ are in reference to the global system of coordinates.

The end forces $\{P\}$ in reference to the local system of coordinates may then be calculated by eq. (7.5). However, it is somewhat more convenient to calculate the element forces (tension or compression) by determining first the axial deformation and then the element axial force. Consider in Figure 7.2 an element of a truss showing the nodal displacements $\bar{\delta}_1$ through $\bar{\delta}_4$ calculated in global coordinates.

The elongation Δ of this element along its longitudinal axis x is given by

$$\Delta = \left(\bar{\delta}_3 - \bar{\delta}_1\right)\cos\theta + \left(\bar{\delta}_4 - \bar{\delta}_2\right)\sin\theta. \qquad (7.11)$$

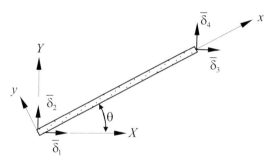

Fig. 7.2   Plane truss element showing the nodal displacements $\overline{\delta}_1$ through $\overline{\delta}_4$ in global coordinates

Its strain is then $\varepsilon = \Delta/L$, the stress $\sigma = E\,\varepsilon = E\,\Delta/L$, and the axial force $P = A\sigma = EA\,\Delta/L$, or using eq. (7.11):

$$P = \frac{EA}{L}\left[\left(\overline{\delta}_3 - \overline{\delta}_1\right)\cos\theta + \left(\overline{\delta}_4 - \overline{\delta}_2\right)\sin\theta\right] \tag{7.12}$$

in which

E = modulus of elasticity
A = cross-sectional area
L = length of the element
θ = angle between global axis X and the local axis x

Illustrative Example 7.1

The plane truss shown in Figure 7.3 (which has only three members) is used to illustrate the application of the stiffness method for trusses. Determine:

(a) Displacements at the joints
(b) Axial forces in the members
(c) Reactions at the supports

Solution:

  1.   Mathematical Model.

       Figure 7.3 shows the model needed for this truss with 3 elements, 3 nodes and 6 nodal coordinates of which the first three are free coordinates and the last three are fixed.

  2.   Element stiffness matrices in global coordinates [eq. (7.9)]:

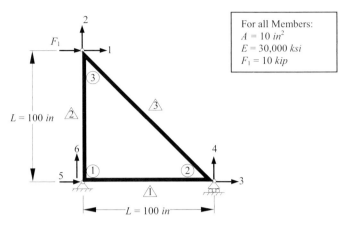

Fig. 7.3 Plane truss for Illustrative Example 7.1

ELEMENT 1 with $\theta = 0$ and $L = 100$

$$
[\bar{k}]_1 = \frac{30E3x10}{100}
\begin{bmatrix}
1 & 0 & -1 & 0 \\
0 & 0 & 0 & 0 \\
-1 & 0 & 1 & 0 \\
0 & 0 & 0 & 0
\end{bmatrix}
= 3000
\begin{array}{cccc}
5 & 6 & 3 & 4 \\
\end{array}
\begin{bmatrix}
1 & 0 & -1 & 0 \\
0 & 0 & 0 & 0 \\
-1 & 0 & 1 & 0 \\
0 & 0 & 0 & 0
\end{bmatrix}
\begin{array}{c}
5 \\ 6 \\ 3 \\ 4
\end{array}
\qquad \text{(a)}
$$

ELEMENT 2: $\theta = 90°$ and $L = 100$

$$
[\bar{k}]_2 = 3000
\begin{array}{cccc}
5 & 6 & 1 & 2 \\
\end{array}
\begin{bmatrix}
0 & 0 & 0 & 0 \\
0 & 1 & 0 & -1 \\
0 & 0 & 0 & 0 \\
0 & -1 & 0 & 1
\end{bmatrix}
\begin{array}{c}
5 \\ 6 \\ 1 \\ 2
\end{array}
\qquad \text{(b)}
$$

ELEMENT 3: $\theta = 135°$ and $L = 100\sqrt{2}$

$$
[\bar{k}]_3 = 1060
\begin{array}{cccc}
3 & 4 & 1 & 2 \\
\end{array}
\begin{bmatrix}
1 & -1 & -1 & 1 \\
-1 & 1 & 1 & -1 \\
-1 & 1 & 1 & -1 \\
1 & -1 & -1 & 1
\end{bmatrix}
\begin{array}{c}
3 \\ 4 \\ 1 \\ 2
\end{array}
\qquad \text{(c)}
$$

3. Assemble reduced system stiffness matrix.

The coefficients in eqs. (a), (b) and (c) corresponding to the first three nodal coordinates indicated at the top and on the right of the element stiffness matrices in these equations are transferred to the reduced system stiffness matrix, $[K]_R$:

$$[K]_R = \begin{bmatrix} 1060 & -1060 & -1060 \\ -1060 & 3000+1060 & 1060 \\ -1060 & 1060 & 3000+1060 \end{bmatrix} = \begin{bmatrix} 1060 & -1060 & -1060 \\ -1060 & 4060 & 1060 \\ -1060 & 1060 & 4060 \end{bmatrix} \quad (d)$$

4. Reduced system force vector.

The system force vector contains only the applied force 10 *kips* at the first nodal coordinate u_1:

$$\{F\}_R = \begin{Bmatrix} 10 \\ 0 \\ 0 \end{Bmatrix} \quad (e)$$

5. System stiffness equation.

$$\begin{Bmatrix} 10 \\ 0 \\ 0 \end{Bmatrix} = \begin{bmatrix} 1060 & -1060 & -1060 \\ -1060 & 4060 & 1060 \\ -1060 & 1060 & 4060 \end{bmatrix} \begin{Bmatrix} u_1 \\ u_2 \\ u_3 \end{Bmatrix} \quad (f)$$

6. Nodal displacements.

Solution of eq. (f) results in

$$u_1 = 0.01611 \ in$$
$$u_2 = 0.00333 \ in$$
$$u_3 = 0.00333 \ in$$

7. Element end forces. Axial forces $P^{(i)}$ on each element of the truss are calculated by eq. (7.12).

ELEMENT   1

$$P^{(1)} = \frac{30 \, E3x10}{100} \Big[(0.00333 - 0) \cos 0° + (0 - 0) \sin 0° \Big]$$

$$= 10 \ kip$$

ELEMENT 2

$$p^{(2)} = \frac{30E3 \times 10}{100} \left[(0.01611 - 0)\cos 90° + (0.00333 - 0)\sin 90° \right]$$

$$= 10 \; kip$$

ELEMENT 3

$$p^{(2)} = \frac{30E3 \times 10}{100} \left[(0.01611 - 0.00333)\cos 135° + (0.00333 - 0)\sin 135° \right]$$

$$= -14.2 \; kip$$

8. Support reactions:

    Support reactions are determined from the element forces as follows:

    At joint 1:
    $$R_5 = -P^{(1)} = -10 \; kips$$
    $$R_6 = -P^{(2)} = -10 \; kips$$

    At joint 2:
    $$R_4 = -P^{(3)} \sin 45° = (14.2)\,(0.707) = 10 \; kips$$

7.7   Analysis of Plane Trusses Using SAP2000

Illustrative Example 7.2
Use SAP2000 to perform the structural analysis of the plane frame shown in Figure 7.4. For all the members of the truss, the modulus of elasticity is $E = 30,000 \; ksi$ and the cross-sectional area $A = 2 \; in^2$.

Solution:

Begin: Open SAP2000.

Enter: OK to close the "Tip of the Day".

            Hint: Maximize both windows for a full view.

Select: In the lower right-hand corner of the window, use the drop-down menu to select "kip-in".

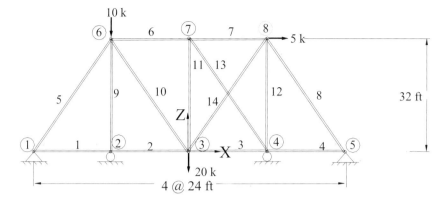

Fig. 7.4   Plane truss of Illustrative Example 7.2

Select: From the main menu select:
FILE>NEW MODEL.
In the Coordinate System Definitions, enter:
 Number of Grid Spaces:
  X direction = 4
  Y direction = 0
  Z direction = 1
 Grid Spacing:
  X direction = 288
  Y direction = 1
  Z direction = 384
Click OK.

Edit: Maximize the 2-D screen. Click the XZ icon to view this plane. Use the PAN icon to center the plot in the window.

Draw: From the main menu select:
DRAW>DRAW FRAME ELEMENT.
Click on the lower left intersection of the grid lines and drag the cursor horizontally to the next grid line; click twice to create an element. Then click again at the last location and proceed horizontally to the next grid line and click twice to create a second element.

Proceed to create additional horizontal elements. Similarly, create all the other elements of the truss shown in Figure 7.4.

Label: From the main menu select:
VIEW>SET ELEMENTS.
Check Labels on Joints and on Frames. Click OK.

Restraints: Select joints 1 and 5 of the truss by clicking at the joints. Then from
the main menu select:
ASSIGN>JOINT>RESTRAINTS.
Restrain the translations 1, 2 and 3. Click OK.

Select joints **2** and **4** at the two rollers. Then from the main menu select:
ASSIGN>JOINT>RESTRAINTS.
Restrain only the translation 3. Click OK.

Select joints 6, 7 and 8. Then from the main menu select:
ASSIGN>JOINT>RESTRAINTS.
Select no restraints by clicking the dot icon.
Click OK.

Material   From the main menu select:
DEFINE>MATERIALS.
Select STEEL and click Modify/Show Material.
Enter Modulus of Elasticity = 30000. Click OK, OK.

Sections: From the main menu select:
DEFINE>FRAME SECTIONS.
Click Add/Wide Flange and scroll down to Add General Section.
Change the cross-section area to 2. Click OK.
Change all other section values to 0 (zero).
Change the label of the section from FSEC2 to TRUSS 7.2 Click OK, OK.

Select all the elements of the truss and from the main menu select:
ASSIGN > FRAME SECTION.
Select TRUSS 7.2. Click OK.

Define: From the main menu select:
DEFINE>STATIC LOAD CASES.
Change DEAD to LIVE and set the Self-Weight Multiplier to 0 (zero).
Click "Change Load". Click OK.

Loads: Select joint 3 at the center of the lower chord of the truss, and then from
the main menu select:
ASSIGN>JOINT STATIC LOADS>FORCES.
Enter Force Global $Z = -20$. Click OK.

Select joint 6 at the left end of the top chord of the truss, and then from the
main menu select:

ASSIGN>JOINT STATIC LOADS>FORCES.
Enter Force Global $Z = -10$. Click OK.
Note: Be sure to change other previous loads to zero.

Select Joint 5 at the right end of the top chord of the truss, and then from the main menu select:
ASSIGN>JOINT STATIC LOADS>FORCES.
Enter Force Global $X = 5$. Click OK.
Note: Be sure to change other previous loads to zero.

Options: From the main menu select:
ANALYZE>SET OPTIONS.
Select available Degrees of Freedom: *UX* and *UZ.* Click OK.

Analysis: From the main menu select:
ANALYZE>RUN.
Enter filename "Example 7.2".
Click SAVE.
When the analysis is concluded, click OK.

Print Input Tables: From the main menu select:
FILE>PRINT INPUT TABLES.
Select Frames, Joints and Reactions. Click OK.

(Table 7.1 provides the edited input data for this Illustrative Example 7.2.)

Table 7.1 Edited Input tables for Illustrative Example 7.2 (units: kips, inches)

JOINT DATA

JOINT	GLOBAL-X	GLOBAL-Y	GLOBAL-Z	RESTR	ANG-A	ANG-B	ANG-C
1	−576.00000	0.00000	0.00000	1 1 1 0 0 0	0.000	0.000	0.000
2	−288.00000	0.00000	0.00000	0 1 1 0 0 0	0.000	0.000	0.000
3	0.00000	0.00000	0.00000	0 0 0 0 0 0	0.000	0.000	0.000
4	288.00000	0.00000	0.00000	0 1 1 0 0 0	0.000	0.000	0.000
5	576.00000	0.00000	0.00000	1 1 1 0 0 0	0.000	0.000	0.000
6	−288.00000	0.00000	384.00000	0 0 0 0 0 0	0.000	0.000	0.000
7	0.00000	0.00000	384.00000	0 0 0 0 0 0	0.000	0.000	0.000
8	288.00000	0.00000	384.00000	0 0 0 0 0 0	0.000	0.000	0.000

FRAME ELEMENT DATA

FRAME	JNT-1	JNT-2	SCTN	ANG	RLS	SGMNT	R1	R2	FCTR	LENGTH
1	1	2	TRUSS72	0.000	000000	4	0.000	0.000	1.000	288.000
2	2	3	TRUSS72	0.000	000000	4	0.000	0.000	1.000	288.000
3	3	4	TRUSS72	0.000	000000	4	0.000	0.000	1.000	288.000
4	4	5	TRUSS72	0.000	000000	4	0.000	0.000	1.000	288.000
5	1	6	TRUSS72	0.000	000000	2	0.000	0.000	1.000	480.000
6	6	7	TRUSS72	0.000	000000	4	0.000	0.000	1.000	288.000
7	7	8	TRUSS72	0.000	000000	4	0.000	0.000	1.000	288.000

Table 7.1 (continued)

FRAME	JNT-1	JNT-2	SCTN	ANG	RLS	SGMNT	R1	R2	FCTR	LENGTH
8	8	5	TRUSS72	0.000	000000	2	0.000	0.000	1.000	480.000
9	2	6	TRUSS72	0.000	000000	2	0.000	0.000	1.000	384.000
10	3	6	TRUSS72	0.000	000000	2	0.000	0.000	1.000	480.000
11	3	7	TRUSS72	0.000	000000	2	0.000	0.000	1.000	384.000
12	4	8	TRUSS72	0.000	000000	2	0.000	0.000	1.000	384.000
13	4	7	TRUSS72	0.000	000000	2	0.000	0.000	1.000	480.000
14	3	8	TRUSS72	0.000	000000	2	0.000	0.000	1.000	480.000

JOINT FORCES Load Case LOAD1

JOINT	GLOBAL-X	GLOBAL-Y	GLOBAL-Z	GLOBAL-XX	GLOBAL-YY	GLOBAL-ZZ
3	0.000	0.000	-20.000	0.000	0.000	0.000
6	0.000	0.000	-10.000	0.000	0.000	0.000
8	5.000	0.000	0.000	0.000	0.000	0.000

Print Output Tables: From the main menu, select:
FILE>PRINT OUTPUT TABLES.
(Table 7.2 provides the edited output tables for Illustrative Example 7.2.)

Table 7.2 Output tables for Illustrative Example 7.2 (Units: kips, inches)

JOINT DISPLACEMENTS

JOINT	LOAD	UX	UY	UZ	RX	RY	RZ
1	LOAD1	0.0000	0.0000	0.0000	0.0000	-4.111E-05	0.0000
2	LOAD1	9.534E-03	0.0000	0.0000	0.0000	3.029E-04	0.0000
3	LOAD1	0.0191	0.0000	-0.1522	0.0000	-2.473E-05	0.0000
4	LOAD1	0.0221	0.0000	0.0000	0.0000	-2.958E-04	0.0000
5	LOAD1	0.0000	0.0000	0.0000	0.0000	9.215E-05	0.0000
6	LOAD1	0.0304	0.0000	-0.0737	0.0000	1.431E-04	0.0000
7	LOAD1	-4.448E-03	0.0000	-0.1074	0.0000	-9.122E-05	0.0000
8	LOAD1	-0.0141	0.0000	-0.0349	0.0000	-1.999E-04	0.0000

JOINT REACTIONS

JOINT	LOAD	F1	F2	F3	M1	M2	M3
1	LOAD1	1.0710	0.0000	4.0761	0.0000	0.0000	0.0000
2	LOAD1	0.0000	0.0000	11.5253	0.0000	0.0000	0.0000
4	LOAD1	0.0000	0.0000	12.4528	0.0000	0.0000	0.0000
5	LOAD1	-6.0710	0.0000	1.9457	0.0000	0.0000	0.0000

FRAME ELEMENT FORCES

FRAME	LOAD	LOC	P	V2	V3	T	M2	M3
1	LOAD1	0.00	1.99	0.00	0.00	0.00	0.00	0.00
2	LOAD1	0.00	1.99	0.00	0.00	0.00	0.00	0.00
3	LOAD1	0.00	-0.638	0.00	0.00	0.00	0.00	0.00
4	LOAD1	0.00	-4.61	0.00	0.00	0.00	0.00	0.00
5	LOAD1	0.00	-5.10	0.00	0.00	0.00	0.00	0.00

Table 7.2 (continued)

FRAME	LOAD	LOC	P	V2	V3	T	M2	M3
6	LOAD1	0.00	−7.26	0.00	0.00	0.00	0.00	0.00
7	LOAD1	0.00	−2.01	0.00	0.00	0.00	0.00	0.00
8	LOAD1	0.00	−2.43	0.00	0.00	0.00	0.00	0.00
9	LOAD1	0.00	−11.52	0.00	0.00	0.00	0.00	0.00
10	LOAD1	0.00	7.00	0.00	0.00	0.00	0.00	0.00
11	LOAD1	0.00	7.00	0.00	0.00	0.00	0.00	0.00
12	LOAD1	0.00	−5.45	0.00	0.00	0.00	0.00	0.00
13	LOAD1	0.00	− 8.75	0.00	0.00	0.00	0.00	0.00
14	LOAD1	0.00	9.25	0.00	0.00	0.00	0.00	0.00

Note: The computer output shows extremely small values for the shear force V2 and the bending moment M3 (in the range of numerical accuracy); however, in editing this table, these values have been set equal to zero.

Plot Displacements: From the main menu select:
DISPLAY>SHOW DEFORMED SHAPE, then select:

FILE>PRINT GRAPHICS.

(The deformed shape is reproduced in Figure 7.5.)

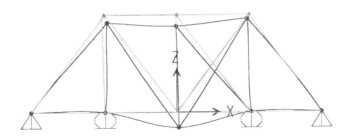

Fig. 7.5 Deformed shape for the plane truss of Illustrative Example 7.2

Plot Axial Forces: From the main menu select:
DISPLAY>SHOW ELEMENT FORCES/STRESSES>FRAMES.
Select Axial Force, click OK, and then select:
FILE>PRINT GRAPHICS.

(The Axial Force diagram is reproduced in Figure 7.6.)

Note: To view plots and values for any member of the truss, right click on the element and a pop-up window will show the axial force of the selected element.

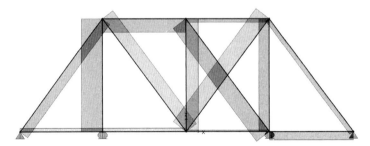

Fig. 7.6   Axial Force diagram for Illustrative Example 7.2 (dark members are in compression and light members in tension)

7.8   Dynamic Analysis of Plane Trusses

The static analysis of trusses whose members are pin-connected reduces to the problem of determining the bar forces due to a set of loads applied at the joints. When the same trusses are subjected to the action of dynamics forces, the simple situation of only axial stresses in the members no longer exists. The inertial forces developed on the members of the truss will, in general, produce flexural bending in addition to axial forces. The bending moments at the ends of the truss members will still remain zero in the absence of external joint moments. The dynamic stiffness method for the analysis of trusses is developed, as in the case of framed structures, by determining the stiffness and mass matrices for an element. Because of the relative simplicity of the transformation of the stiffness matrix from the local to the global coordinate system, conveniently, we obtained explicitly in eq. (7.9) the element stiffness matrix in reference to the global coordinate system. Thus it remains to develop only the mass matrix for such elements.

7.9   Lumped Mass Matrix

The lumped mass matrix for an uniform element of a plane truss requires the allocation of one-half of the total mass of the element. Thus, considering in Figure 7.7 a uniform element having its nodal coordinates labeled as shown, the lumped mass matrix relating the nodal forces {P} with the acceleration is given by the following equation:

$$
\begin{Bmatrix} P_1 \\ P_2 \\ P_3 \\ P_4 \end{Bmatrix} = \frac{\bar{m}L}{2} \begin{bmatrix} 1 & 0 & 0 & 0 \\ 0 & 1 & 0 & 0 \\ 0 & 0 & 1 & 0 \\ 0 & 0 & 0 & 1 \end{bmatrix} \begin{Bmatrix} \ddot{\delta}_1 \\ \ddot{\delta}_2 \\ \ddot{\delta}_3 \\ \ddot{\delta}_4 \end{Bmatrix}
\qquad (7.13)
$$

in which $\overline{m}$ is the distributed mass per unit of length and L the length of the element.

Fig. 7.7 Member of a plane truss showing nodal displacements and forces

7.10 Consistent Mass Matrix

The consistent mass matrix is obtained, as previously demonstrated, using expressions for static displacement functions in the application of the Principle of Virtual Work. These displacement functions, corresponding to a unit deflection at nodal coordinates 2 and 4 shown in Figure 7.8, are given by

$$u_2 = 1 - \frac{x}{L} \qquad (7.14)$$

and

$$u_4 = \frac{x}{L}. \qquad (7.15)$$

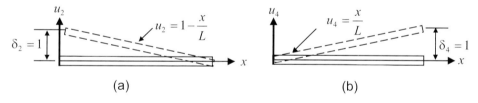

Fig. 7.8   Displacement functions
(a) For a unit displacement $\delta_2 = 1$
(b) For a unit displacement $\delta_4 = 1$

The consistent mass coefficients are given by the general expression. eq. (4.23), which is repeated here for convenience, namely,

$$m_{ij} = \int_0^L \overline{m}(x) u_i(x) u_j(x) dx. \qquad \text{(4.23) Repeated}$$

For a uniform member of mass $\overline{m}$ per unit length, the substitution into eq. (4.23) of eqs. (7.14) and (7.15) yields

$$m_{22} = m_{44} = \frac{\overline{m}L}{3}$$

$$m_{24} = m_{42} = \frac{\overline{m}L}{6}.$$ (7.16)

Finally, assembling the mass coefficients from eqs. (4.26) and (7.16) forms the consistent mass matrix relating forces to accelerations at the nodal coordinates for a uniform member of a plane truss, namely,

$$\begin{Bmatrix} P_1 \\ P_2 \\ P_3 \\ P_4 \end{Bmatrix} = \frac{\overline{m}L}{6} \begin{bmatrix} 2 & 0 & 1 & 0 \\ 0 & 2 & 0 & 1 \\ 1 & 0 & 2 & 0 \\ 0 & 1 & 0 & 2 \end{bmatrix} \begin{Bmatrix} \ddot{\delta}_1 \\ \ddot{\delta}_2 \\ \ddot{\delta}_3 \\ \ddot{\delta}_4 \end{Bmatrix}$$ (7.17)

or in concise notation

$$\{P\} = [m]\,\{\ddot{\delta}\}.$$ (7.18)

The mass matrix, $[m]$, in eq. (7.18) was obtained in reference to the local system of coordinates; however, this mass matrix for an element of a plane truss also refers to the global coordinates system and therefore no transformation is needed. To verify this fact, we substitute the matrices $[m]$ and $[T]$, respectively, from eqs. (7.17) and (7.4) into the following equation, which was used in the previous chapters to obtain the mass matrix $[\overline{m}]$ in global coordinates:

$$[\overline{m}] = [T]^{\mathrm{T}}\,[m]\,[T]$$ (7.19)

to obtain

$$[m] = \frac{\overline{m}L}{6} \begin{bmatrix} c & -s & 0 & 0 \\ s & c & 0 & 0 \\ 0 & 0 & c & -s \\ 0 & 0 & s & c \end{bmatrix} \begin{bmatrix} 2 & 0 & 1 & 0 \\ 0 & 2 & 0 & 1 \\ 1 & 0 & 2 & 0 \\ 0 & 1 & 0 & 2 \end{bmatrix} \begin{bmatrix} c & s & 0 & 0 \\ -s & c & 0 & 0 \\ 0 & 0 & c & s \\ 0 & c & -s & c \end{bmatrix}.$$

Thus,

$$[\overline{m}] = \frac{\overline{m}L}{6} \begin{bmatrix} 2 & 0 & 1 & 0 \\ 0 & 2 & 0 & 1 \\ 1 & 0 & 2 & 0 \\ 0 & 1 & 0 & 2 \end{bmatrix} = [m]$$ (7.20)

in which we use the notation $c = \cos\theta$, $s = \sin\theta$ and the fact that $\cos^2\theta + \sin^2\theta = 1$.

Illustrative Example 7.3

The plane truss shown in Figure 7.9(a), which was analyzed in Illustrative Example 7.1 supporting a static load, is now subjected to the dynamic force shown in Fig 7.9(b). Concentrated masses of magnitude $m = 0.10\ kip\ sec^2/in$ are lumped at each joint of the truss. Determine the three lowest natural frequencies for the truss and the normalized modal matrix.

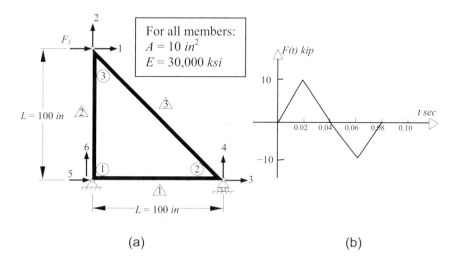

(a)   (b)

Fig. 7.9 (a) Plane truss for Illustrative Examples 7.3 and 7.4, (b) forcing function

Solution:

The reduced stiffness matrix for this truss determined as eq. (d) in Illustrative Example 7.1 is conveniently is reproduced here:

$$[K]_R = \begin{bmatrix} 1060 & -1060 & -1060 \\ -1060 & 4060 & 1060 \\ -1060 & 1060 & 4060 \end{bmatrix} \qquad (a)$$

Lumped mass matrix contains the specified mass $m = 0.1\ kip{\cdot}sec^2/in$ at each location in its diagonal:

$$[M]_R = \begin{bmatrix} 0.1 & 0 & 0 \\ 0 & 0.1 & 0 \\ 0 & 0 & 0.1 \end{bmatrix}. \qquad (b)$$

The differential equation of motion in free vibration is:

$$[M]_R\{\ddot{u}\}+[K]_R\{u\}=0. \qquad (c)$$

Substituting the trial solution $\{u\}=\{a\}\sin\omega t$ into eq. (c) results, after factoring out $\sin\omega t$, in the following eigen problem:

$$[[K]_R-\omega^2[M]_R]\{a\}=\{0\}. \qquad (d)$$

A nontrivial solution of eq. (d) requires

$$\left|\,[K]_R-\omega^2[M]_R\right|=0. \qquad (e)$$

Substituting into eq. (e) $[K]_R$ and $[M]_R$, respectively from eqs. (a) and (b), expanding the determinant and solving the resulting cubic equation in ω^2 yields the following roots:

$$\omega_1^2=5,665\ (rad\,/\sec)^2$$

$$\omega_2^2=30,000\ (rad\,/\sec)^2 \qquad (f)$$

$$\omega_3^2=56,140\ (rad\,/\sec)^2$$

Therefore, the natural frequencies are:

$$\omega_1=75.27\ rad/sec \qquad \omega_2=173.21\ rad/sec \qquad \omega_3=239.04\ rad/sec$$

or in cycles per second $(f=\omega/2\pi)$:

$$f_1=11.98\ cps \qquad f_2=27.57\ cps \qquad f_3=37.71\ cps.$$

Substituting, in turn, into eq. (d) the values of ω_1^2, ω_2^2 and ω_3^2 calculated in eq. (f), setting $a_3=1$, and then solving for a_1 and a_2 results in the corresponding mode shape vectors:

$$\{a\}_1=\begin{Bmatrix}4.2959\\1.0000\\1.0000\end{Bmatrix},\quad \{a\}_2=\begin{Bmatrix}0.0000\\-1.0000\\1.0000\end{Bmatrix},\quad \{a\}_3=\begin{Bmatrix}-0.4655\\1.0000\\1.0000\end{Bmatrix} \qquad (g)$$

which are, respectively, normalized by the following the factors (see eq. IV-5a of Appendix IV):

$$\sqrt{\{a\}_1^T[M]_R\{a\}_1}=1.4302,\ \sqrt{\{a\}_2^T[M]_R\{a\}_2}=0.4472\ \sqrt{\{a\}_3^T\ [M]_R\{a\}_3}=0.4708\,.$$
$$(h)$$

Finally dividing the components of the mode shape vectors in eq. (g) by the factors calculated in eq. (h) yields the following normalized mode shapes:

$$\{\phi\}_1 = \begin{Bmatrix} 3.0037 \\ 0.6992 \\ 0.6992 \end{Bmatrix}, \quad \{\phi\}_2 = \begin{Bmatrix} 0.0000 \\ -2.2360 \\ 2.2360 \end{Bmatrix}, \quad \{\phi\}_3 = \begin{Bmatrix} -0.9888 \\ 2.1239 \\ 2.1239 \end{Bmatrix}.$$

These normalized eigenvectors are then arranged in the columns of the modal matrix:

$$[\Phi] = \begin{bmatrix} 3.0037 & 0.0000 & -0.9888 \\ 0.6992 & -2.2360 & 2.1239 \\ 0.6992 & 2.2360 & 2.1239 \end{bmatrix}.$$

7.11 Dynamic Analysis of Plane Trusses Using SAP2000

Illustrative Example 7.4
For the plane truss in Figure 7.9(a) using SAP2000 determine: (1) the natural frequencies and (2) the response due to the force $F(t)$ depicted in Figure 7.9(b) applied at the nodal coordinate 1.

Solution:
The following commands are implemented in SAP2000:

Begin: Open SAP2000
Hint: Maximize both views for a full window.

Select: In the lower right-hand corner of the window, use the drop-down menu to select "kip-in".

Model: From the main menu select:
FILE>NEW MODEL.
In the Coordinate System Definitions enter:
Number of Grid Spaces:
X direction = 2
Y direction = 0
Z direction = 1
Grid Spacing:
X direction = 100
Y direction = 1
Z direction = 100

Edit: Click the XZ icon and maximize the 2-D view to switch to this plane; then use the PAN icon to drag the figure to the center of the window.

Draw: From main menu select:
DRAW>DRAW FRAME ELEMENT.
Click in the origin of coordinates and drag the mouse horizontally to the next grid line; click twice to create element 1. Press the Esc key.

From main menu select:
DRAW>DRAW FRAME ELEMENT.
Click again in the origin of coordinates and proceed to drag the mouse vertically to the next grid line; click twice to create the second element. Press the Esc key.

From main menu select:
DRAW>DRAW FRAME ELEMENT.
Click on the right end of element 1 and drag the mouse to the upper end of element 2. Click again to create element 3.
Press the Esc key to stop drawing elements.

Label: From the main menu select:
VIEW>SET ELEMENTS.
Click on Joint Labels on Frame Labels; click OK.

Restraints: Select joint 1 of the truss with a click of the mouse.
From the main menu select:
ASSIGN>JOINT>RESTRAINTS.
Select Restraint translations 1 and 3. Click OK.
Select joints 2, and from the main menu select:
ASSIGN>JOINT>RESTRAINTS.
Select Restraint translation 3 only. Click OK.

Sections: From the main menu select:
DEFINE>FRAME SECTIONS.
Click Import/Wide Flange.
Select section properties and click "open".
Scroll to select "General Section".
Enter:
Area = 10 and zero values for other properties of the cross-section.
Click OK, OK.

Assign: Select all members of the truss (click on every member or alternatively, just click at a point outside the truss and drag the mouse to envelope in dotted lines the entire truss).

From the main menu select:
ASSIGN>FRAME>SECTIONS.
Select section SSEC2. Click OK.

Mass: Select all the joints in the truss by clicking on them.
From the main menu select
ASSIGN>JOINT>MASSES.
Enter Masses:
        Direction 1 = 0.1
        Direction 3 = 0.1. Click OK.

Load: From the main menu select:
DEFINE>STATIC LOAD CASES.
        Change Type of load to LIVE.
        Set Self-Weight Multiplier to 0.
        Click Change Load and then OK.

Load Function: From the main menu select:
DEFINE>TIME HISTORY FUNCTIONS.
Click "Add new function".
Click "Add" to register the first point (0, 0).
Enter: Time = 0.02, Value =   10.0, then click ADD.
Enter: Time = 0.06, Value = −10.0, then click ADD.
Enter: Time = 0.08, Value =   0.0, then click ADD.
Enter: Time = 0.10, Value =   0.0 then click ADD and then OK, OK.

Load Case: From the main menu select:
DEFINE>TIME HISTORY CASES.
Click Add New History.
Enter:
        Number of output time steps = 100
        Output time step = 0.001
        Check: Envelopes
Load Assignments:
        Load = LOAD1
        FUNCTION=FUNCI
Accept all of the other default values.
Click ADD; then click OK, OK.

Assign Load: Select joint 3 by a click and from the main menu select:
ASSIGN>JOINT STATIC LOADS>FORCES.
Enter: Force Global X = 1.0; click OK.

Options: From the main menu select:
ANALYZE>SET OPTIONS.

Select available DOFs in the X and Z directions.
Click on "Dynamic Analysis" and on "Set Dynamic Parameters".
Enter: Numbers of Modes = 3; then click OK, OK.

Analyze: From the main menu select:
ANALYZE > RUN.
Enter Filename EXD7.4; then click SAVE.
At the conclusion of the calculations and after checking for errors, click
OK.

Print Input Tables: From the main menu select:
FILES>PRINT INPUT TABLES.
Check "Print to File". Then click OK.
Use NOTEPAD or another editor to retrieve, edit and print
the file C:\SAP2000\EXD7.4.text.

(The edited input table for Illustrative Example 7.4 is reproduced as Table 7.3.)

Table 7.3 Edited input table for Illustrative Example 7.4 (Units: kips, inches)

```
TIME HISTORY CASES
  HISTORY      HISTORY     NUMBER OF      TIME STEP
  CASE          TYPE      TIME STEPS     INCREMENT
  HIST1        LINEAR         20          0.05000
```

JOINT DATA

JOINT	GLOBAL-X	GLOBAL-Y	GLOBAL-Z	RESTRAINTS
1	0.00000	0.00000	0.00000	1 1 1 0 0 0
2	100.00000	0.00000	0.00000	0 0 1 0 0 0
3	0.00000	0.00000	100.00000	0 0 0 0 0 0

JOINT MASS DATA

JOINT	M-U1	M-U2	M-U3	M-R1	M-R2	M-R3
1	0.100	0.000	0.100	0.000	0.000	0.000
2	0.100	0.000	0.100	0.00	0.000	0.000
3	0.100	0.000	0.100	0.000	0.000	0.000

FRAME ELEMENT DATA

FRAME	JNT-1	JNT-2	SECTION	ANGLE	RELEASES	SEGMENTS	LENGTH
1	1	2	FSEC2	0.000	000000	4	100.000
2	1	3	FSEC2	0.000	000000	2	100.000
3	2	3	FSEC2	0.000	000000	2	141.421

JOINT FORCES Load Case LOAD1

JOINT	GLOBAL-X	GLOBAL-Y	GLOBAL-Z	GLOBAL-XX	GLOBAL-YY	GLOBAL-ZZ
3	1.000	0.000	0.00	0.000	0.000	0.000

Plot Displacement Function: From the main menu select:
      DISPLAY>SHOW TIME HISTORY TRACES.
      Click Define Functions.
            Click to "Add Joint Disps/Forces".
                  Joint ID = 3
                  Vector Type = Displacement
                  Component = UX
                  Mode Number = Include All. Click OK.

      Under List of Functions, click once on Joint3.
      Click OK.

      Click again on Define Functions.
            Click to "Add Joint Displ./Forces".
                  Joint ID = 3
                  Vector Type = Displacement
                  Component = UZ
                  Mode Number = Include All. Click OK.

      Under List of Functions, click once on Joint3.
      Click ADD to move Joint3-1 to the Plot Functions column.

      Under List of Functions, click once on Joint3-1.
      Click ADD to move Joint3-1 to Plot Functions column.
            Axis Labels
                  Horizontal = Time (sec)
                  Vertical = Displacements of joint 3 in X and Z directions (in)
Click "Display" to view on screen the time–displacement plot for joint 3 in the X and Z directions. Then, to obtain a printed version, select:
FILE>PRINT GRAPHICS.

(Figure 7.10 reproduces the plot for the Displacement component of joint 3 in the X and Z directions for the truss of Illustrative Example 7.4.)

Alternatively, a table of the displacement function may be obtained by selecting:

FILE>PRINT TABLES TO FILE.

The values in this table may be plotted using Excel or similar software.

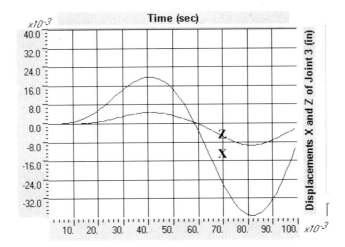

Fig. 7.10   Displacements in the X and Z directions of joint 3 in the truss of
Illustrative Example 7.4

(Table 7.4 provides a portion of the Displacement Function Values in the X and Z
directions for joint 3 of the truss in Illustrative Example 7.4.)

Table 7.4   Displacement Functions in the X and Z directions for Joint 3 of the
truss of Illustrative Example 7.4 (in)

Time (sec)	Displ. X (in)	Displ. Z (in)
0.00000	0.00000	0.00000
0.00100	0.00000	0.00000
0.00200	6.653E-06	0.00000
0.00300	2.239E-05	0.00000
0.00400	5.289E-05	0.00000
0.00500	1.028E-04	1.331E-06
0.00600	1.767E-04	3.260E-06
0.00700	2.787E-04	6.912E-06
0.00800	4.129E-04	1.318E-05
0.00900	5.829E-04	2.318E-05
----	----	----
0.08000	−0.03898	−0.00942
0.08100	−0.03926	−0.00947
0.08200	−0.03932	−0.00945
0.08300	−0.03917	−0.00937
0.08400	−0.03881	−0.00922
0.08500	−0.03822	−0.00901
0.08600	−0.03743	−0.00875
0.08700	−0.03642	−0.00843

Table 7.4 (continued)

Time (sec)	Displ. X (in)	Displ. Z (in)
0.08800	−0.03520	−0.00807
0.08900	−0.03378	−0.00767
0.09000	−0.03217	−0.00723
0.09100	−0.03037	−0.00677
0.09200	−0.02839	−0.00628
0.09300	−0.02624	−0.00577
0.09400	−0.02394	−0.00524
0.09500	−0.02149	−0.00470
0.09600	−0.01892	−0.00415
0.09700	−0.01623	−0.00358
0.09800	−0.01345	−0.00300

Natural Periods and Frequencies:

Using Notepad or other word editor, open the following file:
C:/SAP2000e/EXD 7.4.OUT.
The Natural Period and Frequency table may be found in that file.

(Table 7.5 provides the edited Natural Period and Frequency values for the truss of Illustrative Example 7.4.)

Table 7.5 Natural Periods and Natural Frequencies for Illustrative Example 7.4

MODE	PERIOD (TIME)	FREQUENCY (CYC/TIME)	EIGENVALUE (RAD/TIME)	$(RAD/TIME)^2$
1	0.083468	11.980685	75.276862	5666.606
2	0.036276	27.566445	173.205081	30000.000
3	0.026515	37.714416	236.966663	56153.199

Mode Shapes: From the main menu select:
DISPLAY>SHOW MODE SHAPE.
Modal Shape 1: select "Wire Shadow" and "Cubic Curve" option.
Scale Factor = 1. Click OK.
(To view Modes 2 and 3, click on the + sign on lower right corner of the window.)

Note: Numerical values for the nodal displacements of a mode shape may be viewed on screen by right-clicking on a node of the modal shape.

Illustrative Example 7.5

Use SAP2000 to perform the structural dynamic analysis of the plane truss shown in Figure 7.11(a). For all members of the truss, the cross-sectional area is $A = 2 \ in^2$ and the modulus of elasticity $E = 30,000 \ ksi$. Assign concentrated weights at all joints of the truss, $W = 3.86 \ kip$. The truss is acted on by a time force function shown in Figure 7.11 (b) applied at the center joint on the lower cord in the vertical direction.

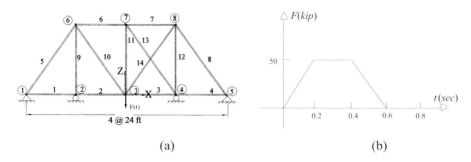

(a)   (b)

Fig. 7.11   (a) Plane truss for Illustrative Example 7.6;   (b) Force function applied at
the center of the truss on the lower cord in the vertical direction

Solution:
The following commands are implemented in SAP2000:

Begin: Open SAP2000
          Hint: Maximize both windows for a full view.

Select: In the lower right-hand corner of the window, use the drop-down menu to
          select "kip-in".

Model: From the main menu select:
          FILE>NEW MODEL.
          In the Coordinate System Definitions, enter:
               Number of Grid Spaces:
                    X direction = 4
                    Y direction = 0
                    Z direction = 1

               Grid Spacing:
                    X direction = 288
                    Y direction = 1
                    Z direction = 384

Edit: Maximize the 2-D view and click the XZ icon to set the view to this plane.
         Then use the PAN icon to drag the figure to the center of the window.

Draw: From main menu select:
         DRAW>DRAW FRAME ELEMENT.
         Click in the lower-left intersection of the grid lines and drag the mouse
         horizontally to the next grid line; click twice to create an element. Then,
         click again at this location and proceed to drag the mouse horizontally to

the next grid line; click twice to create the second element. Repeat this process to create the next two horizontal elements of the truss. Press the Esc key to stop drawing elements.

Using the draw command, proceed to create all the other members of the truss shown in Figure 7.11.

Label: From the main menu select:
VIEW>SET ELEMENTS.
Click Joint Labels and Frame Labels. Click OK.

Restraints: Select joints 1and 5 of the truss by clicking on them, and then select from the main menu:
ASSIGN>JOINT>RESTRAINTS.
Select Restraint translations 1, 2 and 3. Click OK.

Select joints 2 and 4 at the two rollers and from the main menu select:
ASSIGN>JOINT>RESTRAINTS.
Select the Restraint translation for 3 only. Click OK.

Select joints 3, 6, 7, and 8 and from the main menu select:
ASSIGN>JOINT>RESTRAINTS.
Select no restraints by clicking the dot icon. Click OK.

Sections: From the main menu, select:
DEFINE>FRAME SECTIONS.
Click Import/Wide Flange.
Select section properties and click "open".
Scroll on the sections to select "General Section".
Enter:
Area = 2.0
Zero values for other values of the cross-section.
Click OK.
Select: Material Steel.
Click OK.

Assign: Select all members of the truss (click on every member or click at a point outside the truss and drag the pointer to envelope in dotted lines the entire truss). Then from the main menu, select:
ASSIGN>FRAME>SECTIONS.
Select section SSEC2.
Click OK.

Mass: Select all the joints in the truss by clicking on them.
From the main menu select:
ASSIGN>JOINT>MASSES.
Enter Masses:

Direction 1= 0.01
Direction 3= 0.01
Click OK.

Load: From the main menu select:
DEFINE>STATIC LOAD CASES.
Change the Type of load to LIVE.
Set the Self-Weight Multiplier to 0.
Click Change Load.
Click OK.

Load Function: From the main menu select:
DEFINE>TIME HISTORY FUNCTIONS.
Click "Add new function".
Click "Add" to register the first point (0,0).
Enter: Time = 0.2, Value = 50.0, then click ADD.
Enter: Time = 0.4, Value = 50.0, then click ADD.
Enter: Time = 0.6, Value = 0.0, then click ADD.
Enter: Time = 0.8, Value = 0.0, then click ADD.
Click OK, OK.

Load Case: From the main menu select:
DEFINE>TIME HISTORY CASES.
Click Add New History.
Enter:
    Number of output time steps = 80
    Output time step = 0.01
    Check: Envelopes
Load Assignments:
    Load = LOAD1
    FUNCTION = FUNCI
Accept all of the other default values.
Click ADD.
Click OK, OK.

Assign Load: Select joint 3 by a click and then from the main menu select:
ASSIGN>JOINT STATIC LOADS>FORCES.
Enter: Force Global Z = −1.0; click OK.

Analyze: From the main menu select:
ANALYZE>SET OPTIONS.
Select available DOFs in the X and Z directions.
Click on "Dynamic Analysis" and on "Set Dynamic Parameters".
Enter: Numbers of Modes = 3.
        Click OK, OK.

From the main menu select:
ANALYZE>RUN.
Enter Filename EXD7.5.
Click SAVE.
At the conclusion of the calculations and after checking for errors, click
OK.

Print Input Tables: From the main menu select:
FILES>PRINT INPUT TABLES.
Check: "Print to File". Click OK.
Use NOTEPAD or another editor to retrieve, edit and print
the file C:\SAP2000\EXD7.5.text.

(The edited input table for Illustrative Example 7.5 is reproduced as Table 7.6.)

Table 7.6 Edited input table for Illustrative Example 7.5

```
STATIC  LOAD  CASES
  STATIC      CASE    SELF WT
  LOAD1       LIVE    0.0000

TIME  HISTORY  CASES

  HISTORY     HISTORY     NUMBER OF  TIME STEP
   CASE        TYPE       TIME STEPS  INCREMENT
   HIST1      LINEAR         80        0.01000

JOINT  DATA
```

JOINT	GLOBAL-X	GLOBAL-Y	GLOBAL-Z	RESTRAINTS
1	−576.00000	0.00000	0.00000	1 1 1 0 0 0
2	−288.00000	0.00000	0.00000	0 0 1 0 0 0

JOINT	GLOBAL-X	GLOBAL-Y	GLOBAL-Z	RESTRAINTS
3	0.00000	0.00000	0.00000	0 0 0 0 0 0
4	288.00000	0.00000	0.00000	0 0 1 0 0 0
5	576.00000	0.00000	0.00000	1 1 1 0 0 0
6	−288.00000	0.00000	384.00000	0 0 0 0 0 0
7	0.00000	0. 00000	384.00000	0 0 0 0 0 0
8	288.00000	0.00000	384.00000	0 0 0 0 0 0

```
JOINT  MASS  DATA
```

JOINT	M-U1	M-U2	M-U3	M-R1	M-R2	M-R3
1	1.000E-02	1.000E-02	0.000	0.000	0.000	0.000
2	1.000E-02	1.000E-02	0.000	0.000	0.000	0.000
3	1.000E-02	1.000E-02	0.000	0.000	0.000	0.000
4	1.000E-02	1.000E-02	0.000	0.000	0.000	0.000
5	1.000E-02	1.000E-02	0.000	0.000	0.000	0.000
6	1.000E-02	1.000E-02	0.000	0.000	0.000	0.000
7	1.000E-02	1.000E-02	0.000	0.000	0.000	0.000
8	1.000E-02	1.000E-02	0.000	0.000	0.000	0.000

Table 7.6 (continued)

F R A M E   E L E M E N T   D A T A

FRAME	JNT-1	JNT-2	SECTION	RELEASES	SEGMENTS	LENGTH
1	1	2	SSEC2	0.000	4	288.000
2	2	3	SSEC2	0.000	4	288.000
3	3	4	SSEC2	0.000	4	288.000
4	4	5	SSEC2	0.000	4	288.000
5	1	6	SSEC2	0.000	2	480.000
6	6	7	SSEC2	0.000	4	288.000
7	7	8	SSEC2	0.000	4	288.000
8	8	5	SSEC2	0.000	2	480.000
9	2	6	SSEC2	0.000	2	384.000
10	6	3	SSEC2	0.000	2	480.000
11	3	7	SSEC2	0.000	2	384.000
12	8	4	SSEC2	0.000	2	384.000
13	4	7	SSEC2	0.000	2	480.000
14	3	8	SSEC2	0.000	2	480.000

J O I N T   F O R C E S   Load Case LOAD1

JOINT	GLOBAL-X	GLOBAL-Y	GLOBAL-Z	GLOBAL-XX	GLOBAL-YY	GLOBAL-ZZ
3	0.000	0.000	−1.000	0.000	0.000	0.000

Plot Displacement Function: From the main menu select:
  DISPLAY>SHOW TIME HISTORY TRACES.
  Click Define Functions.
  Click to "Add Joint Disps/Forces".
    Joint ID = 3
    Vector Type = Displacement
    Component = UZ
    Mode Number = Include All. Click OK.
  Under List of Functions, click once on Joint3.
  Click ADD to move Joint3 to Plot Functions column.
  Axis Labels
    Horizontal = Time (sec)
    Vertical = Z-Displacements Joint 3 (in)
  Click "Display" to view on screen the time–displacement plot for joint 3 in
  the Z direction. Then, to obtain a printed version, select:
  FILE>PRINT GRAPHICS.

(Figure 7.12 reproduces the plot shown in the screen for the Displacement
in Z direction of Joint 3 for the truss of Illustrative Example 7.5.)

Alternatively, a table of the displacement function may be obtained by
selecting from the main menu:
FILE>PRINT TABLES TO FILE

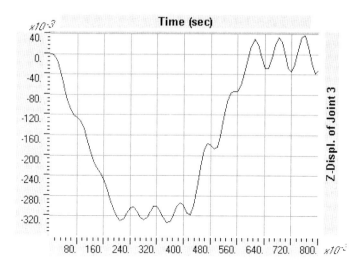

Fig. 7.12   Displacement for Joints 3 in Z direction of the truss of Illustrative Example
7.6

The values in this table may be found in the file EXD75.txt and then
plotted using Excel or similar software.

(Table 7.7 contains a portion of the Displacement Function Values in the Z direction
for joint 3 of the truss in Illustrative Example 7.5.)

Table 7.7   Displacement Function in the Z direction for Joint 3 of the truss of
Illustrative Example 7.5

Time (sec)	Displ. (in)
0.00000	0.00000
0.01000	−0.00183
0.02000	−0.01280
0.03000	−0.03501
0.04000	−0.06330
0.05000	−0.09005
-------------------	-------------------
-------------------	-------------------
0.76000	0.03678
0.77000	0.01447
0.78000	−0.01844
0.79000	−0.03773

Natural Periods and Frequencies:
Using Notepad or other word editor, open the following file:
C:/SAP2000e/EXD 7.5.OUT.
The Natural Period and Frequency Table may be found in this file.

(Table 7.8 contains the edited Natural Period and Frequency values for the truss of Illustrative Example 7.5.)

Table 7.8 Natural Periods and Natural Frequencies for Illustrative Example 7.5

PERIODS AND FREQUENCIES

MODE	PERIOD (SEC)	FREQUENCY (CYC/SEC)	FREQUENCY (RAD/SEC)	EIGENVALUE (RAD/SEC)^2
1	0.095	10.52	65.81	4,331
2	0.072	13.89	86.17	7,426
3	0.047	21.28	131.83	17,379

Mode Shapes: From the main menu, select:
DISPLAY>SHOW MODE SHAPE.
Modal Shape 1: click the "Wire Shadow" and "Cubic Curve" options.
Scale Factor = 1. Click OK.
(To view Modes 2 and 3, click on the "+" icon located in the lower right corner on the screen.)

Note: Numerical values for a mode shape in the X and Y directions may be viewed on screen by right-clicking on a node in the nodal shape.

7.12 Practice Problems

Problem 7.1
For the plane truss shown in Figure P7.1, determine:

(a) Joint displacements
(b) Member axial forces
(c) Support reactions

All Members:
$E = 200\ GP_a$
$A = 48\ cm^2$

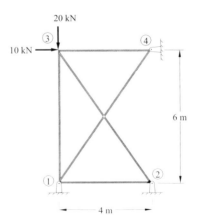

Fig. P7.1

Problem 7.2
Solve Problem 7.1 for the combined effect of the load shown in Fig P7.1 and a settlement of support 1 of 10 *mm* down and 6 *mm* horizontally to the right.

Problem 7.3
For the plane truss shown in Figure P7.3, determine:

(a) Joint displacements
(b) Member end forces
(c) Support reactions

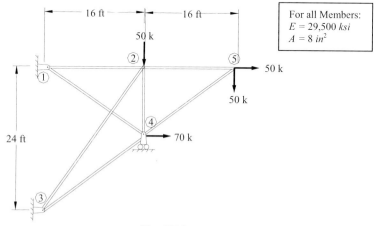

Fig. P7.3

Problem 7.4
Solve Problem 7.3 for the combined effect of the loads shown in Figure P7.3 and a vertical settlement of 1.5 *in* at the support in joint 4.

Problem 7.5
For the plane frame shown in Figure P7.5, determine:

(a) Joint displacements
(b) Axial member forces
(c) Support reactions

All Members:
$E = 29,500\ ksi$
$A = 8\ in^2$

Fig. P7.5

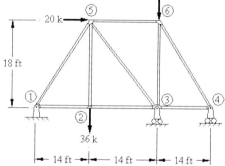

Problem 7.6
Solve Problem 7.5 for the combined effect of the loads shown in Figure P7.5 and a settlement of 1.0 in a downward direction and 2 *in* to the right at the support in joint 1.

Problem 7.7
For the truss shown in Figure P7.7 determine:

(a) Joint displacements
(b) Member axial forces
(c) Support reactions

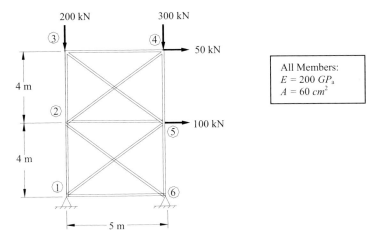

All Members:
$E = 200\ GP_a$
$A = 60\ cm^2$

Fig P7.7

Problem 7.8
Solve Problem 7.7 for the combined effect of the loads shown in Figure P7.7 and settlement of support at joint of 12 *mm* horizontally to the right and 18 *mm* down vertically.

Problem 7.9
For the truss shown in Figure P7.9 determine:

(a) Joint displacements
(b) Axial member forces
(c) Support reactions

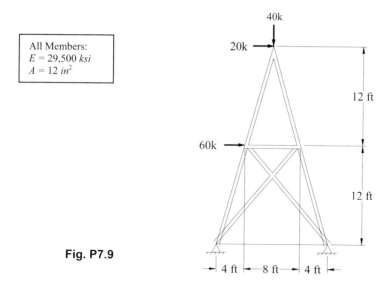

All Members:
$E = 29,500 \ ksi$
$A = 12 \ in^2$

40k

20k

12 ft

60k

12 ft

Fig. P7.9

4 ft   8 ft   4 ft

Problem 7.10

For the plane truss shown in Figure P7.10, determine the system stiffness and mass matrices corresponding to the two nodal coordinates indicated in the figure.

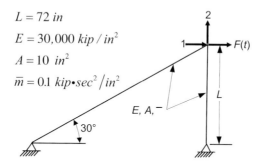

$L = 72 \ in$

$E = 30,000 \ kip / in^2$

$A = 10 \ in^2$

$\bar{m} = 0.1 \ kip{\cdot}sec^2 / in^2$

$E, A, —$

$30°$

2

1   $F(t)$

L

Fig. P7.10

Problem 7.11

Determine the natural frequencies and corresponding normal modes for the truss shown in Figure P7.10.

Problem 7.12

Determine the response of the truss shown in Figure P7.10 when subjected to a force $F(t) = 10$ *kips* suddenly applied for 2 *sec* at nodal coordinate 1. Use the results of Problem 7.11 to obtain the modal equations. Neglect damping in the system.

Problem 7.13

Solve Problem 7.12 assuming 10% damping in all the modes.

Determine the maximum response of the truss shown in Figure P7.10 when subjected to a rectangular pulse of magnitude $F_0 = 10$ *kip* and duration $t_d = 0.1$ *sec*. Neglect damping in the system.

Problem 7.14

Determine the dynamic response of the frame shown in Figure P7.10 when subjected to a harmonic force $F(t) = 10 \sin 10t$ *kips* along nodal coordinate 1. Neglect damping in the system.

Problem 7.15

Repeat Problem 7.14 assuming that the damping in the system is proportional to the stiffness, $[C] = a_0 [K]$ where $a_0 = 0.1$.

Problem 7.16

Determine the response of the truss shown in Figure P7.16 when acted upon by the forces $F_1(t) = 10t$ and $F_2(t) = 5t^2$ during 1 sec. Neglect damping.

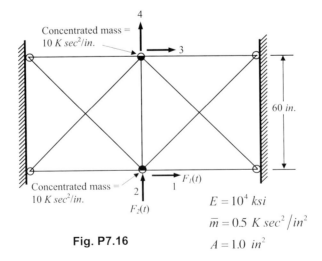

Fig. P7.16

$$E = 10^4 \ ksi$$
$$\bar{m} = 0.5 \ K \ sec^2 / in^2$$
$$A = 1.0 \ in^2$$

Problem 7.17
Determine the response of the truss shown in Figure P7.16 when acted upon by the forces $F_1(t)= 10t$ and $F_2(t)= 5t^2$ during 1 sec. Neglect damping.

Problem 7.18
Solve problem 7.17 assuming 10% modal damping in all the modes.

8 Space Trusses

8.1 Introduction

Space trusses are three-dimensional structures with longitudinal members connected at their ends by assumed-to-be-frictionless hinges. The loads on space trusses are applied only at the nodes or joints, thus the self-weight is allocated for each element at its two ends joining other elements of the truss. The conditions imposed on space trusses are certainly the same as those on plane trusses. Essentially, the only difference in the analysis of space trusses compared with plane trusses is that an element of a space truss has three translatory nodal coordinates at each node while an element of a plane truss has only two.

8.2 Element Stiffness Matrix of a Space Truss – Local Coordinates

The stiffness matrix for an element of a space truss can be obtained as an extension of the corresponding matrix for the plane truss. Figure 8.1 shows the nodal coordinates in the local system (unbarred) and in the global system (barred) for a member of a space truss. The local axis x is directed along the longitudinal axis of the member while the y and z axes are set to agree with the principal directions of the cross-section of the member. The following matrix, in local coordinates, is then written for a uniform member of a space truss:

$$[k] = \frac{AE}{L}\begin{bmatrix} 1 & 0 & 0 & -1 & 0 & 0 \\ 0 & 0 & 0 & 0 & 0 & 0 \\ 0 & 0 & 0 & 0 & 0 & 0 \\ -1 & 0 & 0 & 1 & 0 & 0 \\ 0 & 0 & 0 & 0 & 0 & 0 \\ 0 & 0 & 0 & 0 & 0 & 0 \end{bmatrix} \qquad (8.1)$$

Therefore, the element stiffness equation for an element of a space truss in reference to the local system of coordinate axes is given by

$$\{P\} = [k]\{\delta\} \qquad (8.2)$$

in which $\{\delta\}$ is the element nodal displacement vector in reference to the local coordinates, as shown in Figure 8.1(a), and $\{P\}$ is the corresponding nodal force vector.

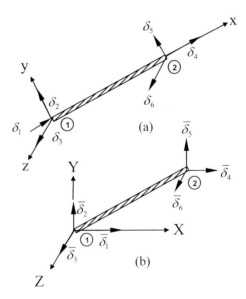

Fig. 8.1 Member of a space truss showing nodal displacement coordinates
(a) in the local system (unbarred), (b) in the global system (barred)

These vectors are, respectively,

$$\{\delta\} = \begin{Bmatrix} \delta_1 \\ \delta_2 \\ \delta_3 \\ \delta_4 \\ \delta_5 \\ \delta_6 \end{Bmatrix} \quad \text{and} \quad \{P\} = \begin{Bmatrix} P_1 \\ P_2 \\ P_3 \\ P_4 \\ P_5 \\ P_6 \end{Bmatrix}. \tag{8.3}$$

8.3   Element Stiffness Matrix in Global Coordinates

The transformation matrix $[T]_1$ corresponding to the three nodal coordinates at a node of a beam as developed in Chapter 6 is given by eq. (6.4) and it is repeated here for convenience.

$$[T]_1 = \begin{bmatrix} \cos xX & \cos xY & \cos xZ \\ \cos yX & \cos yY & \cos yZ \\ \cos zX & \cos zY & \cos zZ \end{bmatrix} \qquad \text{(6.4) Repeated}$$

where $\cos xY$ is the cosine of the angle between the local axis x and the global axis Y and analogous definitions for the other cosine functions in eq. (6.4).

The direction cosines of local axis x in the first row of eq. (6.4), designated as

$$c_1 = \cos xX, \quad c_2 = \cos xY, \text{ and } c_3 = \cos xZ, \tag{8.4}$$

are calculated from the coordinates of the two points $P_1(X_1 \ Y_1, Z_1)$ and $P_2(X_2, Y_2, Z_2)$ at the two ends of the truss element, that is

$$c_1 = \frac{X_2 - X_1}{L}, \quad c_2 = \frac{Y_2 - Y_1}{L} \quad \text{and} \quad c_3 = \frac{Z_2 - Z_1}{L} \tag{8.5}$$

with the length L of the element given by

$$L = \sqrt{(X_2 - X_1)^2 + (Y_2 - Y_1)^2 + (Z_2 - Z_1)^2}. \tag{8.6}$$

The transformation matrix for the nodal coordinates at the two ends of a truss member is then given by

$$[T] = \begin{bmatrix} [T]_1 & [0] \\ [0] & [T]_1 \end{bmatrix} \tag{8.7}$$

in which $[T]_1$ is given by eq. (6.4). It follows that the relationship between the global displacement vector $\{\bar{\delta}\}$ and the local displacement vector $\{\delta\}$ at the two nodes of a truss element is

$$\{\delta\} = [T]\{\bar{\delta}\}.$$ (8.8)

Analogously, the transformation of vector $\{\bar{P}\}$ with components along the global system of coordinates and the element force vector $\{P\}$ in reference to the local coordinates is then given by

$$\{P\} = [T]\{\bar{P}\}.$$ (8.9)

The substitution into the element stiffness eq. (8.2) of $\{\delta\}$ and $\{P\}$, respectively, from eqs. (8.8) and (8.9) results in

$$[T]\{\bar{P}\} = [k][T]\{\bar{\delta}\}.$$ (8.10)

Since the transformation matrix $[T]$ is orthogonal, $\left([T]^{-1} = [T]^{T}\right)$, eq. (8.10) may be written as

$$\{\bar{P}\} = [\bar{k}]\{\bar{\delta}\}$$ (8.11)

in which the element stiffness matrix $[\bar{k}]$ in reference to global coordinates is given by

$$[\bar{k}] = [T]^{T}[k][T].$$ (8.12)

In the evaluation of the element stiffness matrix $[\bar{k}]$ in global coordinates of a space truss, it is only necessary to calculate the direction cosines of the centroidal axis x of the element, which are given by eq. (8.5). The other direction cosines in eq. (6.4) corresponding to the axes y and z do not appear in the final expression for the element stiffness matrix $[\bar{k}]$ as may be verified by substituting eqs. (8.1) and (8.7) into eq. (8.12) and proceeding to multiply the matrices indicated in this last equation. The final result of this operation may be written as follows:

$$[\bar{k}] = \frac{AE}{L} \begin{bmatrix} c_1^2 & c_1c_2 & c_1c_3 & -c_1^2 & -c_1c_2 & -c_1c_3 \\ c_2c_1 & c_2^2 & c_2c_3 & -c_2c_1 & -c_2^2 & -c_2c_3 \\ c_3c_1 & c_3c_2 & c_3^2 & -c_3c_1 & -c_3c_2 & -c_3^2 \\ -c_1^2 & -c_1c_2 & -c_1c_3 & c_1^2 & c_1c_2 & c_1c_3 \\ -c_2c_1 & -c_2^2 & -c_2c_3 & c_2c_1 & c_2^2 & c_2c_3 \\ -c_3c_1 & -c_3c_2 & -c_3^2 & c_3c_1 & c_3c_2 & c_3^2 \end{bmatrix}. \tag{8.13}$$

Consequently, the determination of the stiffness matrix for an element of a space truss, in reference to the global system of coordinates [eq. (8.13)] requires only the evaluation by eq. (8.5) of the direction cosines c_1, c_2, and c_3 of the local axis x along the element.

8.4   Element Axial Force

The expression to determine the axial force (tension or compression) in an element of a space truss may be obtained as an extension of eq. (7.12) for calculating the axial force in an element of a plane truss. Thus, extending eq. (7.12) to consider the three nodal displacements at the two ends of a space truss element, we obtain the expression for the axial force P as

$$P = \frac{EA}{L}\left[\left(\bar{\delta}_4 - \bar{\delta}_1\right)c_1 + \left(\bar{\delta}_5 - \bar{\delta}_2\right)c_2 + \left(\bar{\delta}_6 - \bar{\delta}_3\right)c_3\right] \tag{8.14}$$

where $\bar{\delta}_1$ through $\bar{\delta}_6$ are the element nodal displacements in reference to the global system and $c_1 = \cos xX$, $c_2 = \cos xY$, and $c_3 = \cos xZ$ are the direction cosines of the axial axis x of the element. As stated, these direction cosines are calculated from the coordinates at the two points at the ends of the truss element.

8.5   Assemble the System Stiffness Matrix

The reduced system stiffness matrix in which only the free coordinates are considered, is assembled, as explained in the previous chapters for frame type structures, by transferring to the appropriate location of the system stiffness matrix the coefficients in the element stiffness matrices. The following example illustrates the analysis of a space truss using the matrix stiffness method of analysis.

Illustrative Example 8.1
Determine for the space truss shown in Figure 8.2: (a) the reduced system stiffness matrix, (b) the reduced system force vector, (c) the nodal displacements, and (d) the

element's axial forces. The modulus of elasticity is $E = 30,000\ ksi$ and the cross-sectional area $A = 1.0\ in^2$ for all members of the truss. The truss is supporting a force applied at joint 5 having equal components of magnitude 10 $kips$ in the directions of the axes X, Y and Z as shown in Fig 8.2.

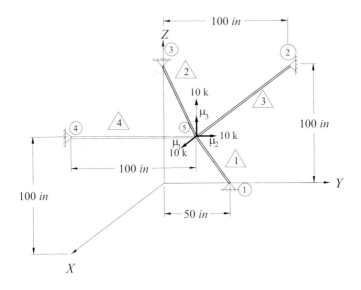

Fig. 8.2 Space truss for Illustrative Example 8.1

Solution:

1. Mathematical Model

    This truss is modeled as having 4 elements and 5 joints, as shown in Figure 8.2. Since only joint 5 is free to displace, the truss has only three free nodal coordinates: u_1, u_2 and u_3, as indicated in Figure 8.2. The coordinates X, Y and Z for the joints of the truss are conveniently arranged in the following table:

    Joint Coordinates (inches)

Joint #	X	Y	Z
1	0	50	0
2	0	100	100
3	0	0	100
4	100	0	100
5	100	100	100

2. Element Stiffness Matrices – Global Coordinates

The stiffness matrix for an element of a space truss is given by eq. (8.13) in which the direction cosines c_1, c_2 and c_3 are calculated by eq. (8.5).

ELEMENT 1 (Joint 5 to Joint 1)

Direction Cosines: $c_1 = -0.667$ $c_2 = -0.333$ $c_3 = -0.667$

$$[\bar{k}]_1 = \begin{bmatrix} 88.888 & 44.444 & 88.888 & -88.888 & -44.444 & -88.888 \\ 44.444 & 22.222 & 44.444 & -44.444 & -22.222 & -44.444 \\ 88.888 & 44.444 & 88.888 & -88.888 & -44.444 & -88.888 \\ -88.888 & -44.444 & -88.888 & 88.888 & 44.444 & 88.888 \\ -44.444 & -22.222 & -44.444 & 44.444 & 22.222 & 44.444 \\ -88.888 & -44.444 & -88.888 & 88.888 & 44.444 & 88.888 \end{bmatrix} \begin{matrix} 1 \\ 2 \\ 3 \\ \\ \\ \end{matrix} \quad (a)$$

ELEMENT 2 (Joint 5 to Joint 3)

Direction Cosines: $c_1 = -0.707$ $c_2 = 0.707$ $c_3 = 0$

$$[\bar{k}]_2 = \begin{bmatrix} 106.066 & 106.066 & 0.000 & -106.066 & -106.066 & 0.000 \\ 106.066 & 106.066 & 0.000 & -106.066 & -106.066 & 0.000 \\ 0.000 & 0.000 & 0.000 & 0.000 & 0.000 & 0.000 \\ -106.066 & -106.066 & 0.000 & 106.066 & 106.066 & 0.000 \\ -106.066 & -106.066 & 0.000 & 106.066 & 106.066 & 0.000 \\ 0.000 & 0.000 & 0.000 & 0.000 & 0.000 & 0.000 \end{bmatrix} \begin{matrix} 1 \\ 2 \\ 3 \\ \\ \\ \end{matrix} \quad (b)$$

ELEMENT 3 (Joint 5 to Joint 2)

Direction Cosines: $c_1 = -1$ $c_2 = 0$ $c_3 = 0$

$$[\bar{k}]_3 = \begin{bmatrix} 300.000 & 0.000 & 0.000 & -300.000 & 0.000 & 0.000 \\ 0.000 & 0.000 & 0.000 & 0.000 & 0.000 & 0.000 \\ 0.000 & 0.000 & 0.000 & 0.000 & 0.000 & 0.000 \\ -300.000 & 0.000 & 0.000 & 300.000 & 0.000 & 0.000 \\ 0.000 & 0.000 & 0.000 & 0.000 & 0.000 & 0.000 \\ 0.000 & 0.000 & 0.000 & 0.000 & 0.000 & 0.000 \end{bmatrix} \begin{matrix} 1 \\ 2 \\ 3 \\ \\ \\ \end{matrix} \quad (c)$$

ELEMENT 4 (Joint 5 to Joint 4)

Direction cosines: $c_1 = 0$ $c_2 = -1$ $c_3 = 0$

$$
[\bar{k}]_4 = \begin{matrix} & 1 & & 2 & 3 & & & \\ \begin{bmatrix} 0.000 & 0.000 & 0.000 & 0.000 & 0.000 & 0.000 \\ 0.000 & 300.000 & 0.000 & 0.000 & -300.000 & 0.000 \\ 0.000 & 0.000 & 0.000 & 0.000 & 0.000 & 0.000 \\ 0.000 & 0.000 & 0.000 & 0.000 & 0.000 & 0.000 \\ 0.000 & -300.000 & 0.000 & 0.000 & 300.000 & 0.000 \\ 0.000 & 0.000 & 0.000 & 0.000 & 0.000 & 0.000 \end{bmatrix} & \begin{matrix} 1 \\ 2 \\ 3 . \\ \\ \\ \end{matrix} \end{matrix}
$$

(d)

3. Reduced System Stiffness Matrix.

The reduced system stiffness matrix $[K]_R$ is assembled by transferring the coefficients in the stiffness matrices of the elements according to the nodal coordinates indicated for the free coordinates 1, 2 and 3 at the top and on the right side of these matrices: (a), (b), (c) and (d). The transfer of these coefficients to the appropriate locations in the reduced system stiffness matrix results in

$$
[K]_R = \begin{matrix} & 1 & 2 & 3 & \\ \begin{bmatrix} 88.89 + 106.07 + 300 & 44.44 + 106.07 & 88.89 \\ 44.44 + 106.07 & 22.22 + 106.07 + 300 & 44.44 \\ 88.89 & 44.44 & 88.89 \end{bmatrix} & \begin{matrix} 1 \\ 2 \\ 3 \end{matrix} \end{matrix}
$$

or

$$
[K]_R = \begin{bmatrix} 494.96 & 150.51 & 88.89 \\ 150.51 & 428.92 & 44.44 \\ 88.89 & 44.44 & 88.89 \end{bmatrix} .
$$

(e)

4. Reduced system force vector

For this Illustrative Example, the reduced system vector $\{F\}_R$ contains the three forces of 10 *kips* applied to joint 5 of the truss, that is

$$
\{F\}_R = \begin{Bmatrix} 10 \\ 10 \\ 10 \end{Bmatrix} .
$$

(f)

5. Reduced System Stiffness Matrix Equation

The reduced system stiffness matrix equation obtained from eqs. (e) and (f) is

$$\begin{Bmatrix} 10 \\ 10 \\ 10 \end{Bmatrix} = \begin{bmatrix} 494.96 & 150.51 & 88.89 \\ 150.51 & 428.92 & 44.44 \\ 88.89 & 44.44 & 88.89 \end{bmatrix} \begin{Bmatrix} u_1 \\ u_2 \\ u_3 \end{Bmatrix}. \qquad (g)$$

6. Displacements at the Free Coordinates

The solution of eq. (g) yields the displacements at the free nodal coordinates as

$$u_1 = -0.0032 \ in$$

$$u_2 = 0.0133 \ in$$

$$u_3 = 0.1080 \ in$$

7. Element Axial Forces

The element axial forces are calculated using eq. (8.14) after identifying the components of the element nodal displacements. For example, for element 1, the nodal displacement vector is identified as

$$\{\bar{\delta}\}_1 = \begin{Bmatrix} u_1 \\ u_2 \\ u_3 \\ 0 \\ 0 \\ 0 \end{Bmatrix} = \begin{Bmatrix} -0.0032 \\ 0.0133 \\ 0.1080 \\ 0 \\ 0 \\ 0 \end{Bmatrix}.$$

The substitution into eqs. (8.5) and (8.6) of the coordinates for the two joints of element 1 ($X_1 = 100$, $Y_1 = 100$, $Z_1 = 100$) and ($X_2 = 0$, $Y_2 = 50$, $Z_2 = 0$) results in

$$L = \sqrt{(0-100)^2 + (50-100)^2 + (0-100)^2} = 150 \ in$$

and

$$c_1 = \frac{0-100}{150} = -0.667 \qquad c_2 = \frac{50-100}{150} = -0.333 \qquad c_3 = \frac{0-100}{150} = -0.667.$$

Then substitution of numerical values into eq. (8.14) gives the axial force $P^{(1)}$ of element 1 as

$$P^{(1)} = \frac{30000 \times 1.0}{150} \left[(0.0032)(-0.667) + (-0.0133)(-0.333) + (-0.1080)(-0.667) \right]$$

$$P^{(1)} = 14.86 \ kips \ (\text{Tension}).$$

Analogously, the axial forces $P^{(2)}$, $P^{(3)}$ and $P^{(4)}$, calculated for the other elements of this truss are:

$$P^{(2)} = 1.51 \ kip \ (\text{Tension})$$
$$P^{(3)} = -0.96 \ kip \ (\text{Compression})$$
$$P^{(4)} = 3.98 \ kip \ (\text{Tension})$$

8. Reaction components at support, node 1:

$$R_{1x} = p^{(1)} \cos xX = (14.86)(-0.667) = -9.91 \ kip$$

$$R_{1y} = p^{(1)} \cos xY = (14.86)(-0.333) = -4.95 \ kip$$

$$R_{1z} = p^{(1)} \cos xZ = (14.86)(-0.667) = -9.91 \ kip$$

8.6   Analysis of Space Trusses Using SAP2000

Illustrative Example 8.2
Use SAP2000 to analyze the space truss of Illustrative Example 8.1. The model of this space truss is reproduced, for convenience, in Figure 8.3.

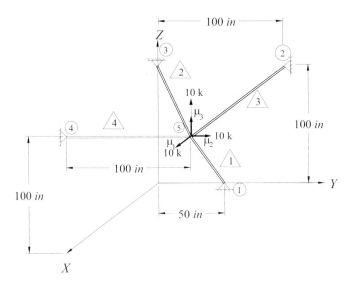

Fig. 8.3 Space Truss for Illustrative Example 8.2

Begin: Open SAP2000.

Enter: "OK" to close the "Tip of the Day".

Units: In the lower right-hand corner of the window, select "kip-in".

Select: From the main menu select:
FILE>NEW MODEL.
Number of Grid Spaces
        X-Direction = 4
        Y-Direction = 4
        Z-Direction = 4
Grid Spacing:
        X-Direction = 50
        Y-Direction = 50
        Z-Direction = 50. Then click OK.

Set: Translate the system of coordinates on each window of the split screen:
Click on the 3-D view to activate this window.
From the main menu select:
VIEW>SET 3-D VIEW.
Change the view direction angle to Plan = 37°, Elevation = 23° and
Aperture = 30°. Then click OK.

From the main menu select:
VIEW>SET 2-D VIEW.
    Set X-Y plane at $Z = 0$
    Set X-Z plane at $Y = 0$
    Set Y-Z plane at $X = 0$. Then click OK.

Select the X-Y plane window, and from the main menu select:
VIEW>SET 2-D VIEW.
    Set X-Y plane at $Z = 0$
    Set X-Z plane at $Y = 0$
    Set Y-Z plane at $X = 0$. Then click OK.

Use the PAN icon to center the coordinate axes in the split screen.

Draw: Click the YZ icon to select the YZ plane.
From the main menu select:
DRAW>ADD SPECIAL JOINT.
Click on the point with coordinates $Y = 50$, $Z = 0$
Click at Point $Y = 100$, $Z = 100$
Click at Point $Y = 0$, $Z = 100$

Draw points in the *X-Z* plane.   Click the *XZ* icon; click on the point with
coordinates $X = 100$, $Z = 100$.
Default to the 3-D window and select or pick any point in the neighbor-
hood of $X = 100$, $Y = 100$, $Z = 100$.
Then right click on this selected point and change coordinates to:
$X = 100$, $Y = 100$, $Z = 100$.

Label: From the main menu select:
VIEW>SET ELEMENT.
Click on "Joint Labels" and on "Frame Labels". Then click OK.

Draw: From the main menu select:
DRAW>DRAW FRAME ELEMENT.
Click on point 5 and drag to point 1, then release and press Enter.
Click on point 5 and drag to point, 2 then release and press Enter.
Click on Point 5 and drag to point, 3 then release and press Enter.
Click on Point 5 and drag to point, 4 then release and press Enter.

Set: Maximize the 3D window.
Use the PAN icon to center the structure in the view.

Label: From the main menu select:
VIEW>SET ELEMENTS.
Click on "Fame Labels". Then click OK.

Material: From the main menu select:
DEFINE >MATERIALS.
Select STEEL and click Modify/Show Material.
     Set $E = 30000$.
     Click OK, OK.

Sections: From the main menu select:
DEFINE>FRAME SECTIONS.
Click on Add/Wide Flange and select "Add General".
Set area = 1.00. Click OK.
For convenience, change the name of the selected section from FSEC2 to
TRUSS 8.2. Click OK.

Assign: Mark or select all the elements, and then from the main menu select:
ASSIGN>FRAME>SECTIONS.
Select TRUSS 8.2. Click OK.

Load: From the main menu select:
DEFINE>STATIC LOAD CASES.
Change DEAD to LIVE.
Set the Self-Weight Multiplier to 0.

Click Change Load. Click OK.

Boundary: Mark Joints 1, 2, 3 and 4; then from the main menu select:
ASSIGN>JOINT>RESTRAINTS.
Select Restraints in translation 1, 3, and 4. Click OK.
Mark Joint 5 and from the main menu select:
ASSIGN>JOINT>RESTRAINTS.
Select no restraints. Then click OK.

Loads: Mark Joint 5 and from the main menu select:
ASSIGN>JOINT STATIC LOADS>FORCES.
For the Load, enter:
FORCE GLOBAL $X = 10$
FORCE GLOBAL $Y = 10$
FORCE GLOBAL $Z = 10$. Then click OK.

Plot: From the main menu select:
DISPLAY>UNDEFORMED SHAPE.
From the main menu select:
DISPLAY>SHOW LOADS>JOINT.
Select FILE>PRINT GRAPHICS.

Analyze: From the main menu select:
ANALYZE>SET OPTIONS.
Select available degrees of freedom translation in X, Y, and Z. Click OK.
From the main menu select:
ANALYZE>RUN.
Respond to filename request by entering "Example 8.2". Click SAVE.
When the analysis is completed click OK.

Print Input Tables: From the main menu select:
FILE>PRINT INPUT TABLES.
(Edited input tables are reproduced as Table 8.1.)

Print Output Tables: From the main menu select:
FILE>PRINT OUTPUT TABLES.
(Edited output tables are reproduced as Table 8.2.)

Table 8.1 Edited input tables for Illustrative Example 8.2 (Units: kips, inches)

```
JOINT  DATA
JOINT GLOBAL-X  GLOBAL-Y  GLOBAL-Z RESTRAINTS ANG-A  ANG-B  ANG-C

  1    0.00000   50.00000    0.00000 1 1 1 0 0 0  0.000  0.000  0.000
  2    0.00000  100.00000  100.00000 1 1 1 0 0 0  0.000  0.000  0.000
  3    0.00000    0.00000  100.00000 1 1 1 0 0 0  0.000  0.000  0.000
```

Table 8.1 (continued)

JOINT	GLOBAL-X	GLOBAL-Y	GLOBAL-Z	RESTRAINTS	ANG-A	ANG-B	ANG-C
4	100.00000	0.00000	100.00000	1 1 1 0 0 0	0.000	0.000	0.000
5	100.00000	100.00000	100.00000	0 0 0 0 0 0	0.000	0.000	0.000

FRAME ELEMENT DATA

FRM	JNT-1	JNT-2	SCTN	ANG	RLS	SGMNTS	R1	R2	FACTOR	LENGTH
1	5	1	TRUSS8-2	0.000	000000	2	0.000	0.000	1.000	150.000
2	5	3	TRUSS8-2	0.000	000000	4	0.000	0.000	1.000	141.421
3	5	2	TRUSS8-2	0.000	000000	4	0.000	0.000	1.000	100.000
4	5	4	TRUSS8-2	0.000	000000	4	0.000	0.000	1.000	100.000

JOINT FORCES   Load Case LOAD1

JOINT	GLOBAL-X	GLOBAL-Y	GLOBAL-Z	GLOBAL-XX	GLOBAL-YY	GLOBAL-ZZ
5	10.000	10.000	10.000	0.000	0.000	0.000

Table 8.2 Edited output results for Illustrative Example 8.2 (Units: kips, inches)

JOINT DISPLACEMENTS

JOINT	LOAD	UX	UY	UZ	RX	RY	RZ
1	LOAD1	0.0000	0.0000	0.0000	0.0000	0.0000	0.0000
2	LOAD1	0.0000	0.0000	0.0000	0.0000	0.0000	0.0000
3	LOAD1	0.0000	0.0000	0.0000	0.0000	0.0000	0.0000
4	LOAD1	0.0000	0.0000	0.0000	0.0000	0.0000	0.0000
5	LOAD1	-3.204E-03	0.0133	0.1080	0.0000	0.0000	0.0000

JOINT REACTIONS

JOINT	LOAD	F1	F2	F3	M1	M2	M3
1	LOAD1	-9.8969	-4.9500	-9.9088	0.0000	0.0000	0.0000
2	LOAD1	0.9612	-4.758E-03	-0.0388	0.0000	0.0000	0.0000
3	LOAD1	-1.0654	-1.0675	-0.0137	0.0000	0.0000	0.0000
4	LOAD1	1.150E-03	-3.9777	-0.0388	0.0000	0.0000	0.0000

FRAME ELEMENT FORCES

FRAME	LOAD	LOC	P	V2	V3	T	M2	M3
1	LOAD1							
		0.00	14.85	0.00	0.00	0.00	0.00	0.00
		75.00	14.85	0.00	0.00	0.00	0.00	0.00
		150.00	14.85	0.00	0.00	0.00	0.00	0.00
2	LOAD1							
		0.00	1.51	0.00	0.00	0.00	0.00	0.00
		35.36	1.51	0.00	0.00	0.00	0.00	0.00
		70.71	1.51	0.00	0.00	0.00	0.00	0.00
		106.07	1.51	0.00	0.00	0.00	0.00	0.00
		141.42	1.51	0.00	0.00	0.00	0.00	0.00
3	LOAD1							
		0.00	-0.96	0.00	0.00	0.00	0.00	0.00
		25.00	-0.96	0.00	0.00	0.00	0.00	0.00
		50.00	-0.96	0.00	0.00	0.00	0.00	0.00
		75.00	-0.96	0.00	0.00	0.00	0.00	0.00
		100.00	-0.96	0.00	0.00	0.00	0.00	0.00

Table 8.2 (continued)

FRAME LOAD	LOC	P	V2	V3	T	M2	M3
4 LOAD1							
	0.00	3.98	0.00	0.00	0.00	0.00	0.00
	25.00	3.98	0.00	0.00	0.00	0.00	0.00
	50.00	3.98	0.00	0.00	0.00	0.00	0.00
	75.00	3.98	0.00	0.00	0.00	0.00	0.00
	100.00	3.98	0.00	0.00	0.00	0.00	0.00

Note: The computer output shows extremely small values for the shear forces V2, V3 and the bending moments M2, M3 (in the range of numerical accuracy). These values have been set equal to zero in Table 8.2.

Plot Displacements: From the main menu select:
DISPLAY>SHOW DEFORMED SHAPE, then select:
FILE>PRINT GRAPHICS.

(Figure 8.4 shows the deformed shape of the truss.)

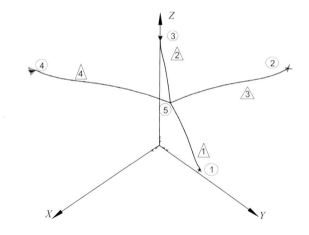

Fig. 8.4 Deformed shape for the space truss of Illustrative Example 8.2

Plot Axial Forces: From the main menu select:
DISPLAY>SHOW ELEMENT FORCES/STRESSES>FRAMES.
In the member force diagram for frames, select:
Axial Force and Fill Diagram. Click OK.

(Figure 8.5 shows the axial force diagram for the members of the truss.)

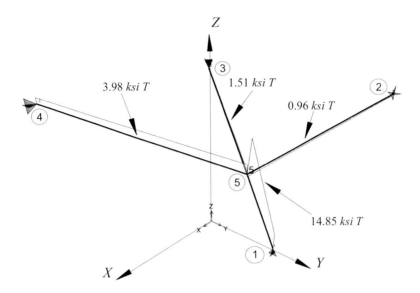

Fig. 8.5 Axial force diagram for the members of the space truss of Illustrative
Example 8.2

8.7   Dynamic Analysis of Space Trusses

The analysis of structures modeled as *Space Trusses* subjected to time dependent
forces requires the development of the system stiffness and the mass matrices. The
stiffness matrix for an element of a space truss in local and in global coordinates is
given, respectively, by eqs. (8.1) and (8.13). Thus it remains only to develop the
mass matrix.

8.7.1   Inertial Properties: Lumped Mass Matrix
The lumped element mass matrix for a uniform element of a space truss is a
diagonal matrix having one-half of the total mass of the element ($\overline{m}L/2$) allocated
to the three nodal coordinates at each end of the member. Using the special notation
for a diagonal matrix, the lumped mass matrix, $[m]_{\mathrm{L}}$, for a uniform element of space
truss may be written as

$$[m]_{\mathrm{L}} = \frac{\overline{m}L}{2}\lceil 1 \quad 1 \quad 1 \quad 1 \quad 1 \quad 1 \rfloor. \tag{8.15}$$

8.7.2   Inertial Properties: Consistent Mass Matrix
The consistent mass matrix $[m]_{\mathrm{C}}$ for a uniform member of a space truss may be
written as an extension of the mass matrix for an element of a plane truss [eq.(7.17)]
as follows:

$$[m]_C = \frac{\overline{m} \, L}{6} \begin{bmatrix} 2 & 0 & 0 & 1 & 0 & 0 \\ 0 & 2 & 0 & 0 & 1 & 0 \\ 0 & 0 & 2 & 0 & 0 & 1 \\ 1 & 0 & 0 & 2 & 0 & 0 \\ 0 & 1 & 0 & 0 & 2 & 0 \\ 0 & 0 & 1 & 0 & 0 & 2 \end{bmatrix}. \tag{8.16}$$

Also, analogously, to eq. (7.20) for an element of a plane truss, the mass matrix $[\overline{m}]$ for an element of a space truss in reference to global coordinates is equal to the mass matrix of the element $[m]$ in local coordinates. Thus, as it may be verified by performing the operations indicated in eq. (7.19), there is no need to transform the element mass matrix since

$$[\overline{m}] = [m]. \tag{8.17}$$

The system mass matrix, $[M]$, is then assembled using the same method employed in assembling the system stiffness matrix, that is, by transferring the coefficients of the element mass matrices to the appropriate locations in the system mass matrix.

8.8   Dynamic Analysis of Space Trusses using SAP2000

The dynamic equilibrium conditions at the nodes of the space truss result in the differential equations of motion that in matrix notation may be written as follows:

$$[M]\{\ddot{u}\} + [C]\{\dot{u}\} + [K]\{u\} = \{F(t)\} \tag{8.18}$$

in which $[M]$, $[C]$ and $[K]$ are, respectively, the system mass, damping, and stiffness matrices; $\{u\}$, $\{\dot{u}\}$ and $\{\ddot{u}\}$ are, respectively, the displacement, velocity, and acceleration vectors at the nodal coordinates; and $\{F(t)\}$ is the vector of external nodal forces.

As discussed in the preceding chapters, the practical way of evaluating damping is to prescribe damping ratios relative to the critical damping for each mode of vibration. Consequently, when eq. (8.18) is solved using the modal superposition method, the specified modal damping ratios are introduced directly into the modal equations. In that case, there is no need for explicitly obtaining the system damping matrix $[C]$. However, this matrix is required when the solution of eq. (8.18) is sought by other methods of solution, such as the step-by-step integration method, for which the system damping matrix $[C]$ can be obtained from the specified modal damping ratios by any of the methods discussed in the reference (Paz 2004).

Illustrative Example 8.3

For the three-dimensional truss shown in Figure 8.6(a), use SAP2000 to determine: (1) the first three natural frequencies and periods; and (2) the response due to the impulsive force plotted in Figure 8.6(b) applied horizontally parallel to the plane XY and bisecting the angle between the axes X and Y.

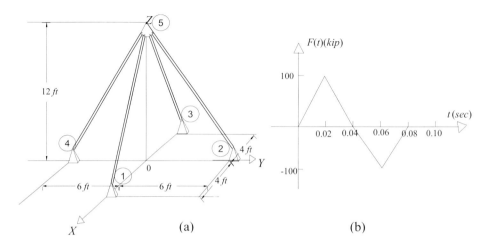

Fig. 8.6 (a) Three-dimensional space truss of Illustrative Example 8.3
(b) Impulsive force applied at joint 5

The cross-sectional area for all the members is 10 in^2, mass density $\rho = 0.01$ $(lb.sec^2/in)/in^3$ and the material modulus of elasticity is 29,500 ksi. In addition, the truss carries a concentrated mass $m = 10$ $lb.sec^2/in$ at the top joint. Assume a damping ratio $\xi = 0.05$ for all the modes.

Solution:

Begin: Open SAP2000
         Hint: Maximize both windows for a full view of all windows.

Units: In the lower right-hand corner of the screen, use the drop-down menu to select "kip-in".

Model: From the main menu select:
         FILE>NEW MODEL
         Number of Grid Spaces:
             X – direction = 2
             Y – direction = 2
             Z – direction = 1

Grid Spacing:
    X – direction = 48
    Y – direction = 72
    Z – direction = 144
Click OK.

Edit: Click the 3-D view.
From the main menu select:
VIEW>3D VIEW.
To set a frontal view of the X axis, rotate the Plan View to 30 degrees by typing 30 in the Plan View box. Click OK.

Draw points: From the main menu select:
DRAW>ADD SPECIAL JOINT.
Click once to draw points at coordinates (48,0,0), (0,72,0), (−48,0,0), (0,−72,0), and (0,0,144).
Note: As the mouse moves over the grid intersections, the coordinates of the mouse's location are shown in the lower right-corner of the screen.
Press the Esc key to exit drawing points.

Label: From the main menu select:
VIEW>SET ELEMENTS.
Click on "Labels Points". Then click OK.

Draw members: From the main menu select:
DRAW>DRAW FRAME ELEMENT.
Click on point 1 and drag the mouse to point 5 and click twice. Then press the Esc key. Now repeat successively these commands to draw frame elements from joints 2, 3 and 4 to joint 5.

Boundary: Select Joints 1, 2, 3 and 4 with a click of the mouse, and the from the main menu select:
ASSIGN>JOINT>RESTRAINTS.
Select the Restraint translations in the 1, 2 and 3 directions. Click OK.

Material: From the main menu select:
DEFINE> MATERIALS.
Select STEEL and click Modify/Show Material.
Enter:
        Mass per Unit Volume = 0.00001
        Weight per Unit Volume = 0
        Modulus of Elasticity = 29500
        Accept remaining default values
Then click OK, OK.

Select Sections: From the main menu select:
  DEFINE>FRAME SECTIONS.
  Click Add/Wide Flange and scroll down to select Add General (Section).
    Enter: Area = 10.0
    Set equal to zero all other cross-sectional properties.
  Then click OK, OK.

Assign Sections: Select all the members of the truss by a click on each member,
  and from the main menu select:
  ASSIGN>FRAME>SECTIONS.
  Select FSEC2. Then click OK.

Concentrated mass: Select joint 5 with a click and then from the main menu select:
  ASSIGN>JOINT>MASSES.
  Enter Mass:
    Direction 1 = 0.01
    Direction 2 = 0.01
    Direction 3 = 0.01. Then click OK.

Load: From the main menu select:
  DEFINE>STATIC LOAD CASES.
    Load = Load1
    Type = Live
    Self-Weight Multiplier = 0
  Click Change Load and then click OK.

Time history Function: From the main menu select:
  DEFINE>TIME HISTORY FUNCTIONS.
  Click on Add New Function.
  Enter: Time = 0 and Value = 0, then click ADD.
  Enter: Time = 0.02 and Value = 1, then click ADD.
  Enter: Time = 0.06 and Value = −1, then click ADD.
  Enter: Time = 0.08 and Value = 0, then click ADD.
  Enter: Time = 0.40 and Value = 0, then click ADD.
  Then click OK, OK.

Time History Case: From the main menu select:
  DEFINE>TIME HISTORY CASES.
  Click Add New History button.
  Accept the default HIST1.
  Select Analysis Type: Linear
  Click for Modal Damping, the Modify/Show button.
    Set damping in all modes = 0.05 and click OK.
  Check "Envelopes".
  Enter Number of Output Time Steps = 400

Output Time Step Size = 0.001
Load Assignment:
    Select LOAD1 and FUNC1
    Then click ADD. Then click OK, OK.

Assign Load: Click on joint 5, and from the main menu select:
    ASSIGN>JOINT STATIC LOADS>FORCES.
    Select LOAD1 and enter:
        Force Global Direction X = 70.7
        Force Global Direction Y = 70.7
    Then click OK.

Options: From the main menu select:
    ANALYZE>SET OPTIONS.
    Click only on UX, UY and UZ for available DOFs.
    Click on "Dynamic Analysis" and on "Set Dynamic Parameters".
    Enter Number of Modes = 3. Then click OK, OK.

Analyze: From the main menu select:
    ANALYZE>RUN.
    Enter Filename "EXD 8.3". Then click SAVE.
    When the calculations are concluded, click OK.

Plot Time History Displacements: From the main menu select:
    DISPLAY>SHOW TIME HISTORY TRACES.
    Click on "Define Functions" and on "Add Input Functions".
    Then select Add Joint Disps/Forces.
    Click OK.
        Time History Joint Function: enter Joint ID = 5
        Vector Type: select Displ
        Component: select UX
        Click OK.
        Vector Type: select Displ
        Component: select UY
        Click OK, OK.
    In the List of Functions column, highlight Joint5 and click "Add" to move joint5 (X displacement component) to the Plot Function column.
    Highlight Joint5-1 to move joint 5 (Y-displacement component) to the Plot Function Column.
    Axis Labels enter:
        Horizontal: Time (sec)
        Vertical: X, Y displ. components of Joint 5 (*in*)
    Click "Display".

To obtain a print of the plot, select:
FILE and "Print Graphics".
(Figure 8.7 reproduces the plot time displacement components of joint 5 in
the X- and the Y-directions.)

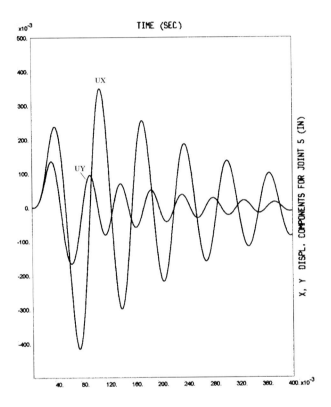

Fig. 8.7 Displacement components of Joint 5 in X, Y directions

Print input tables: From the main menu select:
FILE>INPUT TABLES.
Click "Print to File" and accept the default filename shown on the screen
or enter a new name and click "Save to File".

Use Notepad or other editor to retrieve and edit the file
C:\SAP2000\ > XD 8.3.text containing the input tables.

(Table 8.3 provides the edited Input Tables for Illustrative Example 8.3.)

Table 8.3 Edited input tables for Illustrative Example 8.3 (Units: kip-in)

```
STATIC LOAD CASES
   STATIC      CASE    SELF WT
   CASE        TYPE    FACTOR
   LOAD1       LIVE    0.0000

TIME HISTORY CASES

   HISTORY    HISTORY  NUMBER OF  TIME STEP
   CASE       TYPE   TIME STEPS  INCREMENT
   HIST1      LINEAR       400    0.00100

JOINT DATA

   JOINT    GLOBAL-X      GLOBAL-Y      GLOBAL-Z    RESTRAINTS

     1       48.00000       0.00000       0.00000    1 1 1 0 0 0
     2        0.00000      72.00000       0.00000    1 1 1 0 0 0
     3      -48.00000       0.00000       0.00000    1 1 1 0 0 0
     4        0.00000     -72.00000       0.00000    1 1 1 0 0 0
     5        0.00000       0.00000     144.00000    0 0 0 0 0 0

JOINT MASS DATA

   JOINT   M-U1       M-U2       M-U3       M-R1     M-R2     M-R3

     5   1.000E-02  1.000E-02  1.000E-02   0.000    0.000    0.000

FRAME ELEMENT DATA

FRAME  JNT-1  JNT-2  SECTION   LENGTH

    1     1     5     FSEC2     151.789
    2     2     5     FSEC2     160.997
    3     3     5     FSEC2     151.789
    4     4     5     FSEC2     160.997

JOINT FORCES  Load Case LOAD1
```

JOINT	GLOBAL-X	GLOBAL-Y	GLOBAL-Z	GLOBAL-XX	GLOBAL-YY	GLOBAL-ZZ
5	70.7	70.7	0.000	0.000	0.000	0.000

Natural Periods and Frequencies:

Use Notepad or other word editor to open C:\SAP2000>EXD 8.3.OUT. The natural Period and Frequency values may be found in this file.

(Table 8.4. provides the edited values of the Natural Periods and frequencies.)

Table 8.4 Natural Periods and Frequencies of the truss of Illustrative Example 8.3

MODE	PERIOD (sec)	FREQUENCY (cps)	FREQUENCY (rads/sec)	EIGENVALUE $(rad/sec)^2$
1	0.0647	15.4441	97.038	9416.415
2	0.0471	21.2075	133.250	17755.762
3	0.0159	62.8149	394.678	155770.781

Modes Shapes: From the main menu select:
> DISPLAY>SHOW MODE SHAPE.
> Modal Shape 1: Click the "Wire Shadow" and "Cubic Curve" options.
> Scale Factor = 1. Then click OK.

(To view Modes 2 and 3, repeat the above commands, or alternatively, click on the "+" icon located in the lower-right corner of the window.)

Note: Numerical values for a mode may be viewed directly on the screen by right clicking on a joint in the modal shape.

8.9   Practice Problems

Problem 8.1
For the space truss shown in Figure. P8.1 determine:

(a) Joint displacements
(b) Axial member forces
(c) Support reactions

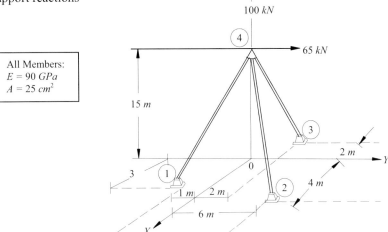

Fig. P8.1

Problem 8.2

For the space truss shown in Figure P8.2 determine:

(a) Joint displacements
(b) Axial member forces
(c) Support reactions

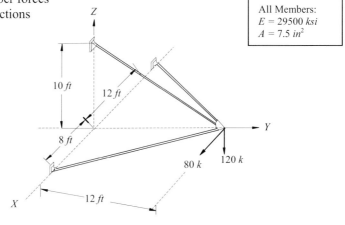

All Members:
$E = 29500\ ksi$
$A = 7.5\ in^2$

Fig. P8.2

Problem 8.3

For the space truss shown in Figure P8.3 determine:

(a) Joint displacements
(b) Member axial forces
(c) Support reactions

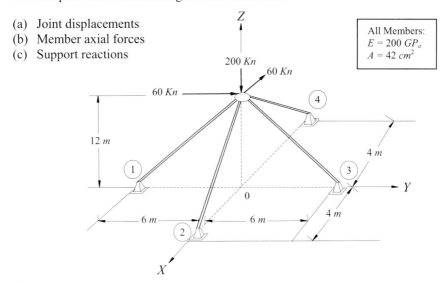

All Members:
$E = 200\ GP_a$
$A = 42\ cm^2$

Fig. P8.3

Problem 8.4
Solve the truss in Problem 8.3 for the additional effect of a vertical settlement of 2 *in.* at the support on joint 2.

Problem 8.5
For the space truss shown in Figure. P8.5 determine:

(a) Joint displacements
(b) Axial member forces
(c) Support reactions

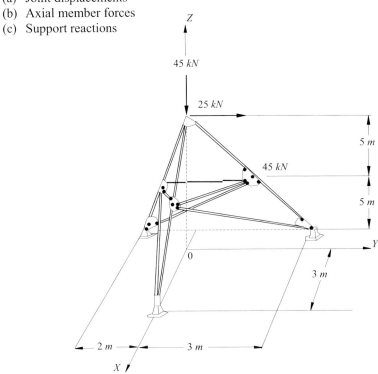

Fig. P8.5

9 Introduction to Finite Element Method

9.1   Introduction

In the preceding chapters, we have considered the matrix analysis of structures modeled as beams, frames, or trusses. The elements of all these types of structures are described by a single coordinate along their longitudinal axis. These are structures with unidirectional elements, called skeletal structures. In general, they consist of individual members or elements connected at points designated as nodal points or joints. For these types of structures, the behavior of each element is considered independently through the calculation of the element stiffness matrices. These matrices are then assembled into the system stiffness matrix in such a way that the equilibrium of forces and the compatibility of displacements are satisfied at each nodal point. The analysis of such structures is commonly known as the Matrix Structural Analysis and could be applied equally to static and dynamic problems.

The structures presented in this chapter are continuous structures that are conveniently idealized as consisting of two-dimensional elements (that most naturally may be extended to three-dimensional structures) connected only at the selected nodal points. For example, Figure 9.1 shows a thin plate idealized with plane triangular elements. The method of analysis for such idealized structures is known as the *Finite Element Method* (*FEM*). This is a powerful method for the analysis of structures with complex geometrical configurations, material properties, or loading conditions. This method is entirely analogous to Matrix Structural Analysis for skeletal structures (beams, frames, and trusses) presented in the preceding chapters. The Finite Element Method differs from the Matrix Structural Method only in two respects: (1) the selection of elements and nodal points is not naturally or clearly established by the geometry as it is for skeletal structures; and (2) the displacements at interior points of an element are expressed by *interpolating functions* and not by an exact analytical relationship as it is in the Matrix Structural Method. Further-more, for skeletal structures, the displacement of an interior point of an element is governed by an ordinary differential equation, while for a continuous two-dimensional element it is governed by a partial differential equation of much greater complexity.

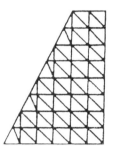

Fig. 9.1 Finite Element Method of a thin plate modeled using triangular elements

9.2   Plane Elasticity Problems

Plane elasticity problems refer to plates that are loaded in their own planes. Out-of-plane displacements are induced when plates are loaded by normal forces that are perpendicular to the plane of the plate, such problems are generally referred to as *plate bending*. (Plate bending is considered in Section 9.4.)

There are two different types of plane elasticity problems: (1) plane stress problems and (2) plane strain problems. In the plane stress problems, the plate is thin relative to the other dimensions and the stresses normal to the plane of the plate are not considered. Figure 9.2 shows a perforated strip-plate in tension as an example of a plane stress problem.

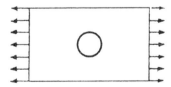

Fig. 9.2   Perforated plate tension element as an example of a structural member loaded in plane stress

For plane strain problems, the strain normal to the plane of loading is suppressed and assumed to be zero. Figure 9.3 shows a transverse slice of a retaining wall as an example of a plane strain problem.

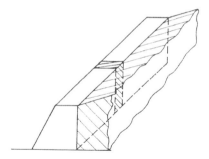

Fig. 9.3   Retaining wall showing a plate slice as an example of plane
                 strain conditions

In the analysis of plane elasticity problems, the continuous plate is idealized as
finite elements interconnected at their nodal points. The displacements at these
nodal points are the basic unknowns as are the displacements at the joints in the
analysis of beams, frames, or trusses. Consequently, the first step in the application
of the FEM is to model the continuous system into discrete elements. The most
common geometric elements used for plane elasticity problems are triangular,
rectangular, or quadrilateral, although other geometrical shapes also could be used.

9.3   Implementation of FEM

The following process is presented to describe the application of the FEM for the
analysis of continuous structural problems:

Step 1: Modeling the structure.

Figure 9.4 shows a thin cantilever plate modeled with triangular plane stress
triangular elements.

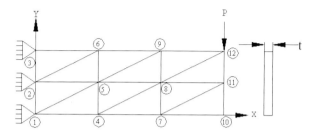

Fig. 9.4 Thin cantilevered plate modeled using plane stress triangular elements

The plate is supporting an external force P at its upper right end in addition to a distributed force on the plane of the plate. This distributed force is generally known as body force and is expressed as a force per unit of volume. Its components on an element along the x and y directions, bx and by, are conveniently arranged as a vector $\{b\}e$.

Consider in Figure 9.5 a triangular element isolated from a plane stress plate. This figure shows the element nodal forces $\{P\}_e$ and corresponding nodal displacements $\{q\}_e$ with components in the x and y directions at the three joints of this triangular element. Figure 9.5 also shows the components $b_x dV$ and $b_y dV$ of the body force applied at an interior location of the plate of volume dV.

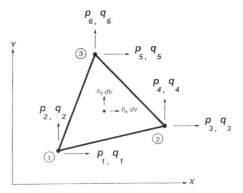

Fig. 9.5   Triangular plate element showing nodal forces P_i and corresponding nodal displacements q_i at the three joints

The triangular element in plane elasticity problems with two nodal coordinates at each of the three joints results in a total of six nodal coordinates. Therefore, for this element, the nodal displacement vector $\{q\}_e$ and corresponding nodal force vector $\{P\}_e$ have six components. These vectors may be written as:

$$\{q\}_e = \begin{Bmatrix} q_1 \\ q_2 \\ q_3 \\ q_4 \\ q_5 \\ q_6 \end{Bmatrix} \qquad \{P\}_e = \begin{Bmatrix} P_1 \\ P_2 \\ P_3 \\ P_4 \\ P_5 \\ P_6 \end{Bmatrix} \qquad (9.1)$$

Consequently, for the plane triangular element with three nodes, the element stiffness matrix $[k]$ relating the nodal forces and the nodal displacements is of dimension 6 x 6.

Step 2: Selection of a suitable displacement function.

The displacements, $u = u(x,y)$ and $v = v(x,y)$, respectively, in the x and y directions at any interior point $P(x,y)$ of the triangular element are expressed approximately by polynomial functions, with a total of six coefficients equal in number to the possible nodal displacements. In this case, the simplest expressions for the displacement functions, $u(x,y)$ and $v(x,y)$, at an interior point of the triangular element are:

$$u(x,y) = c_1 + c_2 x + c_3 y$$
$$v(x,y) = c_4 + c_5 x + c_6 y \tag{9.2}$$

or

$$\{q(x,y)\} = [g(x,y)](c) \tag{9.3}$$

where $\{c\}$ is a vector containing the six coefficients c_i in eq. (9.2); $[g(x,y)]$ is a matrix function of the position of a point in the element having coordinates (x,y); and $\{q(x,y)\}$ is a vector with the displacement components $u(x,y)$ and $v(x,y)$ at an interior point along the x and y directions, respectively.

Step 3: Displacements $\{q(x,y)\}$ at a point in the element are expressed in terms of the nodal displacements $\{q\}_e$.

The evaluation of eq. (9.3) for the displacements of the three nodes of the triangular element results in

$$\begin{Bmatrix} q_1 \\ q_2 \\ q_3 \\ q_4 \\ q_5 \\ q_6 \end{Bmatrix} = \begin{bmatrix} 1 & x_1 & y_1 & 0 & 0 & 0 \\ 0 & 0 & 0 & 1 & x_1 & y_1 \\ 1 & x_2 & y_2 & 0 & 0 & 0 \\ 0 & 0 & 0 & 1 & x_2 & y_2 \\ 1 & x_3 & y_3 & 0 & 0 & 0 \\ 0 & 0 & 0 & 1 & x_3 & y_3 \end{bmatrix} \begin{Bmatrix} c_1 \\ c_2 \\ c_3 \\ c_4 \\ c_5 \\ c_6 \end{Bmatrix}. \tag{9.4}$$

Equation (9.4) is solved for the coefficients c_1 through c_6 in terms of the nodal displacements q_1 through q_6. Then these coefficients are substituted back into eq. (9.2) to obtain in matrix notation:

$$\begin{Bmatrix} u(x,y) \\ v(x,y) \end{Bmatrix} = \begin{bmatrix} N_1 & 0 & N_2 & 0 & N_3 & 0 \\ 0 & N_1 & 0 & N_2 & 0 & N_3 \end{bmatrix} \begin{Bmatrix} q_1 \\ q_2 \\ q_3 \\ q_4 \\ q_5 \\ q_6 \end{Bmatrix} \tag{9.5}$$

where the shape functions N_1, N_2 and N_3 are given by

$$N_1 = \frac{1}{D_0} \left[(x_2 y_3 - x_3 y_2) + (y_2 - y_3)x + (x_3 - x_2)y \right]$$

$$N_2 = \frac{1}{D_0} \left[(x_3 y_1 - x_1 y_3) + (y_3 - y_1)x + (x_1 - x_3)y \right] \tag{9.6}$$

$$N_3 = \frac{1}{D_0} \left[(x_1 y_2 - x_2 y_1) + (y_1 - y_2)x + (x_2 - x_1)y \right]$$

in which

$$D_0 = x_2 y_3 - x_3 y_2 + x_1 \left(y_2 - y_3 \right) + y_1 (x_3 - x_2). \tag{9.7}$$

The expression for D_0 is equal to twice the area of the triangular element. Equation (9.5) may be written in condensed notation as

$$\{ q(x,y) \} = \begin{Bmatrix} u(x,y) \\ v(x,y) \end{Bmatrix} = \left[f(x,y) \right] \{ q \}_e \tag{9.8}$$

where $f(x,y)$ is a function of the coordinates at a point in the plate element.

Step 4: Relationship between strain, $\{\varepsilon(x,y)\}$ at any point within the element to the displacements $\{q(x,y)\}$ and hence to the nodal displacements $\{q\}_e$.

It is shown in the Theory of Elasticity (Timoshenko and Goodier 1970) that for the strain vector $\{\varepsilon(x,y)\}$ with axial strain components ε_x, ε_y, the shearing strain component γ_{xy} is given in terms of derivatives of the displacement function by

$$\{\varepsilon(x,y)\} = \left\{ \begin{array}{c} \varepsilon_x \\ \varepsilon_y \\ \gamma_{xy} \end{array} \right\} = \left\{ \begin{array}{c} \dfrac{\partial u}{\partial x} \\[2mm] \dfrac{\partial v}{\partial y} \\[2mm] \dfrac{\partial u}{\partial y} + \dfrac{\partial v}{\partial x} \end{array} \right\} \tag{9.9}$$

in which $u = u(x,y)$ and $v = v(x,y)$ refer to displacement functions in the x and y directions, respectively.

The strain components may be expressed in terms of the nodal displacements $\{q\}_e$ by substituting into eq. (9.9) the derivatives of $u(x,y)$ and $v(x,y)$ given by eqs. (9.5) and (9.6) to obtain:

$$\left\{ \begin{array}{c} \varepsilon_x \\ \varepsilon_y \\ \gamma_{xy} \end{array} \right\} = \frac{1}{D_0} \begin{bmatrix} (y_2 - y_3) & 0 & (y_3 - y_1) & 0 & (y_1 - y_2) & 0 \\ 0 & (x_3 - x_2) & 0 & (x_1 - x_3) & 0 & (x_2 - x_1) \\ (x_3 - x_2) & (y_2 - y_3) & (x_1 - x_3) & (y_3 - y_1) & (x_2 - x_1) & (y_1 - y_2) \end{bmatrix} \left\{ \begin{array}{c} q_1 \\ q_2 \\ q_3 \\ q_4 \\ q_5 \\ q_6 \end{array} \right\} \tag{9.10}$$

or in condensed notation

$$\{\varepsilon(x, y)\} = [B]\{q\}_e \tag{9.11}$$

in which the matrix $[B]$ is defined in eq. (9.10).

Step 5: Relationship between internal stresses $\{\sigma(x, y)\}$ to strains $\{\varepsilon(x, y)\}$ and hence to the nodal displacements $\{q\}_e$.

For plane elasticity problems, the relationship between the normal stresses σ_x, σ_y and shearing stress τ_{xy} and the corresponding strains $\varepsilon_x, \varepsilon_y$, and γ_{xy} may be expressed, in general, as follows:

$$\left\{ \begin{array}{c} \sigma_x \\ \sigma_y \\ \tau_{xy} \end{array} \right\} = \begin{bmatrix} d_{11} & d_{12} & 0 \\ d_{21} & d_{22} & 0 \\ 0 & 0 & d_{33} \end{bmatrix} \left\{ \begin{array}{c} \varepsilon_x \\ \varepsilon_y \\ \gamma_{xy} \end{array} \right\} \tag{9.12}$$

or in a short matrix notation as:

$$\{\sigma(x, y)\} = [D]\{\varepsilon(x, y)\} \qquad (9.13)$$

where the material matrix $[D]$ is defined in eq. (9.12).

The substitution into eq. (9.13) of $\{\varepsilon(x, y)\}$ from eq. (9.11) gives the desired relationship between stresses $\{\sigma(x, y)\}$ at a point in the element and the displacements $\{q\}_e$ at the nodes as:

$$\{\sigma(x, y)\} = [D][B]\{q\}_e. \qquad (9.14)$$

The coefficients of the matrix $[D]$ in eq. (9.14) have different expressions for plane stress problems and for plane strain problems. These expressions are as follows:

a)   Plane stress problems:

$$[D] = \frac{E}{1-v^2} \begin{bmatrix} 1 & v & 0 \\ v & 1 & 0 \\ 0 & 0 & \dfrac{1-v}{2} \end{bmatrix} \qquad (9.15)$$

b)   Plane strain problems:

$$[D] = \frac{E}{(1+v)(1-2v)} \begin{bmatrix} 1-v & v & 0 \\ v & 1-v & 0 \\ 0 & 0 & \dfrac{1-2v}{2} \end{bmatrix} \qquad (9.16)$$

in which E is the modulus of elasticity and v is the Poisson's ratio.

Step 6: Element stiffness matrix.

Use is made of the *Principle of Virtual Work* to establish the expression for the element stiffness matrix $[k]_e$. This principle states that for structures in equilibrium subjected to small compatible virtual displacements, the virtual work δW_E of the external forces is equal to the virtual work of internal stresses δW_I, that is

$$\delta W_I = \delta W_E \qquad (9.17)$$

In applying this principle, we consider a virtual displacement vector $\delta\{q(x, y)\}$ of the displacement function $\{q(x, y)\}$. Hence, by eq. (9.8)

$$\delta\{q(x, y)\} = [f(x, y)]\delta\{q\}_e. \qquad (9.18)$$

The virtual internal work δW_I during this applied virtual displacement is then given by the product of the virtual strain and the stresses integrated over the volume of the element:

$$\delta\, W_1 = \int_V \delta\{\varepsilon(x,y)\}^T \{\sigma(x,y)\}\, dV. \qquad (9.19)$$

The virtual external work δW_E includes the work of the body forces $\{b\}_e\, dV$ and that of the nodal forces $\{P\}_e$ shown in Figure 9.5. The total external virtual work is then equal to the product of these forces times the corresponding virtual displacements, that is

$$\delta\, W_E = \int_V \delta\{q(x,y)\}^T \{b\}_e\, dV + \delta\{q\}_e^T \{P\}_e. \qquad (9.20)$$

The substitution into eq. (9.17) of δW_I and δW_E, respectively, from eqs. (9.19) and (9.20) results in

$$\int_V \delta\{\varepsilon(x,y)\}^T \{\sigma(x,y)dV = \int_V \delta\{q(x,y)\}^T \{b\}_e\, dV + \delta\{q\}_e^T \{P\}_e \qquad (9.21)$$

Finally, substituting into eq. (9.21) $\{\sigma(x,y)\}$, $\delta\{q(x,y)\}^T$ and $\delta\{\varepsilon(x,y)\}^T$, respectively, from eqs. (9.13), (9.18), and (9.11) we obtain, after cancellation of the common factor $\delta\{q\}_e^T$, the stiffness equation of the element:

$$[k]\{q\}_e = \{P\}_e + \{P_b\}_e \qquad (9.22)$$

in which the element stiffness matrix $[k]$ is given in general by

$$[k] = \int_V [B]^T [D][B] dV \qquad (9.23)$$

and the vector of equivalent nodal forces $\{P_b\}_e$ due to the body forces by

$$\{P_b\}_e = \int_V [f(x,y)]^T \{b\}_e\, dV. \qquad (9.24)$$

For plane elasticity problems, the matrices $[B]$ and $[D]$ result in constant expressions. Consequently, by eq. (9.23), the element stiffness matrix is given by

$$[k] = [B]^T [D][B]\, A\, t \qquad (9.25)$$

where t is the thickness of the plate element and A its area.

Step 7: Assemblage of the reduced system stiffness matrix $[K]_R$ and of the reduced system equivalent nodal force vector $\{F\}_R$ at the free nodal coordinates.

The reduced system stiffness matrix $[K]_R$ is assembled by transferring the coefficients of the element stiffness matrices to appropriate locations in the system stiffness matrix by exactly the same process as was used in the previous chapters to assemble the system stiffness matrix for skeletal structures. The system force vector $\{F\}_R$ is assembled from the element equivalent nodal forces vector in addition to external forces F_n applied directly to the nodal coordinates. Hence, we may symbolically write

$$[K]_R = \Sigma[k] \tag{9.26}$$

and

$$\{F\}_R = \Sigma\{P_b\}_e + \Sigma F_n. \tag{9.27}$$

Step 8: Solution of the system stiffness equation.

The reduced system stiffness equation is then given by

$$[K]_R\{u\} = \{F\}_R \tag{9.28}$$

in which $[K]_R$ is the reduced system stiffness matrix, $\{F\}_R$ is the reduced system force vector, and $\{u\}$ is the vector of the unknown displacements at the free nodal coordinates. Thus, the solution of eq. (9.28) provides the displacement vector $\{u\}$ at the free nodal coordinates.

Step 9: Determination of nodal stresses.

The final step is the calculation of stresses. These stresses can be calculated from the element nodal displacements $\{q\}_e$ identified from the system nodal displacements $\{u\}$ already determined. The element stresses are given by eq. (9.14) as

$$\{\sigma(x, y)\} = [D][B]\{q\}_e. \tag{9.29}$$

For plane elasticity problems, using triangular elements and linear interpolating functions, both matrices $[D]$ and $[B]$ are constants. Therefore, in these problems the stress through the element or at its nodal coordinates is constant.

Illustrative Example 9.1

Consider the deep cantilever beam of rectangular cross-section supporting a load of 3 *kips* as shown in Figure 9.6(a). Model this structure with two plane stress triangular elements as shown in Figure 9.6(b) and perform the finite element analysis.

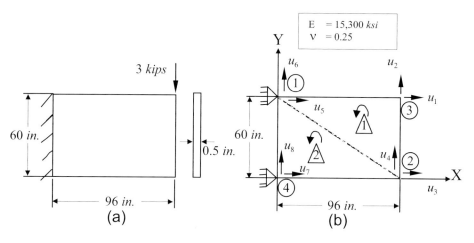

Fig. 9.6   Deep cantilever beam for Illustrative Example 9.1 modeled using two plane stress triangular elements

Solution:
Step 1: Element stiffness matrices.

ELEMENT 1

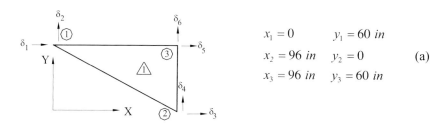

$$x_1 = 0 \qquad y_1 = 60 \ in$$
$$x_2 = 96 \ in \qquad y_2 = 0 \qquad (a)$$
$$x_3 = 96 \ in \qquad y_3 = 60 \ in$$

Fig. 9.7 Element 1 showing nodal coordinates

From eq. (9.7):

$$D_0 = (60)(96) = 5760 \ in^2 = 2 \times \ area \qquad (b)$$

From eqs. (9.10) and (9.11):

$$[B]_1 = \frac{1}{5760} \begin{bmatrix} -60 & 0 & 0 & 0 & 60 & 0 \\ 0 & 0 & 0 & -96 & 0 & 96 \\ 0 & -60 & -96 & 0 & 96 & 60 \end{bmatrix}. \qquad (c)$$

From eq. (9.15):

$$[D]_1 = \frac{15300}{0.9375} \begin{bmatrix} 1 & 0.25 & 0 \\ 0.25 & 1 & 0 \\ 0 & 0 & 0.375 \end{bmatrix}. \tag{d}$$

The element stiffness matrix for element 1, $[k]_1$, is then obtained by substituting into eq. (9.25) the matrices $[B]$ and $[D]$, respectively, from eqs. (c) and (d):

$$[k]_1 = \frac{15300}{5760^2 \times 0.9375} \begin{bmatrix} -60 & 0 & 0 \\ 0 & 0 & -60 \\ 0 & 0 & -96 \\ 0 & -96 & 0 \\ 60 & 0 & 96 \\ 0 & 96 & 60 \end{bmatrix} \begin{bmatrix} 1 & 0.25 & 0 \\ 0.25 & 1 & 0 \\ 0 & 0 & 0.375 \end{bmatrix}$$

$$\begin{bmatrix} -60 & 0 & 0 & 0 & 60 & 0 \\ 0 & 0 & 0 & -96 & 0 & 96 \\ 0 & -60 & -96 & 0 & 96 & 60 \end{bmatrix} (2880)(0.5)$$

or

$$[k]_1 = \begin{matrix} & 5 & 6 & 3 & 4 & 1 & 2 \\ & \begin{bmatrix} 2550 & 0 & 0 & 1020 & -2550 & -1020 \\ 0 & 956 & 1530 & 0 & -1530 & -956 \\ 0 & 1530 & 2448 & 0 & -2448 & -1530 \\ 1020 & 0 & 0 & 6528 & -1020 & -6528 \\ -2550 & -1530 & -2448 & -1020 & 4948 & 2550 \\ -1020 & -956 & -1530 & -6528 & 2550 & 7484 \end{bmatrix} & \begin{matrix} 5 \\ 6 \\ 3 \\ 4 \\ 1 \\ 2 \end{matrix} \end{matrix}. \tag{e}$$

ELEMENT 2

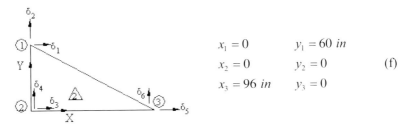

$$x_1 = 0 \qquad y_1 = 60 \; in$$
$$x_2 = 0 \qquad y_2 = 0 \qquad (f)$$
$$x_3 = 96 \; in \qquad y_3 = 0$$

Fig. 9.8 Element 2 showing nodal coordinates

Analogously, we obtain:

$$[B]_2 = \frac{1}{5760} \begin{bmatrix} 0 & 0 & -60 & 0 & 60 & 0 \\ 0 & 96 & 0 & -96 & 0 & 0 \\ 96 & 0 & -96 & -60 & 0 & 60 \end{bmatrix}$$

$$[D]_2 = \frac{15300}{0.975} \begin{bmatrix} 1 & 0.25 & 0 \\ 0.25 & 1 & 0 \\ 0 & 0 & 0.375 \end{bmatrix}.$$

Then substituting into eq. (9.25) results in

$$[k]_2 = [B]_2^T [D]_2 [B]_2 \; A \, t$$

or

$$[k]_2 = \begin{array}{cccccc} 5 & 6 & 7 & 8 & 3 & 4 \\ \begin{bmatrix} 2448 & 0 & -2448 & -1530 & 0 & 1530 \\ 0 & 6528 & -1020 & -6528 & 1020 & 0 \\ -2448 & -1020 & 4998 & 2550 & 2550 & -1530 \\ -1530 & -6528 & 2550 & 7484 & -1020 & -956 \\ 0 & 1020 & -2550 & -1020 & 2550 & 0 \\ 1530 & 0 & -1530 & -956 & 0 & 956 \end{bmatrix} & \begin{array}{c} 5 \\ 6 \\ 7 \\ 8 \\ 3 \\ 4 \end{array} \end{array}. \qquad \text{(g)}$$

1. Reduced system stiffness matrix.

   Transferring to the system stiffness matrix the coefficients from eqs. (e) and (g) corresponding to the free nodal coordinates (1, 2, 3, and 4) as indicated on the top and on the left of these matrices, yields

$$[K]_R = \begin{array}{cccc} 1 & 2 & 3 & 4 \\ \begin{bmatrix} 4948 & 2550 & -2448 & -1020 \\ 2550 & 7484 & -1530 & -6528 \\ -2448 & -1530 & 2448+2550 & 0 \\ -1020 & -6528 & 0 & 6528+956 \end{bmatrix} & \begin{array}{c} 1 \\ 2 \\ 3 \\ 4 \end{array} \end{array}. \qquad \text{(h)}$$

2. Reduced system force vector.

$$\{F\}_R = \begin{Bmatrix} 0 \\ -3 \\ 0 \\ 0 \end{Bmatrix}$$

(i)

3. Reduced system stiffness equation.

$$\{F\}_R = [K]_R \{u\}$$

(j)

Substituting into eq. (j) the reduced system stiffness matrix $[K]_R$ and the reduced system force vector $\{F\}_R$, respectively, from eqs. (h) and (i) gives:

$$\begin{Bmatrix} 0 \\ -3 \\ 0 \\ 0 \end{Bmatrix} = \begin{bmatrix} 4948 & 2550 & -2448 & -1020 \\ 2550 & 7484 & -1530 & -6528 \\ -2448 & -1530 & 4998 & 0 \\ -1020 & -6528 & 0 & 7484 \end{bmatrix} \begin{Bmatrix} u_1 \\ u_2 \\ u_3 \\ u_4 \end{Bmatrix}$$

{k}

4. System nodal displacements.

Solution of eq. (k) yields

$$u_1 = \quad 6.934 \ 10^{-4} \, in$$
$$u_2 = -27.481 \ 10^{-4} \, in$$
$$u_3 = \ -5.016 \ 10^{-4} \, in$$
$$u_4 = -23.025 \ 10^{-4} \, in$$

(l)

5. Element stresses.

Element stresses are given by eq. (9.29) as

$$\sigma(x, y) = [D][B]\{q\}_e$$

(9.29) Repeated

in which the element displacements $\{q\}_e$ are identified from Figure 9.4 with the numerical values of the system nodal displacements obtained from eq. (l) as follows:

ELEMENT 1

$$q_1 = 0 \quad q_2 = 0 \quad q_3 = -5.016 \ 10^{-4} \quad q_4 = -23.025 \ 10^{-4} \quad q_5 = 6.934 \ 10^{-4} \quad q_6 = -27.025 \ 10^{-4}$$

$$\begin{Bmatrix} \sigma_x \\ \sigma_y \\ \tau_{xy} \end{Bmatrix} = \frac{1530}{0.975x5760} \begin{bmatrix} 1 & 0.25 & 0 \\ 0.25 & 1 & 0 \\ 0 & 0 & 0.375 \end{bmatrix} \begin{bmatrix} -60 & 0 & 0 & 0 & 60 & 0 \\ 0 & 0 & 0 & -96 & 0 & 96 \\ 0 & -60 & -96 & 0 & 96 & 60 \end{bmatrix} \begin{Bmatrix} 0 \\ 0 \\ -5.016 \\ -23.025 \\ 6.934 \\ -27.025 \end{Bmatrix} 10^{-4}$$

or

$$\sigma_x = 0.00871 \ ksi$$
$$\sigma_y = -0.00762$$
$$\tau_{xy} = -0.00484$$

ELEMENT 2

$$q_1 = 0 \quad q_2 = 0 \quad q_3 = 0 \quad q_4 = 0 \quad q_5 = -5.016 \ 10^{-4} \quad q_6 = -23.025 \ 10^{-4}$$

$$\begin{Bmatrix} \sigma_x \\ \sigma_y \\ \tau_{xy} \end{Bmatrix} = \frac{1530}{0.975x5760} \begin{bmatrix} 1 & 0.25 & 0 \\ 0.25 & 1 & 0 \\ 0 & 0 & 0.375 \end{bmatrix} \begin{bmatrix} 0 & 0 & -60 & 0 & 60 & 0 \\ 0 & 96 & 0 & -96 & 0 & 0 \\ 96 & 0 & -96 & -60 & 0 & 60 \end{bmatrix} \begin{Bmatrix} 0 \\ 0 \\ 0 \\ 0 \\ -5.016 \\ -23.025 \end{Bmatrix} 10^{-4}$$

or

$$\sigma_x = -0.00819 \ ksi$$
$$\sigma_y = -0.00205$$
$$\tau_{xy} = -0.0141$$

Illustrative Example 9.2
Use SAP 2000 to solve Illustrative Example 9.1.

Solution:
The following commands are implemented in SAP2000:

Begin: Open SAP2000.
Hint: Maximize the window for a full view.

Units: Select "kip-in" in the drop down menu located on the lower right corner of the window.

Model: From the main menu select:
FILE>NEW MODEL.
Note that for this problem, Plane XZ is chosen as the plane of the plate.
Number of grid spaces:
   X direction = 2
   Y direction = 0
   Z direction = 2
Grid Spacing:
   X direction = 96
   Y direction = 1
   Z direction = 60

Edit: Maximize the 2-D view (X-Z screen). Then drag the figure to the center of the window using the PAN icon on the toolbar.

Draw: From the main menu select:
DRAW>DRAW SHELL ELEMENT.
Click at grid intersection $x = 0$ and $z = 60$, drag to the grid intersection $x = 96$ and $z = 0$ and click. Then click again at this intersection and drag to grid intersection $x = 96$ and $z = 60$. Finally, click again at this last point and drag to grid intersection $x = 0$ and $z = 60$ to complete the first shell element.

Now draw the second triangular element by clicking and dragging successively starting at the location of node 4 in Figure 9.6 continuing to node 2, then to node 1, and finally to close at node 1 to complete the second shell element.

Label: From the main menu select:
VIEW>SET ELEMENTS.
Click on Joint Labels and on Shell Labels. Then click OK.

Material: From the main menu select:
DEFINE >MATERIALS>OTHER.
Click Modify/Show Materials.
Select Material Name OTHER
Set: Modulus of elasticity = 15300
      Poisson's Ratio = 0.25. Then click OK, OK.

Shell Sections: From the main menu select:
   DEFINE>SHELL SECTIONS.
   Click Add New Sections.
   Set: Material = OTHER
      Thickness = 0.5
      Type = Shell. Then click OK, OK.

Assign Shells: Select shell 1 and 2 by clicking on the area of these shells.
   From the main menu select:
   ASSIGN>SHELL >SECTIONS.
   Click on SSEC2. Then click OK.

Boundary: Select Joints 1 and 4.
   From the main menu select:
   ASSIGN>JOINT >RESTRAINTS.
   Click to select restraints in all directions. Then click OK.

Load: From the main menu select:
   DEFINE>STATIC LOAD CASES.
   Change Dead Load to LIVE.
   Set the Self Weight Multiplier = 0.
   Click Change Load. Then click OK.

Assign Load: Select joint 3, and from the main menu select:
   ASSIGN>JOINT STATIC LOADS>FORCES.
   Set: Force Global Z = −3.0. Then click OK.

Analyze: From the main menu select:
   ANALYZE>SET OPTIONS.
   Set available degrees of freedom UX, UZ, and RY. Then click OK.

   From the main menu select:
   ANALYZE>RUN.
   Enter Filename "Example 9.2". Then click Save.
   When the calculation is complete, check for errors. Then click OK.

   Note: The displacement value may be viewed on the screen by right-clicking on a joint of the deformed shape diagram.

Plot: Select: FILE>PRINT GRAPHICS.

   (Figure 9.9 reproduces the deformed plot of the plate for Illustrative Example 9.2.)

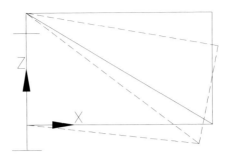

Fig. 9.9 Deformed plate of Illustrative Example 9.2

Input Tables: From the main menu select:
    FILE>PRINT INPUT TABLES.
    Click on Print to File. Then OK.
    Use a word editor (such as Word or Notepad) to edit the input tables and
    then print them.

(Table 9.1 provides the edited input tables for the plate of Illustrative Example 9.2.)

Table 9.1   Edited Input Tables for Illustrative Example 9.2 (Units: kips-inches)

JOINT DATA

JOINT	GLOBAL-X	GLOBAL-Y	GLOBAL-Z	RESTRAINTS
1	0.00000	0.00000	60.00000	1 1 1 1 1 1
2	96.00000	0.00000	0.00000	0 0 0 0 0 0
3	96.00000	0.00000	60.00000	0 0 0 0 0 0
4	0.00000	0.00000	0.00000	1 1 1 1 1 1

SHELL ELEMENT DATA

SHELL	JNT-1	JNT-2	JNT-3	JNT-4	SECTION	ANGLE	AREA
1	2	3	1	1	SSEC2	0.000	2880.000
2	4	2	1	1	SSEC2	0.000	2880.000

JOINT FORCES Load Case LOAD1

JOINT	GLOBAL-X	GLOBAL-Y	GLOBAL-Z	GLOBAL-XX	GLOBAL-YY	GLOBAL-ZZ
3	0.000	0.000	−3.000	0.000	0.000	0.000

Output Tables: From the main menu select:
        FILE>PRINT OUTPUT TABLES
        Click on "Print to File" and on "Append."

(Table 9.2 provides the edited output tables for the plate of Illustrative Example
9.2.)

Table 9.2   Edited output tables for Illustrative Example 9.2 (Units: kips-inches)

JOINT DISPLACEMENTS

JOINT LOAD	UX	UY	UZ	RX	RY	RZ
1 LOAD1	0.0000	0.0000	0.0000	0.0000	0.0000	0.0000
2 LOAD1	−5.014E−04	0.0000	−2.281E−03	0.0000	3.802E−05	0.0000
3 LOAD1	6.774E−04	0.0000	−2.722E−03	0.0000	3.877E−05	0.0000
4 LOAD1	0.0000	0.0000	0.0000	0.0000	0.0000	0.0000

SHELL ELEMENT STRESSES

SHELL/LOAD/JNT			S11-BOT	S22-BOT	S12-BOT	S11-TOP	S22-TOP	S12-TOP
1	1	2	1.371E-02	−1.164E-01	−5.476E-02	−1.371E-02	−1.164E-01	−5.476E-02
		3	1.853E-01	−6.662E-02	−5.247E-02	1.853E-01	− 6.662E-02	−5.247E-02
		1	1.877E-01	−5.687E-02	6.692E-02	1.877E-01	− 5.687E-02	6.692E-02
2	1	4	−8.578E-02	−2.145E-02	−8.783E-02	−8.578E-02	−2.145E-02	−8.783E-02
		2	−8.578E-02	−2.145E-02	−2.049E-01	−8.578E-02	−2.145E-02	−2.049E-01
		1	1.094E-01	2.735E-02	−8.783E-02	1.094E-01	2.735E-0	−8.783E-02

Note: The values given in this example by SAP2000 for Joint Displacements are slightly different from those obtained using hand calculation in Illustrative Example 9.1. This difference is explained by the fact that SAP2000 introduces additional intermediate nodal points in modeling a triangular element, thus allowing the use of a higher order interpolating polynomial. This difference is even greater in the calculation of stress. By using higher interpolating polynomials, SAP2000 provides different values for stresses at the three nodes of the triangular element and not just constant values as those obtained in Illustrative Example 9.1 in which we used linear interpolating polynomials. The use of higher order polynomials improves considerably the accuracy of the solution.

Plot Stresses: From the main menu select:
> DISPLAY>SHOW ELEMENT FORCES/STRESSES/SHELLS.
> Click on Stresses and on S11 (stress parallel to direction X).

> The screen shows a color contour distribution of stresses parallel to the direction X. Right clicking on an element will display on the screen an enlarged stress contour plot for that element showing numerical values as the cursor is moved along the area of the element.

Illustrative Example 9.3

Use SAP2000 to analyze a cantilever rectangular steel plate of dimensions 6 *in* by 4 *in* supporting a concentrated force of 10 *kips* applied at its free end as shown in Figure 9.10. Model the plate using rectangular shell elements.

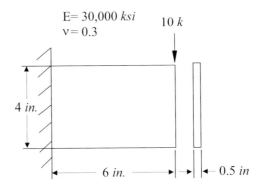

Fig. 9.10 Rectangular steel plate of Illustrative Example 9.3

Solution:
1. Modeling the structure

    The plate is modeled using six shell plate elements of dimensions 2 *in* by 2 *in* as shown in Figure 9.11.

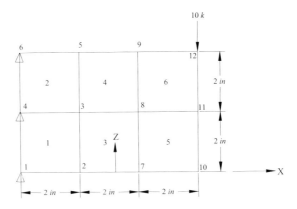

Fig. 9.11   Plate of Illustrative Example 9.3 modeled with squared plate shell elements of dimension 2 (*in*) by 2 (*in*)

The following commands are implemented in SAP2000:

Begin: Open SAP2000.
         Hint: Maximize both screens for a full view of all windows.

Units: Select "kip-in" from the drop down menu located on the lower right corner of the window.

Model: From the main menu select:
FILE>NEW MODEL FROM TEMPLATE.
Select the plate shown with a rectangular grid.
Set:
    Number of spaces along X = 3
    Number of spaces along Z = 2
    Space width along X = 2
    Space width along Z = 2. Then click OK.

Editing: Maximize the 2-D view (X-Z view).
Then drag the figure to the center of the window using the PAN icon on the toolbar.

Material: From the main menu select:
DEFINE>MATERIALS.
Select STEEL, then click Modify/Show Material.
Set: Modulus of Elasticity = 30000
        Poisson's Ratio = 0.3. Then click OK, OK.

Shell Sections: From the main menu select:
DEFINE>SHELL SECTIONS.
Click: Add New Sections.
Select: STEEL.
            Thickness = 0.5. Then click OK, OK.

Click to select the shell elements and from the main menu select:
ASSIGN>SHELL>SECTIONS.
Select: SSEC2. Then click OK.

Label: From the main menu select:
VIEW>SET ELEMENTS.
Click to select Joint Labels and Shell Labels. Then click OK.

Boundary: Select Joints 1, 4, and 6 on the left side of the plate, and from the main menu select:
ASSIGN>JOINT>RESTRAINTS.
Click Restraints in all directions. Then click OK.

Loads: From the main menu select:
DEFINE>STATIC LOAD CASES.
Change DEAD load to LIVE.

Set Self-Weight Multiplier = 0.
Click Change Load. Then click OK.

Assign Load: Select Joint 12, and from the main menu select:
ASSIGN>JOINT STATIC LOADS>FORCES.
Set: Force Global Z = −10.0. Then click OK.

Analyze: From the main menu select:
ANALYZE>OPTIONS.
Set available degrees of freedom UX, UZ and RY. Then click OK.

From the main menu select:
ANALYZE>RUN.
Assign Filename "Example 9.3". Then click SAVE.

When the calculation is completed, check for errors. Then click OK.

Plot deformed shape: From the main menu select:
DISPLAY>SHOW DEFORMED SHAPE.
Check the Wire Shadow option. Then click OK.

Select FILE>PRINT GRAPHICS.

Note: The displacement values may be viewed on screen by right-clicking
on a joint of the deformed plot of the plate.

(Figure 9.12 reproduces the deformed plot of the plate for Illustrative
Example 9.3.)

Fig. 9.12 Deformed Plate of Illustrative Example 9.3

Input Tables: From the main menu select:
  FILE>PRINT INPUT TABLES.
  Click on Print to File. Then click OK.
  Use an editor (such as Word or Notepad) to edit and print the Input Tables.

(Table 9.3 provides the edited input tables for Illustrative Example 9.3.)

Table 9.3   Edited input tables for Illustrative Example 9.3 (Units: kips-inches)

JOINT DATA

JOINT	GLOBAL-X	GLOBAL-Y	GLOBAL-Z	RESTRAINTS
1	−3.00000	0.00000	0.00000	1 1 1 1 1 1
2	−1.00000	0.00000	0.00000	0 0 0 0 0 0
3	−1.00000	0.00000	2.00000	0 0 0 0 0 0
4	−3.00000	0.00000	2.00000	1 1 1 1 1 1
5	−1.00000	0.00000	4.00000	0 0 0 0 0 0
6	−3.00000	0.00000	4.00000	1 1 1 1 1 1
7	1.00000	0.00000	0.00000	0 0 0 0 0 0
8	1.00000	0.00000	2.00000	0 0 0 0 0 0
9	1.00000	0.00000	4.00000	0 0 0 0 0 0
10	3.00000	0.00000	0.00000	0 0 0 0 0 0
11	3.00000	0.00000	2.00000	0 0 0 0 0 0
12	3.00000	0.00000	4.00000	0 0 0 0 0 0

SHELL ELEMENT DATA

SHELL	JNT-1	JNT-2	JNT-3	JNT-4	SECTION	ANGLE	AREA
1	1	2	3	4	SSEC2	0.000	4.000
2	4	3	5	6	SSEC2	0.000	4.000
3	2	7	8	3	SSEC2	0.000	4.000
4	3	8	9	5	SSEC2	0.000	4.000
5	7	10	11	8	SSEC2	0.000	4.000
6	8	11	12	9	SSEC2	0.000	4.000

JOINT FORCES Load Case LOAD1

JOINT	GLOBAL-X	GLOBAL-Y	GLOBAL-Z	GLOBAL-XX	GLOBAL-YY	GLOBAL-ZZ
12	0.000	0.000	−10.000	0.000	0.000	0.000

Output Tables: From the main menu select:
  FILE>PRINT OUTPUT TABLES.
  Check Print to File and Append. Then click

(Table 9.4 provides the edited output tables for the plate of Illustrative Example 9.3.)

Table 9.4   Edited output tables for Illustrative Example 9.3 (Units: kips-inches)

JOINT DISPLACEMENTS

JOINT	LOAD	U1	U2	U3	R1	R2	R3
1	LOAD1	0.0000	0.0000	0.0000	0.0000	0.0000	0.0000
2	LOAD1	−2.440E-03	0.0000	−2.285E-03	0.0000	1.661E-03	0.0000
3	LOAD1	2.265E-05	0.0000	−1.801E-03	0.0000	1.265E-03	0.0000

Table 9.4 (continued)

JOINT	LOAD	U1	U2	U3	R1	R2	R3
4	LOAD1	0.0000	0.0000	0.0000	0.0000	0.0000	0.0000
5	LOAD1	2.381E-03	0.0000	-2.239E-03	0.0000	1.477E-03	0.0000
6	LOAD1	0.0000	0.0000	0.0000	0.0000	0.0000	0.0000
7	LOAD1	-3.801E-03	0.0000	-6.161E-03	0.0000	1.619E-03	0.0000
8	LOAD1	-1.485E-04	0.0000	-6.071E-03	0.0000	2.464E-03	0.0000
9.	LOAD1	3.985E-03	0.0000	-5.877E-03	0.0000	2.012E-03	0.0000
10.	LOAD1	-4.098E-03	0.0000	-0.0104	0.0000	2.358E-03	0.0000
11.	LOAD1	-1.957E-04	0.0000	-0.0108	0.0000	1.869E-03	0.0000
12.	LOAD1	4.879E-03	0.0000	-0.0124	0.0000	3.621E-03	0.0000

JOINT REACTIONS

		F1	F2	F3	M1	M2	M3
JOINT	LOAD						
1	LOAD1	14.3843	0.0000	4.0266	0.0000	-0.6096	0.0000
4	LOAD1	-0.0503	0.0000	1.6783	0.0000	-1.2678	0.0000
6	LOAD1	-14.3339	0.0000	4.2951	0.0000	-0.6862	0.0000

SHELL ELEMENT STRESSES

SHELL LOAD	JOINT	S11-BOT	S22-BOT	S12-BOT	S11-TOP	S22-TOP	S12-TOP
1 LOAD1							
	1	-40.23	-12.07	-3.60	-40.23	-12.07	-3.60
	2	-37.83	-4.08	-6.2	-37.83	-4.08	-6.27
	4	3.733E-01	1.120E-01	-3.09	3.733E-01	1.120E-01	-3.09
	3	2.77	8.10	-5.76	2.77	8.10	-5.76
2 LOAD1							
	4	3.733E-01	1.120E-01	-3.09	3.733E-01	1.120E-01	-3.09
	3	-1.79	-7.11	-5.31	-1.79	-7.11	-5.31
	6	39.25	11.77	-4.40	39.25	11.77	-4.40
	5	37.08	4.55	-6.61	37.08	4.55	-6.61
3 LOAD1							
	2	-20.04	1.25	-6.11	-20.04	1.25	-6.11
	7	-21.99	-5.24	-5.92	-21.99	-5.24	-5.92
	3	-4.262E-01	7.14	-5.79	-4.262E-01	7.14	-5.79
	8	-2.38	6.410E-01	-5.60	-2.38	6.410E-01	-5.60
4 LOAD1							
	3	-4.99	-8.07	-5.33	-4.99	-8.07	-5.33
	8	-1.86	2.36	-5.10	-1.86	2.36	-5.10
	5	24.27	7.110E-01	-3.07	24.27	7.110E-01	-3.07
	9	27.40	11.14	-2.83	27.40	11.14	-2.83
5 LOAD1							
	7	-4.45	1.917E-02	-4.15	-4.45	1.917E-02	-4.15
	10	-6.62	-7.23	-3.55	-6.62	-7.23	-3.55
	8	-3.315E-01	1.25	-4.63	-3.315E-01	1.25	-4.63
	11	-2.51	-6.00	-4.03	-2.51	-6.00	-4.03
6 LOAD1							
	8	1.834E-01	2.97	-4.13	1.834E-01	2.97	-4.13
	11	-8.74	-26.77	-4.55	-8.74	-26.77	-4.55
	9	15.69	7.62	-7.04	15.69	7.62	-7.04
	12	6.77	-22.11	-7.46	6.77	-22.11	-7.46

Note: In order to save space, two additional tables, "Shell Element Principal Stresses" and "Shell Element Resultants", which are given as output tables by SAP2000 are not reproduced as a part of Table 9.4.

Plot Stresses: From the main menu select:
DISPLAY>SHOW ELEMENTS-FORCES/STRESSES/SHELLS.
Click on a Force or Stress and then on a direction.
Then click OK.

For example, click on STRESS and on direction FF1. The screen will then show a color contour plot of the stress distribution on the plate. By right clicking on a shell element, the selected element will be shown in the screen with its contour stresses in a magnified plot. Numerical values for the stresses at any location in the element will be shown on the screen where the cursor is located.

9.4   Plate Bending

The application of the finite element method is now considered for the analysis of plate bending, that is, plates loaded by forces that are perpendicular to the plane of the plate. The presentation that follows is based on two assumptions: (1) the thickness of the plate is assumed to be small compared to other dimensions of the plate; and (2) the deflections of the plate under the load are assumed to be small relative to its thickness. These assumptions are not particular to the application of the finite element method; they are also made in the classical theory of elasticity for bending of thin plates. These two assumptions are necessary because if the thickness of the plate is large, the plate has to be analyzed as a three-dimensional problem. Also, if the deflections are large, then in-plane membrane forces are developed and should be accounted for in the analysis. The analysis of plates can be undertaken by the finite element method without these two assumptions. The program SAP2000 used in this book may be applied for the analysis of thin plates undergoing small deflections or thick plates in which these two assumptions are not required. However, the theoretical presentation in this section is limited to thin plates that undergo small deflections.

9.5   Plate Bending: Rectangular Element

The derivation of the element stiffness matrix as well as the vector of equivalent nodal forces for body forces or any other forces distributed on the plane element is obtained by following the same process as was used for derivation of the triangular element subjected to in-plane loads as presented in Section 9.2.

Step 1: Modeling the structure.

A suitable system of coordinates and node numbering for a rectangular plate is defined in Figure 9.13(a) with the x,y axes along the continuous sides of the rectangular plate element, and the z axis normal to the plane of the plate, completing a right-hand system of Cartesian coordinates. This rectangular element has three nodal displacements at each of its four nodes, a rotation about the x axis (θ_x), a rotation about the y axis (θ_y), and a displacement (w) along the z axis normal to the plane of the plate. These nodal displacements, labeled q_i (i=1,2,…12), are shown in their positive sense in Figure 9.13(b).

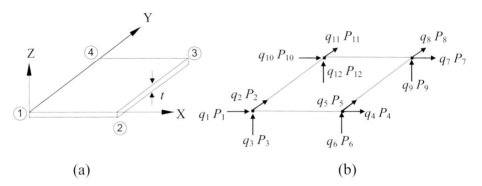

(a)   (b)

Fig. 9.13   Rectangular plate bending element
(a) Coordinate system and node numbering
(b) Nodal displacements q_i and corresponding nodal forces P_i

Corresponding to the three nodal displacements at each node, two moments and a force that are labeled P_i (i = 1, 2,…12), also are shown in this figure. These 12 nodal displacements and 12 nodal forces are conveniently arranged in two vectors of 12 components, $\{q\}_e$ and $\{P\}_e$. Therefore, the stiffness matrix relating the nodal forces and the nodal displacements for this rectangular plate bending element with four nodes is of dimension 12 x 12.

The angular displacements θ_x and θ_y at any point $P(x,y)$ of the plate element are related to the normal displacement w by the following expressions:

$$\theta_x = -\frac{\partial w}{\partial y} \text{ and } \theta_y = \frac{\partial w}{\partial x}. \qquad (9.30)$$

The positive directions of θ_x and θ_y are shown to agree with the angular nodal displacements q_1, q_2, q_4, q_5, etc. as indicated in Figure 9.13. Therefore, after a function $w = w(x,y)$ is chosen for the lateral displacement, the angular displacements are established through the relations in eq. (9.30).

Step 2: Selection of a suitable displacement function.

Since the rectangular element in plate bending has twelve degrees of freedom, the polynomial expression chosen for the normal displacements w must contain 12 constants. A suitable polynomial function is given by

$$w = c_1 + c_2 x + c_3 y + c_4 x^2 + c_5 xy + c_6 y^2 + c_7 x^3 + c_8 x^2 y + c_9 xy^2 + c_{10} y^3$$
$$+ c_{11} x^3 y + c_{12} xy^3 \qquad (9.31)$$

The displacement functions of the rotations θ_x and θ_y are then obtained from eqs. (9.30) and (9.31) as

$$\theta_x = -\frac{\partial w}{\partial y} = -\left(c_3 + c_5 x + 2c_6 y + c_8 x^2 + 2c_9 xy + 3c_{10} y^2 + c_{11} x^3 + 3c_{12} xy^2\right) \quad (9.32)$$

and

$$\theta_y = \frac{\partial w}{\partial x} = c_2 + 2c_4 x + c_5 y + 3c_7 x^2 + 2c_8 xy + c_9 y^2 + 3c_{11} x^2 y + c_{12} y^2 \qquad (9.33)$$

By considering the displacements at the edge of one element, that is, in the boundary between adjacent elements, it may be demonstrated that there is continuity of normal lateral displacements and of the rotational displacement in the direction of the boundary line, but not in the direction transverse to this line as it is shown graphically in Figure 9.14. The displacement function in eq. (9.31) is called *a non-conforming function* because it does not satisfy the condition of continuity at the boundaries between elements for all three displacements w, θ_x, and θ_y.

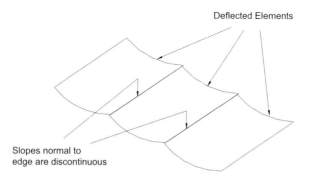

Deflected Elements

Slopes normal to
edge are discontinuous

Fig. 9.14 Deflected continuous rectangular elements

Step 3: Displacements $\{q(x,y)\}$ at a point within the element are expressed in terms of the nodal displacements $\{q\}_e$.

Equations (9.31), (9.32), and (9.33) may be written in matrix notation as

$$\{q(x,y)\} = \begin{Bmatrix} \theta_x \\ \theta_y \\ w \end{Bmatrix} = [g(x,y)]\{c\} \tag{9.34}$$

in which $g(x, y)$ is a function of the coordinates x,y at a point in the element. Equation (9.34) is now used to evaluate the displacements at the nodal coordinates of the element $\{q\}_e$ to obtain:

$$\{q\}_e = [A]\{c\} \tag{9.35}$$

in which the matrix $[A]$ is a function of the coordinates at the nodes of the element. The vector $\{c\}$ containing the constant coefficients is then given by

$$\{c\} = [A]^{-1}\{q\}_e \tag{9.36}$$

where $[A]^{-1}$ is the inverse of the matrix $[A]$ defined in eq. (9.35). Finally, the substitution of the vector of constants $\{c\}$ from eq. (9.36) into eq. (9.34) provides the required relationship for displacements $\{q(x,y)\}$ at any interior point in the rectangular element and the displacements $\{q\}_e$ at the nodes:

$$\{q(x,y)\} = \begin{Bmatrix} -\dfrac{\partial w}{\partial y} \\ \dfrac{\partial w}{\partial x} \\ w \end{Bmatrix} = [f(x,y)]\{q\}_e \tag{9.37}$$

in which $[f(x,y)] = [g(x,y)][A]^{-1}$ is solely a function of the coordinates x,y at a point within the element.

Step 4: Relationship between strains $[\varepsilon(x,y)]$ at a point within the element to displacements $\{q(x,y)\}$ and hence to the nodal displacements $\{q\}_e$.

For plate bending, the state of strain at any point of the element may be represented by three components: (1) the curvature in the x direction; (2) the curvature in the y direction; and (3) a component representing torsion in the plate. The curvature in the

x direction is equal to the rate of change of the slope in that direction, that is, to the derivative of the slope,

$$-\frac{\partial}{\partial x}\left(\frac{\partial w}{\partial x}\right) = -\frac{\partial^2 w}{\partial x^2}.$$ (9.38)

Similarly, the curvature in the y direction is

$$-\frac{\partial}{\partial y}\left(\frac{\partial w}{\partial y}\right) = -\frac{\partial^2 w}{\partial y^2}.$$ (9.39)

Finally, the torsional strain component is equal to the rate of change, with respect to y, of the slope in the x direction, that is

$$\frac{\partial}{\partial y}\left(\frac{\partial w}{\partial x}\right) = -\frac{\partial^2 w}{\partial x \partial y}.$$ (9.40)

The bending moments M_x and M_y and the torsional moments M_{xy} and M_{yx} each act on two opposite sides of the element, but since M_{xy} is numerically equal to M_{yx}, one of these torsional moments, M_{xy}, can be considered to act in all four sides of the element, thus allowing for simply doubling the torsional strain component. Hence, the "strain" vector, $[\varepsilon(x,y)]$, for a plate bending element can be expressed by

$$\{\varepsilon(x,y)\} = \left\{\begin{array}{c} -\dfrac{\partial^2 w}{\partial x^2} \\[2mm] -\dfrac{\partial^2 w}{\partial y^2} \\[2mm] 2\dfrac{\partial^2 w}{\partial x \partial y} \end{array}\right\}.$$ (9.41)

The substitution of the second derivatives obtained by differentiation of eq. (9.37) into eq. (9.41) yields

$$\{\varepsilon(x,y)\} = [B]\{q\}_e$$ (9.42)

in which $[B]$ is a function of the coordinates (x,y) only.

Step 5:   Relationship between internal stresses $\{\sigma(x,y)\}$ to internal strains $[\varepsilon(x,y)]$ and hence to nodal displacements $\{q\}_e$.

In a plate bending, the internal "stresses" are expressed as bending and twisting moments, and the "strains" are the curvatures and the twist. Thus, for plate bending, the state of internal "stresses" can be represented by

$$\{\sigma(x,y)\} = \begin{Bmatrix} M_x \\ M_y \\ M_{xy} \end{Bmatrix} \qquad (9.43)$$

where M_x and M_y are the internal bending moments per unit of length, and M_{xy} is the internal twisting moment per unit of length. For a small rectangular element of the plate bending, these internal moments are shown in Figure 9.15.

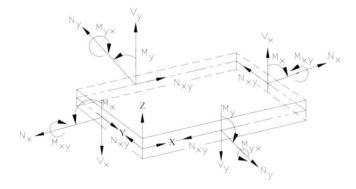

Fig. 9.15 Direction of force and moment components per unit of length as defined for thin plates by SAP2000

The moment-curvature relationships obtained from plate bending theory (Timoshenko and Goodier 1970) are:

$$M_x = -\left(D_x \frac{\partial^2 w}{\partial x^2} + D_1 \frac{\partial^2 w}{\partial y^2} \right)$$

$$M_y = -\left(D_y \frac{\partial^2 w}{\partial y^2} + D_1 \frac{\partial^2 w}{\partial x^2} \right). \qquad (9.44)$$

$$M_{xy} = 2D_{xy} \frac{\partial^2 w}{\partial x \partial y}$$

These relations are written in general for an orthotropic plate, i.e., a plate that has different elastic properties in two perpendicular directions, in which D_x and D_y are flexural rigidities in the x and y directions, respectively, D_1 is a "coupling" rigidity coefficient representing the Poisson's ratio type of effect, and D_{xy} is the torsional rigidity.

For an isotropic plate that has the same properties in all directions, the flexural and twisting rigidities, D_x, D_y and D_{xy}, in eq. (9.44) are given by

$$D_x = D_y = \frac{Et^3}{12(1-v^2)}$$

$$D_1 = v D_x \qquad\qquad (9.45)$$

$$D_{xy} = \frac{(1-v)}{2} D_x$$

in which E is the modulus of elasticity, v the Poisson's Ratio coefficient, and t the thickness of the plate. Equation (9.44) may then be written in matrix notation as

$$\{\sigma(x,y)\} = \begin{Bmatrix} M_x \\ M_y \\ M_{xy} \end{Bmatrix} = \begin{bmatrix} D_x & D_1 & 0 \\ D_1 & D_y & 0 \\ 0 & 0 & D_{xy} \end{bmatrix} \begin{Bmatrix} -\dfrac{\partial^2 w}{\partial x^2} \\ -\dfrac{\partial^2 w}{\partial y^2} \\ 2\dfrac{\partial^2 w}{\partial x \partial y} \end{Bmatrix} \qquad (9.46)$$

or in condensed notation as

$$\{\sigma(x,y)\} = [D]\{\varepsilon(x,y)\} \qquad\qquad (9.47)$$

where the matrix $[D]$ is defined in eq. (9.46). The substitution of $[\varepsilon(x,y)]$ from eq. (9.42) gives the required relationship between the element stresses and nodal displacements as

$$\{\sigma(x,y)\} = [D][B]\{q\}_e . \qquad\qquad (9.48)$$

Step 6: Element stiffness matrix.

The stiffness matrix and the equivalent nodal force vector for an element of plate bending may be obtained by applying the *Principle of Virtual Work,* resulting in the same general equations given by eqs. (9.22), (9.23) and (9.24). For plate bending, the matrices $[f(x,y)]$, $[B]$, and $[D]$ required in these equations are defined, respectively, in eqs. (9.37), (9.42) and (9.47). The calculation of these matrices and also of the integrals indicated in eqs. (9.22) and (9.24), are usually undertaken by numerical methods implemented in the coding of computer programs.

Step 7: Assemblage of the reduced system stiffness matrix $[K]_R$ and of the reduced vector of the external forces $\{F\}_R$.

The system stiffness matrix considering the free nodal coordinates and the system nodal force vector are assembled from the corresponding element stiffness matrices and element force vectors given, respectively, by eqs. (9.25) and (9.27).

Step 8: Solution of the differential equations of motion.

The system stiffness equation is given by eq. (9.28) in which $[K]_R$ is the reduced stiffness matrices and $\{F\}_R$ the reduced system vector of the external forces. The solution of eq. (9.28) thus provides the system nodal displacement vector $\{u\}$.

Step 9: Nodal stresses.

The element nodal stresses $\left\{\sigma\left(x_j, y_j\right)\right\}_e$ for node j of an element are given from eq. (9.48) as

$$\left\{\sigma\left(x_j, y_j\right)\right\}_e = [D]_j [B]_j \{q\}_e \tag{9.49}$$

in which the matrices $[D]_j$ and $[B]_j$ are evaluated for the coordinates of node j of the element.

Illustrative Example 9.4

A square steel plate 40 *in* by 40 *in* assumed to be fixed at the supports on its four sides (Figure 9.16) is loaded with a uniformly distributed force of magnitude $w = 0.01$ *kip/in²* applied downward, normal to the plane of the plate. Use SAP2000 to analyze the plate and determine deflections and stresses.

$E = 29,500$ *ksi*
$v = 0.3$
$t = 0.10$ *in.*
$w = 0.01$ *kip/in²*

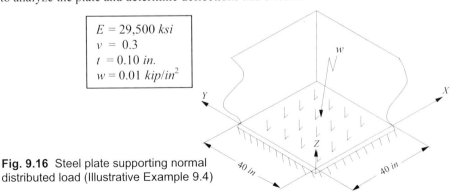

Fig. 9.16 Steel plate supporting normal distributed load (Illustrative Example 9.4)

Solution:

The plate is modeled with sixteen square plate elements of 10 *in* by 10 *in* dimension, as shown in Figure 9.17.

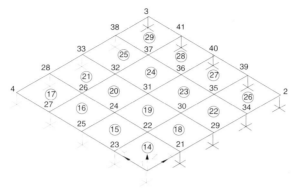

Fig. 9.17 Finite element model for the plate in Illustrative Example 9.4

The following commands are implemented in SAP2000.

Begin: Open SAP2000.
Hint: Maximize both screens for a full view of all windows.

Units: Select kip-in in the drop down menu located in the lower right corner of the window.

Model: From the main menu select:
FILE>NEW MODEL.
Number of grid spaces:
    X direction = 4
    Y direction = 4
    Z direction = 0
Grid Spacing:
    X direction = 10
    Y direction = 10
    Z direction = 1

Select: Click on the top of the right window to default to the *X-Y* plane at $Z = 0$ view.

Draw: From the main menu select:
DRAW>DRAW SHELL ELEMENT.
Click at grid intersection X = −20, Y = −20 and drag the cursor to the X = −20, Y = 20 grid intersection. Then click again at this intersection and

drag the cursor to the X = 20, Y = 20 intersection. Then click again and drag the cursor to the X = 20, Y = −20 intersection. It should be observed that a square plate is drawn.

Mesh Shells: Select (click) the shell. Then from the main menu select:
EDIT>MESH SHELLS.
Change the number of divisions in both entries to 4. Then click OK.
The view will show a mesh of 16 square shells.

Materials: From the main menu select:
DEFINE>MATERIALS>STEEL.
Click Modify/ Show Material.
Set the Modulus of Elasticity = 29500
Set Poisson's Ratio = 0.3. Then click OK, OK.

Sections: From the main menu select:
DEFINE>SHELL SECTIONS.
Add new section.
In the drop-down menu, select STEEL.
Thickness:
Membrane = 0.1
Bending   = 0.1. Then click OK, OK.

Click (or envelope) all 16 shells and from the main menu select:
ASSIGN>SHELL>SECTIONS.
Select SSEC2. Then click OK.

Boundary: Select all joints on the boundary of the plate, and then from the main menu select:
ASSIGN>JOINT>RESTRAINTS.
Restrain in all directions. Then click OK.

Load: From the main menu select:
DEFINE>STATIC LOAD CASES.
Change Type of Load to LIVE load.
Set the Self-Weight Multiplier to 0 (zero).
Click Change Load. Then click OK.

Select all the 16 elements of the plate and from the main menu select:
ASSIGN>SHELL STATIC LOADS>UNIFORM.
Set Load = −0.01.
Select Global Z. Then click OK.

Analyze: From the main menu select:
ANALYZE>SET OPTIONS.

Mark available degrees of freedom = UZ, RX and RY.
Then OK and from the main menu select:
ANALYZE>RUN.
Enter Filename "Example 9.4". Then click SAVE.
At the conclusion of the calculations, click OK.

Note: Displacement values may be viewed by right-clicking on a node of the deformed plate.

Plot: From the main menu select:
DISPLAY>SHOW DEFORMED SHAPE.
Click the Wire Shadow option.
Select FILE>PRINT GRAPHICS.

(Figure 9.18 shows the deformed shape for the plate of Illustrative Example 9.4.)

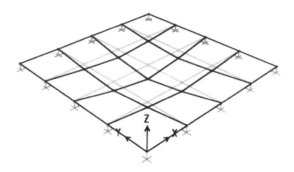

Fig. 9.18   Deformed shape for the plate of Illustrative Example 9.4.

Input Tables: From the main menu select:
FILE>PRINT INPUT TABLES.
Click Print to File. Then click OK.

Use an editor (such as Word or Notepad) to edit the input tables and then print them.

(Table 9.5 provides the edited input tables for the plate of Illustrative Example 9.4.)

Table 9.5 Edited Input Tables for Illustrative Example 9.4 (Units: kips-inches)

JOINT DATA

JOINT	GLOBAL-X	GLOBAL-Y	GLOBAL-Z	RESTRAINTS
1	−20.00000	−20.00000	0.00000	1 1 1 1 1 1
2	20.00000	−20.00000	0.00000	1 1 1 1 1 1
3	20.00000	20.00000	0.00000	1 1 1 1 1 1
4	−20.00000	20.00000	0.00000	1 1 1 1 1 1
21	−10.00000	−20.00000	0.00000	1 1 1 1 1 1
22	−10.00000	−10.00000	0.00000	0 0 0 0 0 0
23	−20.00000	−10.00000	0.00000	1 1 1 1 1 1
24	−10.00000	0.00000	0.00000	0 0 0 0 0 0
25	−20.00000	0.00000	0.00000	1 1 1 1 1 1
26	−10.00000	10.00000	0.00000	0 0 0 0 0 0
27	−20.00000	10.00000	0.00000	1 1 1 1 1 1
28	−10.00000	20.00000	0.00000	1 1 1 1 1 1
29	0.00000	−20.00000	0.00000	1 1 1 1 1 1
30	0.00000	−10.00000	0.00000	0 0 0 0 0 0
31	0.00000	0.00000	0.00000	0 0 0 0 0 0
32	0.00000	10.00000	0.00000	0 0 0 0 0 0
33	0.00000	20.00000	0.00000	1 1 1 1 1 1
34	10.00000	−20.00000	0.00000	1 1 1 1 1 1
35	10.00000	−10.00000	0.00000	0 0 0 0 0 0
36	10.00000	0.00000	0.00000	0 0 0 0 0 0
37	10.00000	10.00000	0.00000	0 0 0 0 0 0
38	10.00000	20.00000	0.00000	1 1 1 1 1 1
39	20.00000	−10.00000	0.00000	1 1 1 1 1 1

JOINT	GLOBAL-X	GLOBAL-Y	GLOBAL-Z	RESTRAINTS
40	20.00000	0.00000	0.00000	1 1 1 1 1 1
41	20.00000	10.00000	0.00000	1 1 1 1 1 1

SHELL ELEMENT DATA

SHELL	JNT-1	JNT-2	JNT-3	JNT-4	SECTION	ANGLE	AREA
14	1	21	22	23	SSEC2	0.000	100.000
15	23	22	24	25	SSEC2	0.000	100.000
16	25	24	26	27	SSEC2	0.000	100.000
17	27	26	28	4	SSEC2	0.000	100.000
18	21	29	30	22	SSEC2	0.000	100.000
19	22	30	31	24	SSEC2	0.000	100.000
20	24	31	32	26	SSEC2	0.000	100.000
21	26	32	33	28	SSEC2	0.000	100.000
22	29	34	35	30	SSEC2	0.000	100.000
23	30	35	36	31	SSEC2	0.000	100.000
24	31	36	37	32	SSEC2	0.000	100.000
25	32	37	38	33	SSEC2	0.000	100.000
26	34	2	39	35	SSEC2	0.000	100.000
27	35	39	40	36	SSEC2	0.000	100.000
28	36	40	41	37	SSEC2	0.000	100.000
29	37	41	3	38	SSEC2	0.000	100.000

Output Tables: From the main menu select:
FILE>OUTPUT TABLES.
Click on Print to File and Append.

(Table 9.6 provides the edited output tables for the plate of Illustrative Example 9.4.)

Table 9.6   Edited output tables for Illustrative Example 9.4 (Units: kips-inches)

JOINT DISPLACEMENTS

JOINT	LOAD	UX	UY	UZ	RX	RY	RZ
1	LOAD1	0.0000	0.0000	0.0000	0.0000	0.0000	0.0000
2	LOAD1	0.0000	0.0000	0.0000	0.0000	0.0000	0.0000
3	LOAD1	0.0000	0.0000	0.0000	0.0000	0.0000	0.0000
4	LOAD1	0.0000	0.0000	0.0000	0.0000	0.0000	0.0000
5	LOAD1	0.0000	0.0000	0.0000	0.0000	0.0000	0.0000
6	LOAD1	0.0000	0.0000	−4.8274	−0.6069	0.6069	0.0000
7	LOAD1	0.0000	0.0000	0.0000	0.0000	0.0000	0.0000
8	LOAD1	0.0000	0.0000	−8.0922	0.0000	1.0120	0.0000
9	LOAD1	0.0000	0.0000	0.0000	0.0000	0.0000	0.0000
10	LOAD1	0.0000	0.0000	−4.8274	0.6069	0.6069	0.0000
11	LOAD1	0.0000	0.0000	0.0000	0.0000	0.0000	0.0000
12	LOAD1	0.0000	0.0000	0.0000	0.0000	0.0000	0.0000
13	LOAD1	0.0000	0.0000	0.0000	0.0000	0.0000	0.0000
14	LOAD1	0.0000	0.0000	−8.0922	−1.0120	0.0000	0.0000
15	LOAD1	0.0000	0.0000	−13.6109	0.0000	0.0000	0.0000
16	LOAD1	0.0000	0.0000	−8.0922	1.0120	0.0000	0.0000
17	LOAD1	0.0000	0.0000	0.0000	0.0000	0.0000	0.0000
18	LOAD1	0.0000	0.0000	0.0000	0.0000	0.0000	0.0000

JOINT	LOAD	UX	UY	UZ	RX	RY	RZ
19	LOAD1	0.0000	0.0000	−4.8274	−0.6069	−0.6069	0.0000
20	LOAD1	0.0000	0.0000	−8.0922	0.0000	−1.0120	0.0000
21	LOAD1	0.0000	0.0000	−4.8274	0.6069	−0.6069	0.0000
22	LOAD1	0.0000	0.0000	0.0000	0.0000	0.0000	0.0000
23	LOAD1	0.0000	0.0000	0.0000	0.0000	0.0000	0.0000
24	LOAD1	0.0000	0.0000	0.0000	0.0000	0.0000	0.0000
25	LOAD1	0.0000	0.0000	0.0000	0.0000	0.0000	0.0000

JOINT REACTIONS

JOINT	LOAD	F1	F2	F3	M1	M2	M3
1	LOAD1	0.0000	0.0000	0.1818	0.7218	−0.7218	0.0000
2	LOAD1	0.0000	0.0000	0.1818	−0.7218	−0.7218	0.0000
3	LOAD1	0.0000	0.0000	0.1818	−0.7218	0.7218	0.0000
4	LOAD1	0.0000	0.0000	0.1818	0.7218	0.7218	0.0000
5	LOAD1	0.0000	0.0000	1.0606	4.3943	−0.0972	0.0000
7	LOAD1	0.0000	0.0000	1.0606	0.0972	−4.3943	0.0000
9	LOAD1	0.0000	0.0000	1.6970	0.0000	−6.7915	0.0000
11	LOAD1	0.0000	0.0000	1.0606	−0.0972	−4.3943	0.0000
12	LOAD1	0.0000	0.0000	1.0606	−4.3943	−0.0972	0.0000
13	LOAD1	0.0000	0.0000	1.6970	6.7915	0.0000	0.0000
17	LOAD1	0.0000	0.0000	1.6970	−6.7915	0.0000	0.0000
18	LOAD1	0.0000	0.0000	1.0606	4.3943	0.0972	0.0000
22	LOAD1	0.0000	0.0000	1.0606	−4.3943	0.0972	0.0000
23	LOAD1	0.0000	0.0000	1.0606	0.0972	4.3943	0.0000
24	LOAD1	0.0000	0.0000	1.6970	0.0000	6.7915	0.0000
25	LOAD1	0.0000	0.0000	1.0606	−0.0972	4.3943	0.0000

Note: The output of SAP2000 also includes the following tables:
        Shell Element Resultant
        Shell Element Stresses
        Shell Element Principal Stresses
        These additional tables are not reproduced as part of Table 9.6.

Plot Stresses: From the main menu select:
        DISPLAY>SHOW ELEMENT FORCES/STRESSES>SHELLS.

Note: The next screen shown allows for selection of forces or stresses to display in color contour plots of the forces or stress components as well as maximum or principal values. Any element on this plot may be selected by a right click for a detailed diagram where numerical values are shown as the cursor is moved over the contoured element.

9.6   Finite Element Method: Structural Dynamics

The application of the Finite Element Method for structures subjected to time-dependent forces requires the consideration of the inertial forces to establish the dynamic equilibrium of an element. The consideration of these forces leads to the calculation of the *element mass matrix* followed by the assemblage of the *system mass matrix* and the solution of the resultant differential equations.

The Principle of Virtual Work as used in step 6 of Section 9.3 will now result in the following element stiffness equation:

$$[m]_e \{\ddot{q}\} + [k]_e \{q\} = \{P\}_e + \{P_b\}_e \qquad\qquad (9.50)$$

in which the element mass matrix $[m]_e$ is given by

$$[m]_e = \int_v \rho [f(x,y)]^T [f(x,y)] dV. \qquad\qquad (9.51)$$

The other quantities in eqs. (9.50) and (9.51) have already been defined:

$[k]_e$ the element stiffness matrix [eq. (9.23)]

$\{P_b\}_e$ the equivalent nodal forces due to the body forces [eq. (9.24)]

$\{P\}_e$ the element nodal forces

$\{q\}$ the element nodal displacements

$\{\ddot{q}\}$ the element nodal acceleration

ρ the mass density

$f(x,y)$ the function defined in eq. (9.8) for plane elasticity problems

Illustrative Example 9.5

A simple supported deep steel beam (Modulus of Elasticity, $E = 29E06$ psi) shown in Figure 9.19(a), with a distributed mass, $\overline{m} = 0.03$ ($lb.sec^2/in/$)/in is loaded at its center by the impulsive force $F(t)$ depicted in Figure 9.19(b). Determine the first five natural periods and the time-displacement response at the center of the beam. Use the computer program SAP2000 and rectangular shell elements to model this structure.

Solution:

Begin: Open SAP2000
       Hint: Maximize both windows for a full view of all windows.

Select: In the lower right-hand corner of the window, use the drop-down menu to select "lb-in."

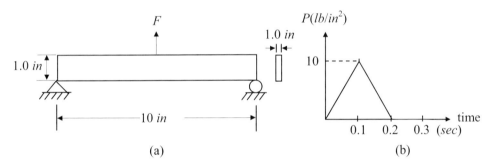

Fig. 9.19 (a) Simple supported deep beam; (b) Impulsive load time function

Model: From the main menu select:
       FILE>NEW MODEL.
       Number of Grid Spaces:
           x direction = 10
           y direction = 1
           z direction = 0
       Grid Spacing:
           x direction = 1
           y direction = 1
           z direction = 1

Edit: Click on the XY icon.
       Maximize the XY window and use the PAN icon in the tool bar to drag the plot into the center of the window.

Draw: From the main menu select:
DRAW>QUICK DRAW SHELL ELEMENT.
Click on each of the square areas.
Then press the Esc key.

Label From the main menu select:
VIEW>SET ELEMENTS.
Click Joint Labels. Then click OK.
Use the PAN icon to drag the figure to the center of the window.

(Figure 9.20 reproduces the modeled beam.)

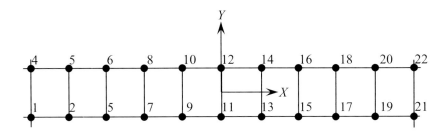

Fig. 9.20 Deep beam of Illustrative Example 9.5 modeled with ten shell elements

Boundary: Select Joints 4, 21, and 22 by clicking on them; then from the main menu select:
ASSIGN>JOINT>RESTRAINTS.
Select Restraint only for Translation 2. Then click OK.

Select joint 1 and from the main menu select:
ASSIGN>JOINT>RESTRAINTS.
Select Restraint in directions 1 and 2. Then click OK.

Section: From the main menu select:
DEFINE>SHELL SECTIONS.
Click Add New Section.
Select Material: STEEL
Enter thickness:
        Membrane = 0.1
        Bending = 0.1
Click on Type: Shell. Then click OK, OK.

Assign: Select the 10 shell elements by clicking on them, and then from the main menu select:
ASSIGN>SHELL>SECTIONS.
Select shell section SSEC2. Then click OK.

Material: From the main menu select:
DEFINE>MATERIALS.
Select: Steel.
Click Modify/Show Materials. Then click OK.
Enter:
    Modulus of Elasticity = 29E6
    Poisson's Ratio = 0.3
Mass per unit of volume = 0.3. [This value is calculated from the mass per unit of length, 0.03 ($lb \cdot sec^2/in$)/in on a thickness of $t = 0.1$ in.] Click OK, OK.

Load: From the main menu select:
DEFINE>STATIC LOAD CASES.
Select Type: Live Load and set Self-Weight Multiplier = 0
Click Change Load. Then click OK, OK.

Load History Function: From the main menu select:
DEFINE>TIME HISTORY FUNCTIONS.
Click Add New Function. Click Add (to add the point 0, 0).
Type 0.1 in the Time box and 1000 in the Value box. Click Add.
Type 0.2 in the Time box and 0 in the Value box. Click Add.
Type 0.4 in the Time box and 0 in the Value box . Click Add.
Then click OK, OK.

Load History Cases: From the main menu select:
DEFINE>TIME HISTORY CASES.
Click "Add New History".
Accept the default HIST1.
Select Analysis Type: Linear.
Click Modify/Show Values.
Type 0.0 for the damping in all modes.
Then click OK.
Click "Envelopes".
Type 40 for the Number of Output Time Steps.
Type 0.01 for Output Time Step Size.
In Load Assignment, select LOAD1.
For Function, select FUNC1.
Click Add. Then click OK, OK.

Assign Load: Click on Joint 12 to select this joint, and then from the main menu select:
ASSIGN>JOINT LOAD STATIC LOADS>FORCES.
Enter Force Global Y = 1.0. Then click OK.

Options: From the main menu select:
ANALYZE>SET OPTIONS.
Click UX, UY and RZ for available DOFs.
Click on "Dynamic Analysis" and on "Set Dynamic Parameters".
Enter Number of Modes = 5. Then click OK, OK.

Analyze: From the main menu select:
ANALYZE>RUN.
Enter Filename "EXD 9.5". Then click SAVE.
When the calculations are concluded, click OK.

Print Input Tables: From the main menu select:
FILE>PRINT INPUT TABLES.
Click "print to File" and accept all other defaults values. Then click OK.
Use Notepad or other word editor to retrieve and edit the Input File "EXD 9.5.txt".

(Input tables of Illustrative Example 9.5 are reproduced in Table 9.7.)

Table 9.7 Edited Input Tables for Illustrative Example 9.5 (Units: lb, in, sec)

TIME HISTORY CASES

HISTORY CASE	HISTORY TYPE	NUMBER OF INCREMENT	TIME STEP
HIST1	LINEAR	40	0.01000

JOINT DATA

JOINT	GLOBAL-X	GLOBAL-Y	GLOBAL-Z	RESTRAINTS
1	−5.00000	−0.50000	0.00000	1 1 0 0 0 0
2	−4.00000	−0.50000	0.00000	0 0 0 0 0 0
3	−4.00000	0.50000	0.00000	0 0 0 0 0 0
4	−5.00000	0.50000	0.00000	0 1 0 0 0 0
5	−3.00000	−0.50000	0.00000	0 0 0 0 0 0
6	−3.00000	0.50000	0.00000	0 0 0 0 0 0
7	−2.00000	−0.50000	0.00000	0 0 0 0 0 0
8	−2.00000	0.50000	0.00000	0 0 0 0 0 0
9	−1.00000	−0.50000	0.00000	0 0 0 0 0 0
10	−1.00000	0.50000	0.00000	0 0 0 0 0 0
11	0.00000	−0.50000	0.00000	0 0 0 0 0 0
12	0.00000	0.50000	0.00000	0 0 0 0 0 0
13	1.00000	−0.50000	0.00000	0 0 0 0 0 0
14	1.00000	0.50000	0.00000	0 0 0 0 0 0
15	2.00000	−0.50000	0.00000	0 0 0 0 0 0
16	2.00000	0.50000	0.00000	0 0 0 0 0 0
17	3.00000	−0.50000	0.00000	0 0 0 0 0 0
18	3.00000	0.50000	0.00000	0 0 0 0 0 0

Table 9.7 (continued)

JOINT	GLOBAL-X	GLOBAL-Y	GLOBAL-Z	RESTRAINTS
19	4.00000	−0.50000	0.00000	0 0 0 0 0 0
20	4.00000	0.50000	0.00000	0 0 0 0 0 0
21	5.00000	−0.50000	0.00000	0 1 0 0 0 0
22	5.00000	0.50000	0.00000	0 1 0 0 0 0

SHELL ELEMENT DATA

SHELL	JNT-1	JNT-2	JNT-3	JNT-4	SECTION	ANGLE	AREA
1	1	2	3	4	SSEC2	0.000	1.000
2	2	5	6	3	SSEC2	0.000	1.000
3	5	7	8	6	SSEC2	0.000	1.000
4	7	9	10	8	SSEC2	0.000	1.000
5	9	11	12	10	SSEC2	0.000	1.000
6	11	13	14	12	SSEC2	0.000	1.000
7	13	15	16	14	SSEC2	0.000	1.000
8	15	17	18	16	SSEC2	0.000	1.000
9	17	19	20	18	SSEC2	0.000	1.000
10	19	21	22	20	SSEC2	0.000	1.000

JOINT FORCES Load Case LOAD1

JOINT	GLOBAL-X	GLOBAL-Y	GLOBAL-Z	ROT-XX	ROT-YY	ROT-ZZ
12	0.000	1.000	0.000	0.000	0.000	0.000

Plot Displacement: From the main menu select:

DISPLAY>SHOW TIME HISTORY TRACES.
Click "Define Functions" and "Add Input Functions".
Select "Add Joint Displs/Forces". Then click OK.
  Time History Joint Function: Joint ID = 12
  Vector Type: select Displ
  Component: UY
  Click OK, OK.

Highlight "Joint 12" in the "List of Functions" column.
Click "ADD" to move joint 12 to the "Plot Functions" column.
Enter time range from 0 to 0.4.
Axis Labels:
  Horizontal: TIME (sec)
  Vertical: Y-DISPLACEMENT JOINT 12 (in)
Click Display.
The window will show the displacement-time function for joint 12.
Click "Done" to exit.

Plot Displacement: From the main menu select:

FILE>PRINT GRAPHICS.
(Fig 9.21 reproduces the Y-Displacement-Time plot for joint 12.)

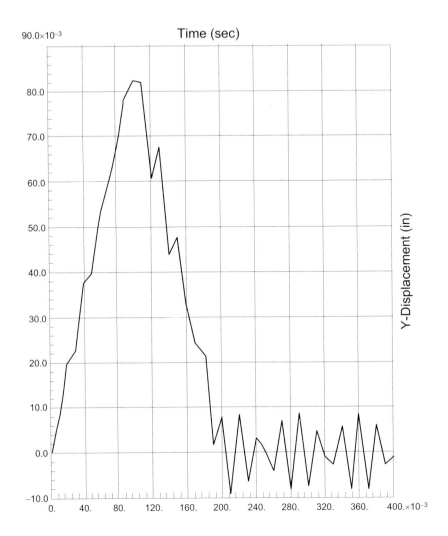

Fig. 9.21 Y Displacement Time plot for Joint 12 of Illustrative Example 9.5

Modal Periods and Natural Frequencies:
>   Using Notepad or other word editor, open C:/SAP2000/EXD 9.5.OUT.
>   The Natural Period and Frequency Table is stored in that file.

(Table 9.8 provides the edited values for the natural periods and frequencies of the structure in Illustrative Example 9.5.)

Table 9.8 Modal periods and natural frequencies of Illustrative Example 9.5

MODAL PERIODS AND FREQUENCIES

MODE	PERIOD (sec)	FREQUENCY (cps)	FREQUENCY (Rad/sec)	EIGENVALUE (Rad/sec)^2
1	0.023117	43.257298	271.793	73871.772
2	0.006832	146.371249	919.677	84507.042
3	0.004458	224.303194	1409.339	1.9862E+06
4	0.002705	369.676234	2322.744	5.3951E+06
5	0.001800	555.537934	3490.548	1.2184`+07

Illustrative Example 9.6

A square steel plate 40 *in* by 40 *in* and thickness 0.10 *in*, assumed to be fixed at the supports on its four sides (Figure 9.22), is acted upon by a harmonic force $F(t) = 0.1$ sin 5.3t *kip* applied normally at its center.

Determine:
(a) the first five natural frequencies
(b) the time-displacement function at the center of the plate
(c) the time-stress function at the center of plate. Use time step $\Delta t = 0.01$ *sec*. Neglect damping. [$E = 29,500$ *ksi*, $\upsilon = 0.3$, $\rho = 0.1$ (k sec^2 /in)/in^3]

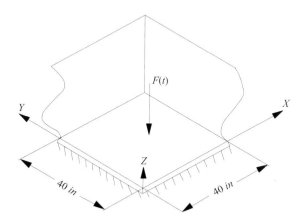

Fig 9.22   For Illustrative Example 9.6, a square plate supporting a normal harmonic force at the center

Solution:

Begin: Open SAP2000
Hint: Maximize both windows for a full view of all windows.
Select: In the lower right-hand corner of the window, use the drop-down menu to select "kip-in."

Model: From the main menu select:
FILE>NEW MODEL.
Number of Grid Spaces:
 x direction = 4
 y direction = 4
 z. direction = 0
Grid Spacing:
 x direction = 10
 y direction = 10
 z direction = 1
Then click OK.

Edit: Click on the XY icon. Maximize the XY window and use the PAN icon to drag the plot into the center of the window.

Draw: From the main menu select:
DRAW>QUICK DRAW SHELL ELEMENT.
Click on each of the square areas.
Then press the Esc key.

Label: From the main menu select:
VIEW>SET ELEMENTS.
Click Joint Labels and Shell Labels. Then click OK.
Use the PAN icon on the tool bar to drag the plot to the center of the window.

Boundary: Select all joints on the boundary of the plate.
From the main menu select:
ASSIGN>JOINT>RESTRAINTS.
Select restraints in all directions. Then click OK.
(Figure 9.23 reproduces the modeled plate for Illustrative Example 9.6.)

Material: From the main menu select:
DEFINE>MATERIALS.
Select: Steel
Click Modify/Show Materials.
Enter:
  Modulus of Elasticity = $29.5E3$

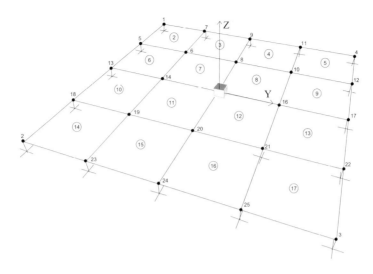

Fig. 9.23 Deep beam of Illustrative Example 9.6 modeled with shell elements

Poisson's Ratio = 0.3
Mass per unit of volume ρ = 0.1
Then click OK, OK.

Section: From the main menu select:
DEFINE>SHELL SECTIONS.
Click on Add New Section.
Select Material: STEEL
Enter thickness:
  Membrane = 0.1
  Bending = 0.1
Click Type: Shell. Then click OK, OK.
Click on the 3-D icon to default to a three-dimensional view.

(Figure 9.23 reproduces the modeled structure shown in the screen.)

Assign: Select all shell elements by clicking on them, and then from the main menu select:
ASSIGN>SHELL>SECTIONS.
Select shell section: SSEC2. Then click OK.

Load: From the main menu select:
DEFINE>STATIC LOAD CASES.

Select Type Live Load and set Self Weight Multiplier = 0.
Click "Change Load." Then click OK.

Load History Function: From the main menu select:
DEFINE>TIME HISTORY FUNCTION.
Click Add New Function. Then enter the values in Table 9.9.
Click Add after entering each line of the table.

Table 9.9 Harmonic Function

$t\ (sec)$	$F(t) = \sin 5.3t$
0	0
0.0593	0.3090
0.1186	0.5878
0.1778	0.8090
0.2371	0.9511
0.2964	1
0.3557	0.9511
0.4149	0.8090
0.4742	0.5878
0.5335	0.3090
0.5928	0
0.6520	−0.3090
0.7113	−0.5878
0.7706	−0.8090
0.8299	−0.9511
0.8891	−1
0.9484	−0.9511
1.0077	−0.8090
1.0670	−0.5878
1.1262	−0.3090
1.1855	0

Then click OK, OK.

Load History Cases: From the main menu select:
DEFINE>TIME HISTORY CASES.
Click "Add New History".
Accept the default HIST1.
Select Analysis Type: Periodic.
Click Modify/Show Values.
Enter 0.0 for damping in all modes.
Then click OK.

Check "Envelopes".
Type 120 in the Number of Output Time Steps.

Type 0.01 for Output Time Step Size.
For Load Assignment, select LOAD1.
For Function, select FUNC1.
Click Add.
Then click OK, OK.

Assign Load: Click on Joint 15 to select this joint.
From the main menu select:
ASSIGN>JOINT LOAD STATIC LOADS>FORCES.
Enter: Force Global Z = −0.1 Then click OK.

Options: From the main menu select:
ANALYZE>SET OPTIONS.
Click UZ, RX, and RY for available DOFs.
Click on "Dynamic Analysis" and on "Set Dynamic Parameters".
Enter Number of Modes = 5. Then click OK, OK.

Analyze: From the main menu select:
ANALYZE>RUN.
Enter Filename "EXD 9.6".
Then click SAVE.
When the calculations are concluded, click OK.

Print Input Tables: From the main menu select:
FILE>PRINT INPUT TABLES.
Click "Print to File" and accept all other defaults values.
Then click OK.
Use Notepad or other word editor to retrieve and edit the Input File stored
in file "C:\SAP2000\EXD 9.6.txt".

(The edited input tables of Illustrative Example 9.6 are reproduced in Table 9.10.)

Table 9.10 Edited Input Tables for Illustrative Example 9.6 (Units: kips, inches)

STATIC LOAD CASES

STATIC CASE	CASE TYPE	SELF WT FACTOR
LOAD1	LIVE	0.0000

TIME HISTORY CASES

HISTORY CASE	HISTORY TYPE	NUMBER OF TIME STEPS	TIME STEP INCREMENT
HIST1	LINEAR	120	0.01000

JOINT DATA

JOINT	GLOBAL-X	GLOBAL-Y	GLOBAL-Z	RESTRAINTS
1	−20.00000	−20.00000	0.00000	1 1 1 1 1 1
2	20.00000	−20.00000	0.00000	1 1 1 1 1 1

Table 9.10 (continued)

JOINT	GLOBAL-X	GLOBAL-Y	GLOBAL-Z	RESTRAINTS
3	20.00000	20.00000	0.00000	1 1 1 1 1 1
4	-20.00000	20.00000	0.00000	1 1 1 1 1 1
5	-10.00000	-20.00000	0.00000	1 1 1 1 1 1
6	-10.00000	-10.00000	0.00000	0 0 0 0 0 0
7	-20.00000	-10.00000	0.00000	1 1 1 1 1 1
8	-10.00000	0.00000	0.00000	0 0 0 0 0 0
9	-20.00000	0.00000	0.00000	1 1 1 1 1 1
10	-10.00000	10.00000	0.00000	0 0 0 0 0 0
11	-20.00000	10.00000	0.00000	1 1 1 1 1 1
12	-10.00000	20.00000	0.00000	1 1 1 1 1 1
13	0.00000	-20.00000	0.00000	1 1 1 1 1 1
14	0.00000	-10.00000	0.00000	0 0 0 0 0 0
15	0.00000	0.00000	0.00000	0 0 0 0 0 0
16	0.00000	10.00000	0.00000	0 0 0 0 0 0
17	0.00000	20.00000	0.00000	1 1 1 1 1 1
18	10.00000	-20.00000	0.00000	1 1 1 1 1 1
19	10.00000	-10.00000	0.00000	0 0 0 0 0 0
20	10.00000	0.00000	0.00000	0 0 0 0 0 0
21	10.00000	10.00000	0.00000	1 1 1 1 1 1
22	10.00000	20.00000	0.00000	1 1 1 1 1 1
23	20.00000	-10.00000	0.00000	1 1 1 1 1 1
24	20.00000	0.00000	0.00000	1 1 1 1 1 1
25	20.00000	10.00000	0.00000	1 1 1 1 1 1

SHELL ELEMENT DATA

SHELL	JNT-1	JNT-2	JNT-3	JNT-4	SECTION	ANGLE	AREA
2	1	5	6	7	SSEC2	0.000	100.000
3	7	6	8	9	SSEC2	0.000	100.000
4	9	8	10	11	SSEC2	0.000	100.000
5	11	10	12	4	SSEC2	0.000	100.000
6	5	13	14	6	SSEC2	0.000	100.000
7	6	14	15	8	SSEC2	0.000	100.000
8	8	15	16	10	SSEC2	0.000	100.000
9	10	16	17	12	SSEC2	0.000	100.000
10	13	18	19	14	SSEC2	0.000	100.000
11	14	19	20	15	SSEC2	0.000	100.000
12	15	20	21	16	SSEC2	0.000	100.000
13	16	21	22	17	SSEC2	0.000	100.000
14	18	2	23	19	SSEC2	0.000	100.000
15	19	23	24	20	SSEC2	0.000	100.000
16	20	24	25	21	SSEC2	0.000	100.000
17	21	25	3	22	SSEC2	0.000	100.000

JOINT FORCES Load Case LOAD1

JOINT	GLOBAL-X	GLOBAL-Y	GLOBAL-Z	GLOBAL-XX	GLOBAL-YY	GLOBAL-ZZ
15	0.000	0.000	-0.100	0.000	0.000	0.000

Modal Periods and Natural Frequencies:

Using Notepad or other editor to open the following file:
C:/SAP2000/EXD 9.2.OUT.

The Natural Period and Frequency Table are stored in this file and reproduced as Table 9.11.

TABLE 9.11 Model Periods and Frequencies

MODE	PERIOD (sec)	FREQUENCY (cps)	FREQUENCY (Rd/sec)	EIGENVALUE $(Rd/sec)^2$
1	0.163827	6.104004	38.352588	1470.921
2	0.093938	10.645303	66.886414	4473.792
3	0.084367	11.852935	74.474189	5546.405
4	0.065660	9.229916	95.692386	9157.033
5	0.061039	6.382842	102.936430	10595.909

Plot Displacement: From the main menu select:
DISPLAY>SHOW TIME HISTORY TRACES.

Click "Define Functions" and "Add Input Functions".
Select "Add Joint Displs/Forces".
Click OK.
    Time History Joint Function: Joint ID = 15
    Vector Type: select Displ
    Component: select UZ
    Then click OK, OK.

Highlight "Joint 15" in the "list of Functions" column.
Click "ADD" to move "Joint 15" to the "Plot Functions" column.
Enter time range from 0 to 0.4.
Axis Labels:
        Horizontal: TIME (sec)
        Vertical: Z-DISPLACEMENT JOINT 15 (in)

Click Display.
The window will show the time-displacement function for Joint 15.

Select FILE>PRINT GRAPHICS.
To exit, click Done.

(Fig 9.24 reproduces the Z-Displacement-Time plot for joint 15.)

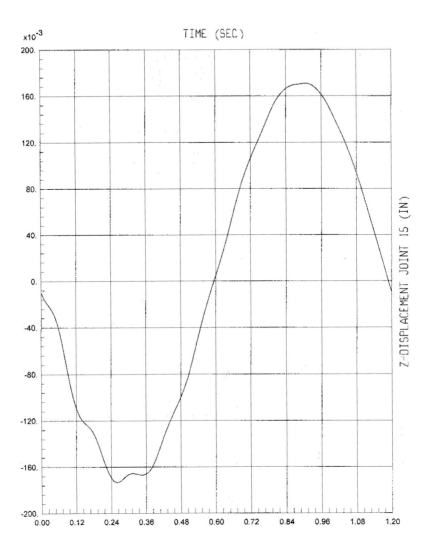

Fig. 9.24 Z Displacement time plot for Joint 15 of Illustrative Example 9.6

9.7   Practice Problems

Problem 9.1
Determine the displacements and stresses for the triangular plate shown in Figure P9.1. Assume a state of plane stress. Use triangular shell elements.

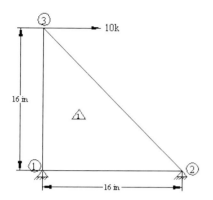

$E = 29,000 \ ksi$
$t \ = 0.3 \ in$
$\rho = 0.3 \ k \ sec^2/in/in^3$

Fig. P9.1

Problem 9.2

The plate shown in Figure P9.2 is modeled with two triangular elements. Determine the displacements and stresses.

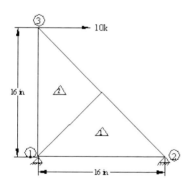

$E = 29,000 \ ksi$
$t = 0.3 \ in$
$\rho = 0.3 \ k \ sec^2/in/in^3$

Fig. P9.2

Analysis of Space Trusses Using SAP2000

Problem 9.3
The plate shown in Figure P9.3 is modeled with two triangular plate elements. Determine displacements and stresses. Assume a state of plane stress.

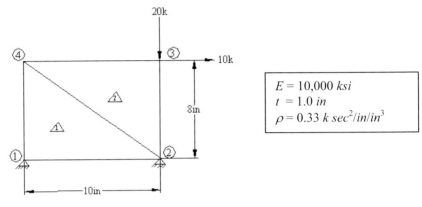

$$E = 10,000 \ ksi$$
$$t = 1.0 \ in$$
$$\rho = 0.33 \ k \ sec^2/in/in^3$$

Fig. P9.3

Problem 9.4
Use SAP2000 to solve Illustrative Example 9.1, modeling the plate with four rectangular plane stress elements as shown in Figure P9.4.

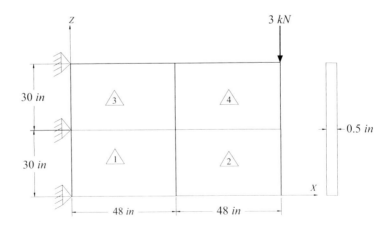

Fig. P9.4

Problem 9.5
The deep cantilever plate shown in Figure P9.5 carries a distributed force per unit of area of magnitude $w = 0.10 \ lb/in^2$ applied in the opposite direction of the axis Z. Model this plate with six rectangular plane stress elements and use SAP2000 to perform the structural analysis.

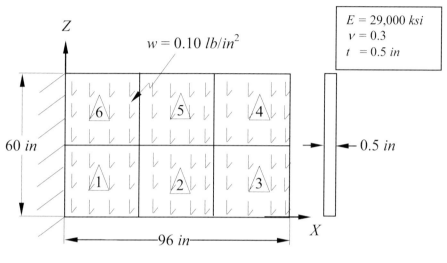

Fig. P9.5

Problem 9.6
Consider the rectangular steel plate of Illustrative Example 9.4 supporting a concentrated vertical force of magnitude $P = 8.0 \ kip$ applied at the center of the plate downward in the vertical direction. Model this plate in bending with 6 square elements as in Illustrative Example 9.5 and use SAP2000 to determine deflections and stresses.

Problem 9.7
Solve Problem 9.6 for the distributed normal load of $0.01 \ kip/in^2$ in addition to the concentrated force of 8 $kips$ prescribed in Problem 9.6.

Note: The solution of Illustrative Example 9.4 (with load $w = 0.01 \ kip/in^2$) together with solutions of Problems 9.6 (with load $P = 8 \ kips$) and 9.7 (with loads $w = 0.01 \ kip/in^2$ and $P = 8 \ kips$) would allow the reader to verify the principle of superposition.

Problem 9.8

The steel plate shown in Figure P9.8 of dimensions 20 *in* × 20 *in* and thickness 0.10 *in* with a circular hole of radius $r = 50$ *in* is subjected to a suddenly applied in-plane lateral compressive pressure along the edges *AD* and *BC* of magnitude $p = 100$ *psi*. Model the plate with quadrilateral elements in each quarter section of the plate. Obtain the first five natural frequencies and plots of the response for displacements and stresses on the elements.

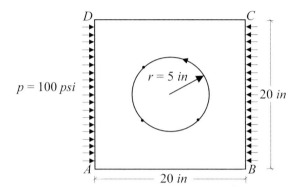

Fig. P9.8

Problem 9.9

Solve Problem 9.8 assuming that applied pressure reaches its maximum value in 0.1 *sec* and then decays to zero in the next 0.1 *sec*, as shown in Figure P9.9.

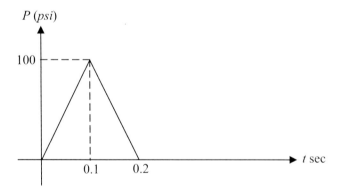

Fig. P9.9. Pressure load for Problem 9.9

Appendix I
Equivalent Nodal Forces

This appendix provides, for some common loads, the values for the equivalent nodal forces. These forces are defined as those forces that, applied at the nodal coordinates of a uniform beam element, would produce the same displacements at the nodal coordinates as those displacements resulting from the loads applied on the element.

Case (a): Concentrated Force

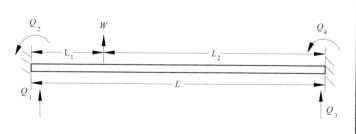

$$Q_1 = \frac{WL_2^2}{L^3}\left(3L_1 + L_2\right)$$

$$Q_2 = \frac{WL_1L_2^2}{L^2}$$

$$Q_3 = \frac{WL_1^2}{L^3}\left(L_1 + 3L_2\right)$$

$$Q_4 = -\frac{WL_1^2L_2}{L^2}$$

Case (b): Concentrated Moment

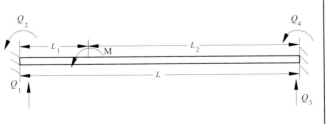

$$Q_1 = -\frac{6\,M L_1 L_2}{L^3}$$

$$Q_2 = \frac{M L_2}{L^2}\left(L_2 - 2L_1\right)$$

$$Q_3 = \frac{6\,M L_1 L_2}{L^3}$$

$$Q_4 = -\frac{M L_1}{L^2}\left(L_1 - 2L_2\right)$$

Case (c): Uniformly Distributed Force

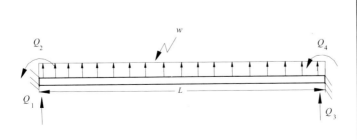

$$Q_1 = \frac{wL}{2}$$

$$Q_2 = \frac{wL^2}{12}$$

$$Q_3 = \frac{wL}{2}$$

$$Q_4 = -\frac{wL^2}{12}$$

Case (d): Trapezoidal Distributed Force

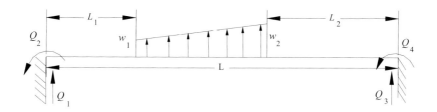

$$Q_1 = \frac{w_1(L-L_1)^3}{20L^3}\left\{(7L+8L_1)-\frac{L_2(3L+2L_1)}{L-L_1}\; x\; \left[\;1+\frac{L_2}{L-L_1}+\frac{L_2^2}{(L-L_1)^2}\right]\;+\;\frac{2L_2^4}{(L-L_1)^3}\right\}$$

$$+\frac{w_2(L-L_1)^3}{20L^3}\left\{(3L+2L_1)\left[1+\frac{L_2}{L-L_1}\;\frac{L_2^2}{(L-L_1)^2}\right]-\frac{L_2^3}{(L-L_1)^2}\left[2+\frac{15L-8L_2}{L-L_1}\right]\right\}$$

$$Q_2 = \frac{w_1(L-L_1)^3}{60L^2}\left\{3(L+4L_1)-\frac{L_2(2L+3L_1)}{L-L_1}\left[1+\frac{L_2}{L-L_1}+\frac{L_2^2}{(L-L_1)^2}\right]+\frac{3L_2^4}{(L-L_1)^3}\right\}$$

$$+\frac{w_2(L-L_1)^3}{60L^2}\left\{(2L+3L_1)\left[1+\frac{L_2}{L-L_1}+\frac{L_2^2}{(L-L_1)^2}\right]-\frac{3L_2^4}{(L-L_1)^3}\right\}+\frac{w_2(L-L_1)^3}{60L^2}$$

$$\left\{(2L+3L_1)\left[1+\frac{L_2}{L-L_1}+\frac{L_2^2}{(L-L_1)^2}\right]-\frac{3L_2^3}{(L-L_1)^2}\left[1+\frac{5L-4L_2}{L-L_1}\right]\right\}$$

$$Q_3 = \left(\frac{w_1+w_2}{2}\right)(L-L_1-L_2)-Q_1$$

$$Q_4 = \frac{L-L_1-L_2}{6}\left[w_1-(2L+2L_1-L_2)-w_2(L-L_1+2L_2)\right]+Q_1L-Q_2$$

Case (e): Axial Concentrated Force

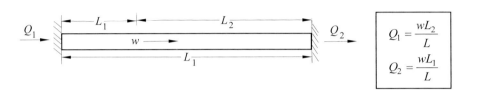

$$Q_1 = \frac{wL_2}{L}$$

$$Q_2 = \frac{wL_1}{L}$$

Case (f): Distributed Axial Force

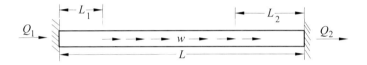

$$Q_1 = \frac{w}{2L}\left(L - L_1 - L_2\right)\left(L - L_1 + L_2\right)$$

$$Q_2 = \frac{w}{2L}\left(L - L_1 - L_2\right)\left(L + L_1 - L_2\right)$$

Case (g): Concentrated Torsional Moment:

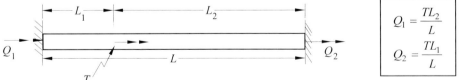

$$Q_1 = \frac{TL_2}{L}$$

$$Q_2 = \frac{TL_1}{L}$$

Appendix II
Displacement Functions for Fixed-End Beams

*Case (a): Concentrated Force

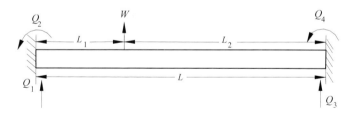

$$y_L(x) = \frac{1}{EI}[-\frac{1}{6}Q_1 x^3 + \frac{1}{2}Q_2 x^2] \qquad\qquad 0 \le x \le L_1$$

$$y_L(x) = \frac{1}{EI}\{\frac{1}{6}(w-Q_1)x^3 + \frac{1}{2}(Q_2 - wL)x^2 - [\frac{1}{2}(w-Q_1)L^2 + (Q_2 - wL_1)L]x$$

$$+ \frac{1}{3}(w-Q_1)L^3 + \frac{1}{2}(Q_2 - wL_1)L^2\} \qquad L_1 \le x \le L$$

*Case (b): Concentrated moment

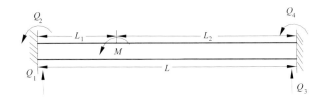

$$y_L(x) = \frac{1}{EI}\left[-\frac{1}{6}Q_1 x^3 + \frac{1}{2}Q_2 x^2\right] \qquad\qquad 0 \le x \le L_1$$

$$y_L(x) = \frac{1}{EI}\left[-\frac{1}{6}Q_1 x^3 + \frac{1}{2}(Q_2 - M)\, x^2\right] \qquad L_1 \le x \le L$$

*Case (c): Uniformly Distributed Force

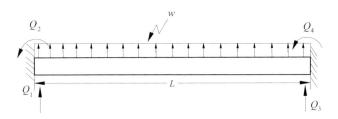

$$y_L(x) = \frac{1}{EI}\left[-\frac{1}{6}Q_1 x^3 + \frac{1}{2}Q_2 x^2 + \frac{1}{24}wx^4\right]$$

*Case (d): Trapezoidal Distributed Load

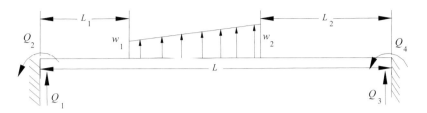

$$y_L(x) = \frac{1}{EI}\left[-\frac{1}{6}Q_1 x^3 + \frac{1}{2}Q_2 x^2\right] \qquad 0 \le x \le L_1$$

$$Y_L(x) = \frac{1}{EI}\left\{-\frac{1}{6}Q_1 x^3 + \frac{1}{2}Q_2 x^2 + \frac{w_1}{2}\left[\frac{x^4}{12} - \frac{L_1 x^3}{3} + \frac{L_1^2 x^2}{2}\right] + \frac{1}{6}\left[\frac{w_2 - w_1}{L - L_1 - L_2}\right]\right.$$

$$\left[\frac{x^5}{20} - \frac{x^4 L_1}{4} + \frac{x^3 L_1^2}{2} - \frac{x^2 L_1^3}{2}\right] - \frac{w_1 L_1^3 x}{6} + \left(\frac{w_2 - w_1}{L - L_1 - L_2}\right)\frac{L_1^4 x}{24} + \frac{w_1 L_1^4}{24} - \frac{l_1^5}{120}\left[\frac{w_2 - w_1}{L - L_1 - L_2}\right]\right\}$$

$$L_1 \le x \le L - L_2$$

$$y_L(x) = \frac{1}{EI}\left\{-\frac{Q_1 x^3}{6} + \frac{Q_2 x^2}{2} + w_1(L - L_1)\left[\left(\frac{L - L_1 - L_2}{4}\right)x^2 + \frac{x^5}{20}\frac{(-L + L_2)}{2}x^2\right]\right.$$

$$\frac{1}{2}(L - L_1 - L_2)(w_2 - w_1)\left[\left(\frac{L - L_1 - L_2}{6}\right)x^2 + \frac{x^5}{20} + \frac{(-L - L_2)}{2}x^2\right] + C_1 x + C_2\right\}$$

$$L - L_2 \le x \le L$$

where

$$C_1^* = \frac{Q_1 L^2}{2} - Q_2 L - w_1(L - L_1)\left[\frac{L^2 - LL_1 - L_{1L_2}}{2} - \frac{L^2}{2} + LL_2\right] - \frac{1}{2}(L - L_1 - L_2)(w_2 - w_1)$$

$$\left[\frac{L^2 - LL_1 - LL_2}{3} - \frac{L^2}{2} + LL_2\right]$$

$$C_2^* = C_1 L + \frac{Q_1 L^2}{2} - \frac{Q_2 L^2}{2} - w_1(L - L_1)\left[\left(\frac{L - L_1 - L_2}{4}\right)L^2 + \frac{L^5}{5} + \frac{\left(L_1^3 - L^2 L_2\right)}{2}\right]$$

$$-\frac{1}{2}(L - L_1 - L_2)(w_2 - w_1)\left[\left(\frac{L - L_1 - L_2}{6}\right)L^2 + \frac{L^5}{20} + \left(\frac{-L^3 - L^2 L_2}{2}\right)\right]$$

* Expressions for the Equivalent Nodal Forces Q_1, Q_2, Q_3, and Q_4 are given in Appendix I.

Appendix III
Structural Dynamics:
Single Degree of Freedom Systems

III-1   Basic Concepts and Definitions:

1. The *analytical* or *mathematical model* of a structure is an idealized representation for its analysis.

2. The number of *degrees of freedom* of a structural system is equal to the number of independent coordinates necessary to describe its position.

3. *A single degree of freedom system* is a structure modeled with one coordinate (see Figure III-1).

4. The *Free Body Diagram* (FBD) for dynamic equilibrium (to allow application of *D'Alembert's Principle*) is a diagram of the system isolated from all other bodies, showing all the external forces on the system, including the inertial force. This force is equal to the product of the mass and the acceleration and it is directed in the opposite sense to the positive direction (see Figure III-2).

5. The *stiffness* or *spring constant* of a linear system is the force necessary to produce a unit displacement. Therefore the relationship between the force, F_s, applied to a linear spring, with spring constant k, and the resulting displacement, u, is given by

$$F_s = ku \qquad\qquad \text{(III-1)}$$

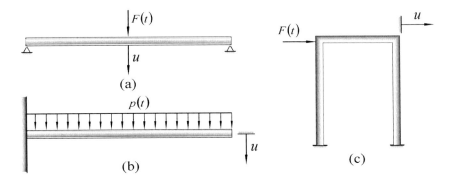

Fig. III-1   Examples of structures modeled as single degree of freedom systems

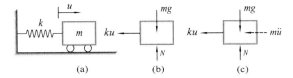

Fig. III-2   Alternate free body diagrams: (a) Single degree of freedom system
(b) Showing only external forces (c) Showing external and inertial forces

6.   The *equivalent spring constant* k_e for two *springs in parallel* (see Figure III-3) is calculated by

$$k_e = k_1 + k_2 \qquad\qquad\text{(III-2)}$$

and in general for *n* springs in parallel by

$$k_e = \sum_{i=1}^{n} k_i \qquad\qquad\text{(III-3)}$$

7.   The *equivalent spring constant* k_e for two *springs in series* (see Figure III-3) is calculated from

$$\frac{1}{k_e} = \frac{1}{k_1} + \frac{1}{k_2} \qquad\qquad\text{(III-4)}$$

and in general for *n springs in series* by

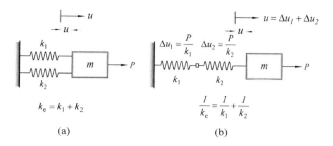

Fig. III-3   Combination of springs: (a) Springs in parallel; (b) Springs in series

$$\frac{1}{k_e} = \sum_{i=1}^{n} \frac{1}{k_i}$$ (III-5)

8. *Newton's Law of Motion* for a particle in *vector notation*:

$$\boldsymbol{F} = m\boldsymbol{a}$$ (III-6)

where $\boldsymbol{F}$ is the resultant force acting on a particle of mass m and $\boldsymbol{a}$ the resultant acceleration.

9. *Newton's Law of Motion* for a particle in *Cartesian components*:

$$\sum F_x = ma_x$$ (III-7a)

$$\sum F_y = ma_y$$ (III-7b)

$$\sum F_z = ma_z$$ (III-7c)

where a_x, a_y, and a_z are the components of the acceleration along the coordinate axes x, y, and z, respectively.

10. *Newton's Law of Motion for plane motion of a rigid body* that is symmetric with respect to the reference plane of motion (x-y plane):

$$\sum F_x = m(a_G)_x$$ (III-8a)

$$\sum F_y = m(a_G)_y$$ (III-8b)

$$\sum M_G = I_G \alpha \qquad\qquad\qquad \text{(III-8c)}$$

In the preceding equations $(a_G)_x$ and $(a_G)_y$ are the acceleration components, respectively, along the x and y axes, of the center of mass G of the body; α is the angular acceleration; I_G is the mass moment of inertia of the body with respect to an axis through G, the center of mass; and $\sum M_G$ is the sum of the moments of all the forces acting on the body with respect to an axis through G, perpendicular to the x-y plane.

Equations (III-8) are certainly also applicable to the motion of a rigid body in pure rotation about a fixed axis; however, for this particular type of plane motion, eq. (III-8c) may be substituted by

$$\sum M_0 = I_0 \alpha \qquad\qquad\qquad \text{(III-8d)}$$

in which the mass moment of inertia I_0 and the moment of the forces M_0 are determined with respect to the fixed axis of rotation.

11. An *undamped simple oscillator* is a structure modeled as a spring-mass system in which the damping or frictional forces has been "neglected" (see Figure III-2).

12. The differential equation of the undamped simple oscillator in free motion (no external forces applied) is given by the following differential equation

$$m\ddot{u} + ku = 0 \qquad\qquad\qquad \text{(III-9)}$$

where the acceleration has been indicated by $\ddot{u}$. In this notation, double over-dots denote the second derivative of the displacement with respect to time (acceleration) and obviously a single over-dot denotes the first derivative with respect to time, that is, the velocity.

13. The general solution of eq. (III-9) is

$$u = A \cos \omega\, t + B \sin \omega\, t. \qquad\qquad\qquad \text{(III-10)}$$

14. The constants of integration A and B are determined from initial conditions of the displacement u_0 and of the velocity υ_0 :

$$A = u_0$$
$$B = \upsilon_0 / \omega$$

15. The frequency ω in *rad/sec* or in *cycle/sec* (f) and the period (T) of vibration are given, respectively by the following,

$$\omega = \sqrt{k/m} \quad \text{is the natural frequency in } rad/sec$$

$$f = \frac{\omega}{2\pi} \quad \text{is the natural frequency in } cps$$

$$T = \frac{1}{f} \quad \text{is the natural period in } seconds$$

Substituting into eq. (III-10) the values of A and B results in

$$u = u_0 \cos \omega\, t + \frac{\upsilon_0}{\omega} \sin \omega\, t. \tag{III-11}$$

16. The natural frequency ω also may be expressed in terms of the static displacement resulting from the weight $W = mg$ applied to the spring:

$$\omega = \sqrt{\frac{g}{u_{st}}} \tag{III-12}$$

where $u_{st} = W/k$ is the static displacement of the spring due to the weight W.

17. The equation of motion, eq. (III-11), also may be written in the alternate forms:

$$u = C \sin(\omega\, t + \alpha) \tag{III-13}$$

or

$$u = C \cos(\omega\, t - \beta) \tag{III-14}$$

where C is the *amplitude of motion* given by

$$C = \sqrt{u_0^2 + (\upsilon_0 / \omega)^2} \tag{III-15}$$

and

$$\tan \alpha = \frac{u_0}{v_0 / \omega}$$

$$\tan \beta = \frac{v_0 / \omega}{u_0} \qquad \text{(III-16)}$$

III-2   Damped Single Degree of Freedom System

Real structures dissipate energy while undergoing vibratory motion. The most common and practical method for considering this dissipation of energy is to assume that it is due to *viscous damping forces*. These forces are assumed to be proportional to the magnitude of the velocity but acting in the direction opposite to the motion.

The factor of proportionality, c, between the frictional force and the velocity is called the *viscous damping coefficient*. It is expedient to express this coefficient as a fraction of the *critical damping* ($c_{cr} = 2\sqrt{km}$) in the system, namely

$$\xi = c / c_{cr} \qquad \text{(III-17)}$$

where ξ is damping ratio coefficient.

Critical damping may be defined as the least value of the damping coefficient for which the system will not oscillate when disturbed initially but will simply return to the equilibrium position.

The differential equation of motion for the free vibration of a damped structural system modeled by a damped single degree of freedom system (oscillator) as shown in Figure III-4 is given by

$$m\ddot{u} + c\dot{u} + ku = 0 \qquad \text{(III-18)}$$

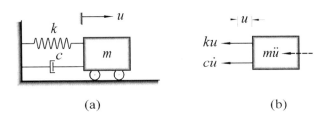

(a)   (b)

Fig. III-4 (a) Viscous damped oscillator   (b) Free body diagram

The analytical expression for the solution of eq. (III-18) depends on the magnitude of the damping ratio coefficient. Three cases are possible:

1. Critically damped system ($\xi = 1$).
2. Overdamped system ($\xi > 1$).
3. Underdamped system ($\xi < 1$).

For the underdamped system ($\xi < 1$), the solution of the differential equation of motion (III-18) may be written as

$$u(t) = e^{-\xi \omega t} \left[u_0 \cos \omega_D t + \frac{v_0 + u_0 \xi \omega}{\omega_D} \sin \omega_D t \right] \tag{III-19}$$

in which

$$\omega = \sqrt{k/m} \quad \text{is the undamped frequency}$$

$$\omega_D = \omega\sqrt{1-\xi^2} \quad \text{is the damped frequency}$$

$$\xi = \frac{c}{c_{cr}} \quad \text{is the damping ratio coefficient} \tag{III-20}$$

$$c_{cr} = 2\sqrt{km} \quad \text{is the critical damping}$$

$$u_0, \; v_o \quad \text{are the initial displacement and the initial velocity}$$

Alternatively, eq.(III-19) can be written as

$$u(t) = Ce^{-\xi \omega t} \cos(\omega_D t - \alpha) \tag{III-21}$$

where

$$C = \sqrt{u_0^2 + \frac{(v_0 + u_0 \xi \omega)^2}{\omega_D^2}} \tag{III-22}$$

and

$$\tan \alpha = \frac{(v_0 + u_0 \xi \omega)}{\omega_D u_0}.$$

III-3   Logarithmic Decrement

A practical method of determining the damping present in a system is to evaluate experimentally the *logarithmic decrement*, δ, which is defined as the natural logarithm of the ratio of two consecutive peaks for the displacement or acceleration, in free vibration as shown in the plot for the displacement reproduced in Figure III-5. Thus,

$$\delta = \ln\frac{u_1}{u_2} \quad \text{or} \quad \delta = \ln\frac{\ddot{u}_1}{\ddot{u}_2} \tag{III-23}$$

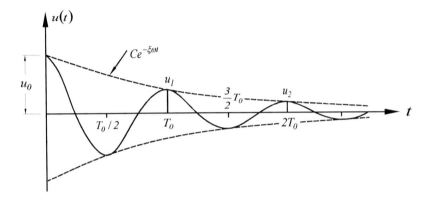

Fig. III-5 Free-vibration response for underdamped system

The value of the damping coefficient for real structures is much less than the critical damping coefficient and usually ranges between 2% and 10% of the critical damping value. Substituting for the extreme value $\xi = 0.10$ into eq. (III-20) results in

$$\omega_D = 0.995\omega.$$

It can be observed that the frequency of vibration for a system with as much as a 10% damping ratio is essentially equal to the undamped natural frequency. Thus, in practice, the natural frequency for a damped system may be taken to be equal to the undamped natural frequency.

III-4   Response to Harmonic Loading

Consider in Figure III-6 a structure modeled as the damped simple oscillator subjected to a harmonic force given by $F_0 \sin \overline{\omega} t$, where F_0 is the amplitude of the force and $\overline{\omega}$ is its frequency expressed in *rad/sec*.

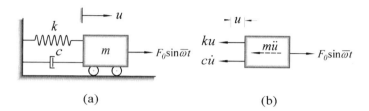

(a)   (b)

Fig. III-6   (a) Damped oscillator harmonically excited; (b) Free body diagram

The differential equation of motion for a linear single degree of freedom system may be expressed by the second-order differential equation:

$$\ddot{u} + 2\xi\omega\, \dot{u} + \omega^2 u = \frac{F_0}{m} \sin \overline{\omega} t \tag{III-24}$$

in which

$\overline{\omega}$ is the forced frequency,

$\xi = \dfrac{c}{c_{cr}}$ is the damping ratio coefficent,

$\omega = \sqrt{\dfrac{k}{m}}$ is the natural frequency.

The general solution of eq. (III-24) is obtained as the superposition of the *complementary* (*transient*) and the *particular* (*steady-state*) *solutions*, namely

$$u = e^{-\xi\omega t}\left(A\cos\omega_D t + B\sin\omega_D t\right) + \frac{F_0/k \sin(\overline{\omega} t - \theta)}{\sqrt{\left(1-r^2\right)^2 + \left(2r\xi\right)^2}} \tag{III-25}$$

$\underbrace{\phantom{e^{-\xi\omega t}\left(A\cos\omega_D t + B\sin\omega_D t\right)}}$ $\underbrace{\phantom{\dfrac{F_0/k \sin(\overline{\omega} t - \theta)}{\sqrt{\left(1-r^2\right)^2}}}}$

      transient solution   steady-state solution

in which

$$r = \frac{\overline{\omega}}{\omega}$$ is the frequency ratio

$$\omega_s = \omega\sqrt{1 - r^2}$$ is the damped natural frequency

$$\theta = \tan^{-1}\left(\frac{2r\xi}{1 - r^2}\right)$$ is the phase angle

and A and B are constants of integration, which can be determined from the initial conditions.

The transient part of the solution vanishes rapidly to zero because of the negative exponential factor, thus leaving only the steady-state solution, u_p. From eq. (III-25), the steady-state solution is

$$u_p = \frac{F_0 / k \, \sin(\overline{\omega}t - \theta)}{\sqrt{\left(1 - r^2\right)^2 + \left(2r\xi\right)^2}} \qquad \text{(III-26)}$$

The maximum value or amplitude, U, divided by the static displacement $u_{st} = F_0/k$ is defined as the *Dynamic Magnification Factor D*, that is,

$$D = \frac{U}{u_{st}} = \frac{1}{\sqrt{\left(1 - r^2\right)^2 + \left(2r\xi\right)^2}} \qquad \text{(III-27)}$$

The parametric plot (Figure III-7) of eq. (III-27) depicting the Dynamic Magnification Factor, D, as a function of the frequency ratio, r, and the parameter ξ (damping ratio) is referred to as the *frequency response*. Of particular significance is the condition of resonance $(r = \overline{\omega} / \omega = 1)$ for which the amplitude of motion becomes very large for the lightly damped system and tends to become infinity for the undamped system.

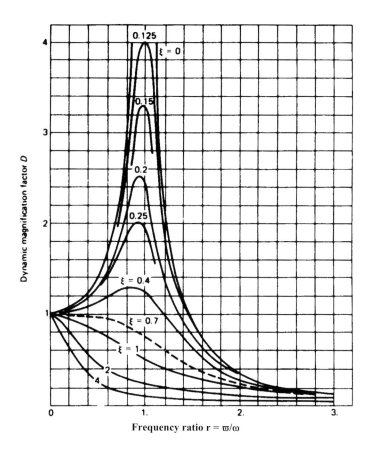

Fig. III-7   Dynamic magnification factor as a function of the frequency ratio for various values of the damping ratio

III-5   Response to Support Motion

The response of the structure to support or foundation motion can be obtained in terms of the absolute motion of the mass or in terms of its motion relative to the motion of the support, $u_s(t)$. In this latter case, the equation assumes a simpler and more convenient form, namely

$$m\ddot{u}_r + c\dot{u}_r + ku_r = F_{eff}(t)$$

(III-28)

in which

$$u_r = u - u_s \quad \text{is the relative displacement} \tag{III-29}$$

and

$$F_{eff}(t) = -m\ddot{u}_s(t) \quad \text{is the effective force.}$$

For harmonic excitation at the support expressed as a displacement function

$$u_s(t) = u_0 \sin \overline{\omega} t \tag{III.30}$$

or as an acceleration function

$$\ddot{u}_s(t) = -u_0 \overline{\omega}^2 \sin \overline{\omega} t. \tag{III.31}$$

The solution in terms of the relative motion u_r is of the same mathematical form as the solution given by eq. (III-26) when the force is applied to the mass, namely

$$u_r(t) = \frac{u_0 \, r^2 \sin (\overline{\omega} t - \theta)}{\sqrt{(1 - r^2)^2 + (2r\xi)^2}} \tag{III-32}$$

in which all the symbols have already been defined.

$$r = \frac{\overline{\omega}}{\omega} \qquad\qquad \text{is the frequency ratio}$$

$$\omega_D = \omega\sqrt{1 - r^2} \qquad \text{is the damped natural frequency}$$

$$\theta = \tan^{-1}\left(\frac{2r\xi}{1 - r^2}\right) \text{ is the phase angle}$$

$$\xi = \frac{c}{c_{cr}} \qquad\qquad \text{is the damping ratio.}$$

III-6   Equivalent Viscous Damping

The *equivalent viscous damping* in a structural system may be evaluated experimentally either from the *peak amplitude* or from the *bandwidth* obtained from a plot of the amplitude-frequency curve when the system is forced to *harmonic vibration*. A typical response curve for a moderately damped structure is shown in Figure III-8.

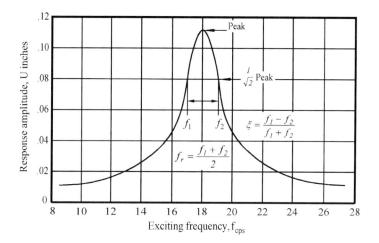

Fig. III-8 Experimental frequency response curve

It is can be shown from eq. (III-27) that, at resonance ($r = 1$), the damping ratio is given by

$$\xi = \frac{1}{2D(r = 1)}$$ (III-33)

where $D\,(r = 1)$ is the dynamic magnification factor evaluated at resonance ($r = 1$).

Alternatively, the damping ratio coefficient may be estimated experimentally from the bandwidth, i.e., the distance between the two branches of the frequency response curve, as shown in Figure III-8. Conveniently, the bandwidth is determined at the response amplitude equal to $1/\sqrt{2}$ of the peak amplitude.

In the typical experimental frequency response curve (Figure III-8), f_1 and f_2 are the frequencies for such bandwidth. The resulting expression to estimate the damping ratio coefficient is

$$\xi = \frac{f_2 - f_1}{f_2 + f_1}$$ (III-34)

Most commonly, *equivalent viscous damping* is evaluated by equating the experimentally measured energy dissipated in the system during a cycle of harmonic vibration at the resonant frequency to the theoretically calculated energy that the system, assumed viscously damped, would dissipate in a cycle (see Figure III-9).

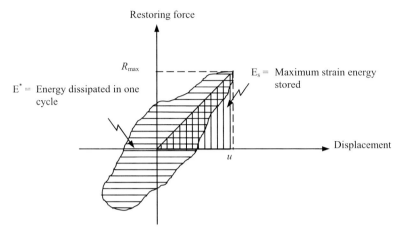

Fig. III-9   Restoring force vs. displacement during a cycle of vibration showing the
energy dissipated E^* (area within the loop) and the maximum energy
stored, assuming elastic behavior (triangular area under maximum
displacement)

This approach leads to the following expression for the equivalent viscous damping
ratio:

$$\xi_{eq} = \frac{1}{4\pi r} \frac{E^*}{E_s} \qquad \text{(III-35)}$$

in which

   E^* = energy dissipated in the system during a cycle of harmonic vibration at
         resonance

   E_s = strain energy stored at maximum displacement if the system were
         elastic

   r = ratio of forced frequency to the natural frequency of the system

III-7   Vibration Isolation

Two related problems are of great importance, in *vibration isolation*: (1) the *motion
transmissibility*, that is, the relative motion transmitted from the foundation to the
structure ($T_r = U/U_s$); and (2) the *force transmissibility*, that is, the relative

magnitude of the force transmitted from the structure to the foundation ($T_r = F_T/F_0$). For both of these problems, the transmissibility T_r is given by

$$T_r = \sqrt{\frac{1+(2r\xi)^2}{(1-r^2)^2+(2r\xi)^2}} \qquad \text{(III-36)}$$

in which

U = maximum harmonic amplitude transmitted to the mass
U_s = maximum harmonic amplitude of the support vibration
F_T = maximum harmonic force transmitted to the support
F_0 = maximum harmonic force applied to the mass

A plot of transmissibility as a function of the frequency ratio and damping ratio is shown in Figure III-10. The curves in this figure are similar to the curves in Figure III-7, representing the frequency response of the damped oscillator. The major difference between these two sets of curves is that all of the curves in Figure III-10 pass through the same point at a frequency ratio $r = \sqrt{2}$.

It may be observed in Figure III-10 that an increase in damping helps to decrease the transmissibility, but only for values of the frequency ratio $r < \sqrt{2}$; however, for the ratio r greater than $\sqrt{2}$, an increase of damping tends to reduce the effectiveness of vibration isolation.

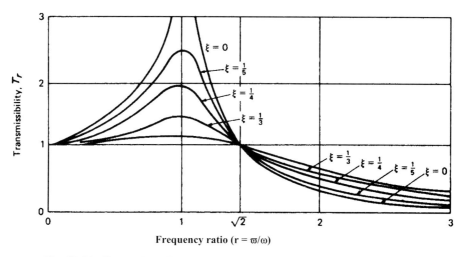

Fig. III-10 Transmissibility versus frequency ratio for vibration isolation

III-8   Response to General Dynamic Loading: Duhamel's Integral

The differential equation of motion for a single degree of freedom linear system can be solved for any forcing function of time τ, $F(\tau)$ as the force depicted in Figure III-11, in terms of *Duhamel's integral,* which for the undamped system is given by

$$u(t) = \frac{1}{m\omega} \int_0^t F(\tau)\sin\omega\left(t - \tau\right) d\tau \qquad \text{(III-37)}$$

and for a damped system by

$$u(t) = \frac{1}{m\omega_D} \int_0^t F(\tau)e^{-\xi\,\omega(t-\tau)} \sin\omega_D\left(t - \tau\right) d\tau \qquad \text{(III-38)}$$

where

$\omega = \sqrt{\dfrac{k}{m}}$ is the (undamped) natural frequency

$\omega_D = \omega\sqrt{1-\xi^2}$ is the damped frequency

$\xi = \dfrac{c}{c_{cr}}$ is the damping ratio and

τ is the dummy variable of integration.

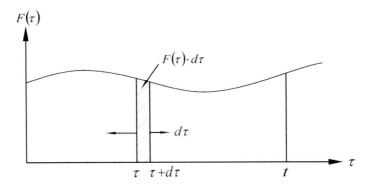

Fig. III-11 General load function as impulsive loading

The expression for the displacement, *u(t)*, calculated by *Duhamel's integral* includes the transient motion corresponding to initial displacement zero and initial velocity also zero. If the initial conditions are of different values, their effect should be added to eq. (III-37) or to eq. (III-38); thus given the total solution $u(t)$ in eq. (III-39) for the undamped system or $u(t)$ in eq. (III-40) for the damped system

$$u(t) = u_0 \cos \omega t + \frac{\upsilon_0}{\omega} \sin \omega t + \frac{1}{m\omega} \int_0^t F(\tau) \sin \omega(t - \tau) d\tau \qquad \text{(III-39)}$$

and

$$u(t) = e^{-\xi\omega t} \left(u_0 \cos \omega_D t + \frac{\upsilon_0 + u_0 \xi\omega}{\omega_D} \sin \omega_D t \right)$$
$$+ \frac{1}{m\omega_D} \int_0^t F(\tau) e^{-\xi\omega(t-\tau)} \sin \omega_D (t - \tau) d\tau. \qquad \text{(III-40)}$$

The application of Duhamel's integral for specific types of forces follows.

1. Constant force

Consider the case of a constant force of magnitude F_0 applied suddenly to the undamped oscillator at time $t = 0$, as shown in Figure III-12.

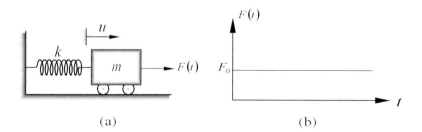

(a)   (b)

Fig. III-12   Undamped oscillator acted upon by constant force

For both initial displacement and initial velocity equal to zero, the application of eq.(III-37) to this case results in

$$u(t) = \frac{F_0}{k} (1 - \cos \omega t) = u_{st} (1 - \cos \omega t) \qquad \text{(III-41)}$$

where $u_{st} = \dfrac{F_0}{k}$ is the static displacement due to a force of magnitude F_0. The response for such a suddenly applied constant force is plotted in Figure III-13.

It can be observed that this solution is very similar to the solution for the free vibration of the undamped oscillator [see eq. (III-13)].

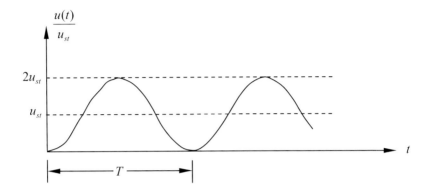

Fig. III-13 Response of an undamped single degree of freedom system to a suddenly applied constant force

The major difference is that the coordinate axis t has been shifted by an amount equal to $u_{st} = F_o/k$. Also, it should be noted that the maximum displacement $2u_{st}$ is exactly twice the displacement that the force F_0 would produce if it were applied statically. This is an important result; the maximum displacement of a linear elastic system for a constant force applied suddenly is twice the displacement caused by the same force applied statically (slowly).

2. Rectangular Load

Consider a constant force F_0 suddenly applied for only a limited time duration, t_d, that is, $F = F_0$ for the time intervals $t \leq t_d$ and $F = 0$ for $t \geq t_d$, as shown in Figure III-14. In this case, the displacement function may be conveniently expressed in terms of the *dynamic load factor* (DLF) (defined as the displacement at any time t divided by the static displacement), namely,

$$DLF = 1 - \cos 2\pi \frac{t}{T}, \qquad\qquad t \leq t_d \qquad\qquad \text{(III-42)}$$

and

$$DLF = \left(2 \sin \frac{\pi t_d}{T}\right) \sin\left[2\pi\left(\frac{t}{T} - \frac{t_d}{2T}\right)\right], \qquad t \geq t_d \qquad\qquad \text{(III-43)}$$

Figure III-14 is a plot of the maximum value of the Dynamic Load Factor $(DLF)_{\text{max}}$ as a function of the ratio of the duration of the force to the natural period of the system, t_d/T. It may be observed in Figure III-14 that for the relatively long duration of the applied force, the $(DLF)_{max}$ is equal to 2, the same value for the force applied indefinitely.

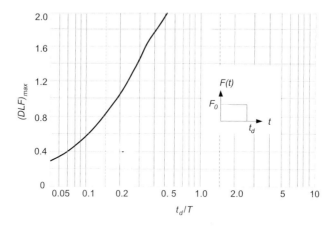

Fig. III-14 Maximum dynamic load factor for the undamped oscillator acted on by a rectangular force of duration t_d.

3. Triangular Load

For the undamped oscillator, initially at rest and subjected to a force $F(t)$ that has an initial value F_0 and that decreases linearly to zero at time t_d, as shown in the Figure III-15, the response is conveniently expressed for the two intervals $t \le t_d$ and $t \ge t_d$ by the following equations:

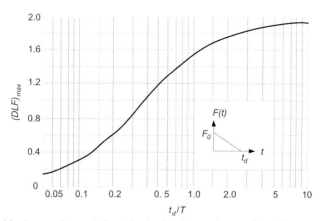

Fig. III-15 Maximum dynamic load factor for the undamped oscillator acted upon by a triangular force for a time duration t_d.

$$DLF = \frac{u}{u_{st}} = 1 - \cos\left(\frac{2\pi t}{T}\right) + \frac{\sin\left(\frac{2\pi t}{T}\right)}{\left(\frac{2\pi t_d}{T}\right)} - \frac{t}{t_d} \qquad 0 \leq t \leq t_d \qquad \text{(III-44)}$$

and

$$DLF = \frac{1}{\dfrac{2\pi t_d}{T}}\left\{\sin 2\pi \frac{t}{T} - \sin 2\pi\left(\frac{t}{T} - \frac{t_d}{T}\right)\right\} - \cos 2\pi \frac{t}{T} \qquad t \geq t_d . \qquad \text{(III-45)}$$

The plot of the maximum dynamic load factor, for this triangular load, as a function of the relative time duration t_d/T for the undamped oscillator is shown in Figure III-15. As it would be expected, the maximum value of the dynamic load factor approaches 2 as t_d/T becomes larger; that is, the effect of the decay of the force is negligible, in this case, for the time required for the system to reach the maximum peak.

III-9   Response to General Dynamic Loading: The Direct Method

The solution of the differential equation of motion also may be obtained by application of the *Direct Method*. In this method, it is assumed that the forcing function is given by a segmental linear function between defining points, as shown in Figure III-16. On the basis of this assumption, the solution obtained is exact. The response is calculated at each time increment for the conditions existent at the end of the preceding time interval (initial conditions for the new time interval) and the action of the excitation force applied during the time interval, which is assumed to be linear. In this case, excitation function $F(t)$, approximated by a piecewise linear function as shown in Figure III-16, is given by

$$F(t) = \left(1 - \frac{t - t_i}{\Delta t}\right)F_i + \left(\frac{t - t_i}{\Delta t}\right)F_{i+1}, \qquad t_i \leq t \leq t_{i+1} \qquad \text{(III-46)}$$

in which $t_i = i \cdot \Delta t$ for equal intervals of duration Δt and $i = 1, 2, 3,...N$. The differential equation of motion, for the interval $t_i \leq t \leq t_{i+1}$, is then given by

$$m\ddot{u} + c\dot{u} + ku = \left(1 - \frac{t - t_i}{\Delta t}\right)F_i + \left(\frac{t - t_i}{\Delta t}\right)F_{i+1}, \qquad t_i \leq t \leq t_{i+1} \qquad \text{(III-47)}$$

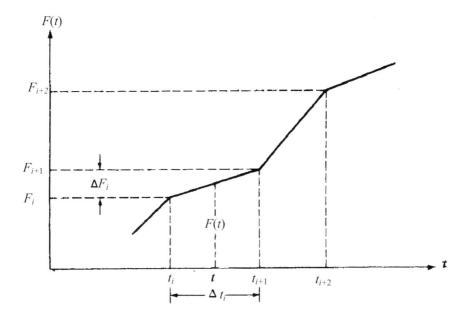

Fig. III-16 Segmental linear loading function.

The solution of eq. (III-47) results in the following formulas to calculate the displacement, velocity and acceleration at the next time step, $t_{i+1} = t_i + \Delta t$:

$$u_{i+1} = A' u_i + B' \dot{u}_i + C' F_i + D'_i F_{i+1} \tag{III-48}$$

$$\dot{u}_{i+1} = A'' u_i + B'' \dot{u}_i + C'' F_i + D'' F_{i+1} \tag{III-49}$$

$$\ddot{u}_{i+1} = -\omega^2 u_{i+1} - 2\xi \omega \dot{u}_{i+1} + \frac{F_{i+1}}{m} \tag{III-50}$$

Equations (III-48), (III-49), and (III-50) are recurrence formulas to calculate, respectively, the displacement, velocity, and acceleration at the next time step $t_{i+1} = t_i + \Delta t$ from the calculated values for these quantities at the preceding time step t_i. Because these recurrence formulas are exact, for a segmental linear force, the only restriction in selecting the length of the time step, Δt, is that it allows a close approximation to the excitation function and that equally spaced time intervals do not miss the peaks of this function. This numerical procedure is highly efficient because the coefficients in eqs. (III-48) and (III-49) need to be calculated only once.

The final expressions to calculate the coefficients A', B',...D'' are given in the Box III.1.

$$A' = e^{-\xi\omega\Delta t}\left(\frac{\xi\omega}{\omega_D}\sin\omega_D\Delta t + \cos\omega_D\Delta t\right)$$

$$B' = e^{-\xi\omega\Delta t}\left(\frac{1}{\omega_D}\sin\omega_D\Delta t\right)$$

$$C' = \frac{1}{k}\left\{e^{-\xi\omega\Delta t}\left[\left(\frac{1-2\xi^2}{\omega_D\Delta t}-\frac{\xi\omega}{\omega_D}\right)\sin\omega_D\Delta t - \left(1+\frac{2\xi}{\omega\Delta t}\right)\cos\omega_D\Delta t\right]+\frac{2\xi}{\omega\Delta t}\right\}$$

$$D' = \frac{1}{k}\left\{e^{-\xi\omega\Delta t}\left[\left(\frac{2\xi-1}{\omega_D\Delta t}\sin\omega_D\Delta t+\frac{2\xi}{\omega\Delta t}\cos\omega_D\Delta t\right)\right]+\left(1-\frac{2\xi}{\omega\Delta t}\right)\right\}$$

$$A'' = -e^{-\xi\omega\Delta t}\left(\frac{\omega^2}{\omega_D}\sin\omega_D\Delta t\right)$$

$$B'' = e^{-\xi\omega\Delta t}\left(\cos\omega_D\Delta t\right)-\frac{\xi\omega}{\omega_D}\sin\omega_D\Delta t$$

$$C'' = \frac{1}{k}\left\{-e^{-\xi\omega\Delta t}\left[\left(\frac{\omega^2}{\omega_D}+\frac{\omega\xi}{\Delta t\omega_D}\right)\sin\omega_D\Delta t+\frac{1}{\Delta t}\cos\omega_D\Delta t\right]-\frac{1}{\Delta t}\right\}$$

$$D'' = \frac{1}{k\Delta t}\left\{-e^{-\xi\omega\Delta t}\left(\frac{\omega\xi}{\omega_D}\sin\omega_D\Delta t+\cos\omega_D\Delta t\right)+1\right\}$$

Box III.1 Coefficients in eqs. (III-48) and (III-49)

III-10   Response Spectrum

The response spectrum method to obtain the response of a single degree of freedom system has gained in recent years wide acceptance in structural dynamic practice, particularly in earthquake engineering analysis and design. Stated briefly, the response spectrum is a plot of the maximum response (maximum displacement, velocity, acceleration, or any other quantity of interest) to a specified load function for all possible single degree of freedom systems. The abscissa of the spectrum is the natural frequency (or period) of the system and the ordinate of the maximum response.

Dynamic response spectra for a single degree of freedom elastic system have been computed for a number of past earthquake registered motions. A typical example of displacement response spectrum for a single degree of freedom system subjected to support motion is shown in Figure 111-17. This plot is the response for the input motion given by the recorded ground acceleration of the North-South component of the 1940 El Centro earthquake. The acceleration record of this earthquake has been used extensively in earthquake engineering investigations. A plot of the acceleration record of El Centro earthquake 1940 is shown in Figure 111-18. Until the time of the San Fernando, California earthquake of 1971, the El Centro record was one of the few records for long and strong earthquake motion.

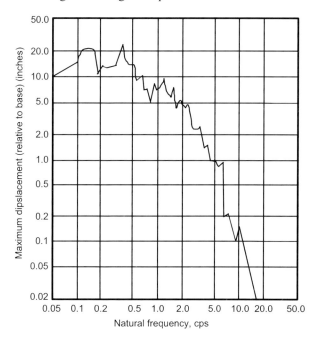

Fig. III-17 Displacement response spectrum for elastic system, subjected to the ground motion of the N-S component of the 1940 El Centro earthquake. (from *Design of Multistory Reinforced Building for Earthquake Motions* by J. A. Blum, N. M. Newmark, and L. H. Corning, Portland Cement Association 1961)

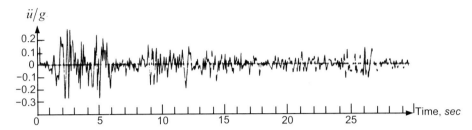

Fig. III-18 Ground acceleration record for El Centro, California earthquake of
May 18, 1940, North–South component

III-11   Tripartite Response Spectra

It is possible to plot in a single chart using logarithmic scales the maximum response in terms of the absolute acceleration, the relative displacement, and a third quantity known as the relative pseudo-velocity. The pseudo-velocity is not exactly the same as the actual velocity, but it is closely related and provides for a convenient substitute for the true velocity. These three quantities—the maximum absolute acceleration, the maximum relative displacement, and the maximum relative pseudo-velocity—are known, respectively, as the *Spectral Acceleration S_a, Spectral Displacement S_D,* and *Spectral Velocity S_V* It is significant that the spectral displacement S_D, that is, the maximum relative displacement, for an undamped system is proportional to the spectral acceleration S_a, the maximum absolute acceleration, namely

$$S_a = -\omega^2 S_D \qquad\qquad\qquad \text{(III-51)}$$

where $\omega = \sqrt{k/m}$ is the natural frequency, $S_a = \ddot{u}_{\max}$, and $S_D = u_{r,\,max}$.

However, for a damped system, the spectral acceleration S_a [eq. (III-51)] is not exactly equal to the maximum acceleration, although in general, it provides a good approximation. Equation (III-51) is by mere coincidence the same as the relationship between acceleration and displacement for a simple harmonic motion. The fictitious velocity associated with the apparent harmonic motion is known as the *pseudo-velocity* and, its maximum value S_V, is defined as the *Spectral Velocity*, thus providing the following relationships:

$$S_v = \omega S_D = \frac{S_a}{\omega} \qquad\qquad\qquad \text{(III-52)}$$

The same type of data that were used to obtain the displacement response spectrum in Figure III-17 is now plotted in Figure III-19 in terms of the spectral velocity, for several values of the damping coefficient, with the difference that the abscissa as well as the ordinate are in this figure plotted on a logarithmic scale. In this type of plot, because of eqs. (III-51) and (III-52), it is possible to draw diagonal scales for the spectral displacement sloping 135° with the abscissa, and for the spectral acceleration 45°, so from a single plot, values of spectral acceleration, spectral velocity, and spectral displacements may be read (Figure III-19). This type of plot is known as a *Tripartite Response Spectra*

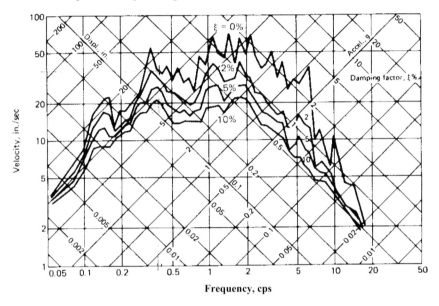

Fig. III-19 Tripartite Response Spectra for elastic system subjected to the 1940 El Centro earthquake (from Blume, et al. 1961)

III-12   Response Spectra for Elastic Design

In general, response spectral charts are prepared by calculating the response to a specified excitation of single degree of freedom systems with various damping values. Numerical integration with short time intervals is applied to calculate the response of the system. The step-by-step process is continued until the total earthquake record has been completed. The greatest value of the function of interest is recorded and becomes the maximum response of that system to the specific excitation. Changing the parameters of the system to change the natural frequency, the process is repeated and a new maximum response is recorded. This process is continued until all frequencies in the range of interest have been covered and the results plotted. Since no two earthquakes are alike, this process may need to be repeated for all earthquakes of interest.

As already stated, until the San Fernando, California earthquake of 1971, there were few recorded strong earthquake motions because there were few accelerometers emplaced to measure them. The El Centro, California earthquake of 1940 was the most severe earthquake recorded and was used as the basis for much analytical work. Since that year, however, many other strong earthquakes have been recorded. Maximum values of ground motion of about 0.32 g for the El Centro earthquake to values of more than 0.5 g for other earthquakes have been recorded. It can be expected that even larger values will be recorded as more instruments are placed closer to the epicenters of future earthquakes.

Earthquakes consist of a series of essentially random ground motions. Usually the north-south, east-west, and vertical components of the ground acceleration are measured. Currently, no accurate method is available to predict the motion that a particular site can be expected to experience in future earthquakes. Thus it is reasonable to use a *design response spectrum* that incorporates the spectra for several earthquakes and that represents a kind of "average" response spectrum for design. Such a design response spectrum is shown in Figure III-20 normalized for a *maximum ground acceleration of 1.0 g.*

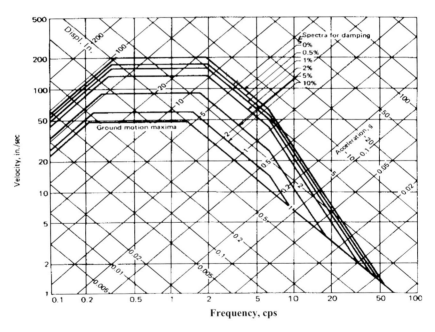

Fig. III-20 Design Spectra normalized to peak ground acceleration of 1.0 g
(From Newark and Hall 1973)

III-13   Influence of Local Soil Conditions

Before the San Fernando earthquake of 1971, earthquake accelerograms were limited in number, and the majority had been recorded on alluvium soil. Therefore, it is only natural that the design spectra based on those data, such as those suggested by Housner (1959) and Newmark-Hall (1973), mainly represent alluvial sites. Since 1971, the wealth of information obtained from earthquakes worldwide and subsequent studies have shown the very significant effect that the local site conditions have on spectral shapes.

An example of a conservative design spectrum is shown in Figure III-21 This figure shows four spectral acceleration curves representing the average of normalized spectral values corresponding to several sets of earthquake records registered on four types of soils. The dashed line through the points A, B, C, and D defines a possible conservative design spectrum for rock and stiff soil sites. Another design spectral chart based on such simplifications is reproduced in Figure III-22 from the 1994 edition of the Uniform Building Code (ICBO 1994).

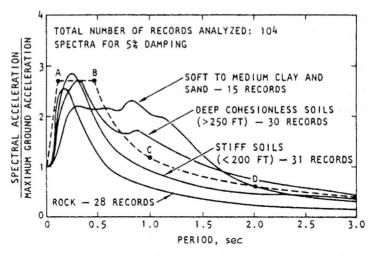

Fig. III-21 Average acceleration spectra for different soil conditions (from Seed and Idriss 1982)

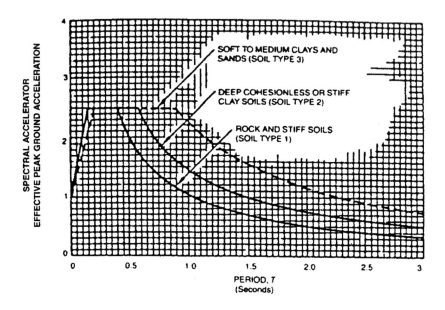

Fig. III-22 Normalized Design Spectra Chart reproduced from the 1994 edition of the Uniform Building Code (ICBO 1994)

Appendix IV
Structural Dynamics:
Multi Degree of Freedom Systems

IV.1 Free Vibration

The motion of an undamped structure (losses of mechanical energy due to friction or other sources are neglected) subjected to dynamic excitation may be described by a discrete number of *nodal coordinates* (displacements at selected locations in the structure). In free vibration, that is when the structure is in motion due only to *initial conditions* (displacement and velocity at the initiation of the motion). The motion of the structure is then governed by a homogenous system of differential equations that in matrix notation may be expressed as follows:

$$[M]\{\ddot{u}\} + [K]\{u\} = 0 \tag{IV-1}$$

where
  $[M]$ is the mass matrix of the system
  $[K]$ is the stiffness matrix, and
  $\{u\}$ and $\{\ddot{u}\}$ are, respectively, the displacement and acceleration vectors at the
      nodal coordinates of the structure.

The process of solving this system of equations leads to a homogenous system of linear algebraic equations of the form

$$[[K] - \omega^2 [M]]\{a\} = \{0\} \tag{IV-2}$$

Mathematically eq. (IV-2) is known as an eigenproblem in which the components of the vector $\{a\}$ are the unknowns and ω^2 is a parameter. For a *nontrivial solution* of this problem, (not all of the components in the vector $\{a\}$ are equal to zero) it is required that the determinant of the coefficients of the unknowns of $\{a\}$ be equal to zero, that is,

$$\left| [K] - \omega^2 [M] \right| = 0 \qquad \qquad \text{(IV-3)}$$

The roots of this polynomial equation provide the squared values of the natural *frequencies of the multi degree of freedom system*, ω_i ($i = 1, 2,...N$), in which N is the number of nodal coordinates. The first frequency (lowest value) is also referred to as the *fundamental frequency*. For each value of ω_i^2 obtained in the solution of eq. (IV-3), it is then possible to solve for the unknowns in $\{a\}$ in terms of relative values. The vector $\{a\}_i$ ($i = 1, 2,...N$) corresponding to the root ω_i^2 is the ith *modal shape* (i^{th}-eigenvector) of the dynamic system. The arrangement of the modal shapes in the columns of a matrix constitutes the modal matrix $[\Phi]$ of the system. It is particularly convenient to normalize the eigenvectors to satisfy the following condition:

$$\{\phi\}_i^T [M] \{\phi\}_i = 1 \qquad i = 1, \ 2,...N \qquad \qquad \text{(IV-4)}$$

where the normalized modal vectors $\{\phi\}_i$ are in the general case obtained by dividing the components of the vector $\{a\}_i$ by

$$\sqrt{\{a\}_i^T [M] \{a\}_i} \qquad \qquad \text{(IV-5a)}$$

and, for the special case in which the mass matrix $[M]$ is a diagonal matrix, by

$$\sqrt{\sum_{i=1}^{N} m_i a_{ij}^2} \qquad \qquad \text{(IV-5b)}$$

in which m_i is the mass allocated to the nodal coordinate "i" and a_{ij} is the 'i' component of the vector $\{a\}_j$ in the mode "j".

The normalized modal vectors satisfy the following important conditions of orthogonality:

$$\{\phi\}_i^T [M] \{\phi\}_j = 0 \qquad \text{for j} \neq i$$
$$\{\phi\}_i [M] \{\phi\}_j = 1 \qquad \text{for j} = i \qquad \qquad \text{(IV-6)}$$

and

$$\{\phi\}_i^T[K]\{\phi\}_j = 0 \quad \text{for } j \neq i$$
$$\{\phi\}_i[K]\{\phi\}_j = \omega_i^2 \quad \text{for } j = i$$

(IV-7)

The preceding relations are equivalent to

$$[\Phi]^T[M][\Phi] = [I]$$

(IV-8)

and

$$[\Phi]^T[K][\Phi] = [\Omega]$$

(IV-9)

in which $[\Phi]$ is the normalized modal matrix of the system and $[\Omega]$ is a diagonal matrix containing the eigenvalues ω_i^2 in the main diagonal.

For a dynamic system with only a few degrees of freedom, the natural frequencies and modal shapes may be determined by expanding the determinant in eq. (IV-3) and calculating the roots of the resulting polynomial equation (*characteristic equation*). However, for a system with a large number of degrees of freedom, this direct method of solution becomes impractical. In general, it is then necessary to resort to iterative numerical methods, such as the Jacobi Method, the Subspace Iteration Method, or other methods.

IV.2 Forced Motion

The motion of an undamped structure acted by external forces is governed by the following differential equation:

$$[M]\{\ddot{u}\} + [K]\{u\} = \{F(t)\}$$

(IV-10)

in which the vector $\{F(t)\}$ contains the external forces applied at the nodal coordinates of the structure. Distributed forces and concentrated forces not applied at the nodal coordinates are included in the force vector $\{F(t)\}$ as *equivalent nodal forces*. The equivalent nodal forces are determined by establishing the condition at which the solution of unknowns in the vector $\{u\}$ of eq. (IV-10) remains unaffected.

The solution of the equations of motion (IV-10) may be obtained either by the *modal superposition method* of dynamic analysis or by a *step-by-step numerical integration* procedure. The modal superposition method is restricted to the analysis of structures governed by a linear system of differential equations, whereas the step-by-step methods of numerical integration are equally applicable to systems with linear or nonlinear behavior.

IV.3   The Modal Superposition Method

The modal superposition method requires determining the solution of the eigenproblem corresponding to the homogeneous equation [right-hand side of eq. (IV-10) equated to zero] as a first step to obtaining the eigenvalues ω_i^2 and corresponding eigenvectors $\{a\}_i$, which conveniently may be normalized and arranged in the columns of the modal matrix $[\Phi]$. Then the process of solution continues by introducing into eq. (IV-10) the following linear transformation:

$$\{u\} = [\Phi]\,\{z\}. \tag{IV-11}$$

This transformation leads to a set of N *uncoupled differential* equations from the original system of coupled equations [eq. (IV-10)], in which each equation is a function of a single unknown variable, z_j, namely

$$\ddot{z}_j + \omega_j^2 z_j = P_j(t) \qquad (j = 1, 2, 3,...N) \tag{IV-12}$$

where

$$P_j = \sum_{i=1}^{N}(\phi_{i\,j}F_i(t)). \tag{IV-13}$$

Damping may be considered in the analysis by simply introducing a damping term into eq. (IV-12) to obtain:

$$\ddot{z}_j + 2\xi_j\omega_j\dot{z}_j + \omega_j^2 z_j = P_j(t). \tag{IV-14}$$

The solution of eq. (IV-14) may then be obtained by any of the methods presented in Appendix III for the single degree of freedom system.

In eq. (IV-14) ξ_j is the relative damping coefficient for mode "j" and is defined as

$$\xi_i = \frac{c_j}{c_{rc\,j}} \tag{IV-15}$$

where c_i is the absolute value of the modal damping coefficient, which is defined as the factor be to applied to the velocity to determine the damping force ($F_D = c\,\dot{z}$) and c_{cr} is the *critical damping coefficient*, which may be defined as the maximum value of the damping coefficient to have vibration.

Numerical values for the absolute coefficient c_i are generally not known; in contradistinction, the value for the relative modal damping coefficient ξ_i in the structure can be estimated. Its value fluctuates around 5% of the critical damping coefficient ($\xi_i \sim 0.05$). The precise value of ξ_i to be used in eq. (VI-14) is estimated by considering the structural materials (steel, reinforced concrete, etc.) and the presence of partitions or other non-structural elements.

IV.4   Combining Maximum Values of Modal Response

When use is made of response spectra to determine maximum values for modal response, these maximum values are usually combined by the *Square Root of the Sum of Squares (SRSS) method*. The SRSS method to estimate the total response calculated from maximum modal values may be expressed in general as

$$R_k = \sqrt{\sum_{j=1}^{N} R_{kj}^2} \qquad \text{(IV-16)}$$

where R_k is the estimated response (force, displacement, etc.) at a specified nodal coordinate k, and R_{kj} is the maximum response at that coordinate in the jth mode.

Application of the SRSS method for combining the maximum modal values generally provides an acceptable estimation of the total maximum response. However, when some of the modes are closely spaced, the use of the SRSS method may result in grossly underestimating or overestimating the maximum response. In particular, large errors have been found in the analysis of three-dimensional structures in which torsional effects are significant. The term "closely spaced" when referring to modes may be arbitrarily defined as the case when the difference between two natural frequencies is within 10% of the smaller of the two frequencies.

A formulation known as the Complete Quadratic Combinations (CQC), which is based on the theory of random vibrations, provides results more accurate than the application of the SRSS method. The CQC method, which may be considered as an extension of the SRSS method, is given by the following equation:

$$R_k = \sqrt{\sum_{i=1}^{N} \sum_{j=1}^{N} R_{ki}\, \rho_{ij} R_{kj}} \qquad \text{(IV-17)}$$

in which the cross-modal coefficient ρ_{ij} may be approximated by

$$\rho_{ij} = \frac{8\left(\xi_i \xi_j\right)^{1/2}\left(\xi_i + r\xi_j\right)r^{3/2}}{\left(1-r^2\right)^2 + 4\xi_i \xi_j r\left(1+r^2\right) + 4\left(\xi_i^2 + \xi_j^2\right)r^2} \qquad \text{(IV-18)}$$

where $r = \omega_j / \omega_i$ is the ratio of the natural frequencies of order i and j, and ξ_i and ξ_j the corresponding damping ratios for modes i and j.

For constant modal damping, ξ, for all the modes, eq. (IV-18) reduces to:

$$\rho_{ij} = \frac{8\xi^2(1+r)r^{3/2}}{\left(1-r^2\right)^2 + 4\xi^2 r\left(1+r\right)^2} \qquad \text{(IV-19)}$$

It is important to note that, for $i = j$, eq.(VI-18) or eq. (VI-19) yields $\rho_{ij} = 1$ for any value of the damping ratio, including $\xi = 0$. Thus, for an undamped structure, the CQC method [eq. (IV-17)] is identical to the SRSS method [eq. (IV-16)].

Glossary

Analytical Model (Mathematical Model) – (Section III-1) – The idealized representation of the structure to be analyzed.

Amplitude – (Section III-4) – Maximum value of a function as it varies with time. If the variation of time is described by a sine or cosine function, it is said to vary harmonically.

Angle of Rolling – (Section 6.3) – It is the angle by which the local axis y appears to have been rotated from the "standard" orientation. This standard orientation exists when the local plane, formed by the local axes x (centroidal axis of the beam element) and y (minor principal axis of the cross-section), is vertical, that is, parallel to the vertical global axis Z.

Angular Frequency (Circular Frequency) – (Section 2.3) – The frequency of a periodic function expressed in radians per second (rd/sec), and it is equal to the frequency in cycles per seconds (Hertz) multiplied by 2π.

Assemble the System Stiffness Matrix – (Section 1.5) – The process of transferring the coefficients in the element stiffness matrices to the appropriate locations in the system or the global stiffness matrix. The appropriate locations are dictated by the nodal coordinates in the structure assigned to an element.

Beam – (Section 1.1) – A longitudinal structure supported at selected locations and carrying loads that are for the most part applied in a direction normal to the longitudinal direction of the beam.

Body Force – (Section 9.3) – Distributed force per unit volume at an interior point of a structural element.

Boundary Condition – (Section 1.3) – A constraint applied to the structure independent of time.

Characteristic Equation – (Section IV-1) – An equation whose roots are the natural frequencies.

Circular Frequency – See Angular Frequency.

Complementary Solution – (Section III-4) – The solution of a homogeneous differential equation in free vibration (no external excitation).

Complete Quadratic Combination (CQC) – (Section IV-4) – A method of combining maximum values of modal contributions that is based on random vibration theory and includes cross-correlation terms.

Consistent Mass – (Section 2.2.2) – Mass influence coefficients determined by assuming that the dynamic displacement functions are equal to the static displacement functions.

Coupled Equations – (Section IV-3) – A system of differential equations in which the equations are not independent from each other.

Critical Damping – (Section III-2) – Minimum amount of viscous damping for which the system will not vibrate.

D'Alembert Principle – (Section III-1) – This principle states that a dynamic system may be assumed to be in equilibrium, provided that the inertial forces are considered as external forces.

Damped Frequency – (Section III-2) – The frequency at which a viscously damped system oscillates in free vibration.

Damping – (Section III-2) – The factor, that during motion of the structure, dissipates mechanical energy.

Damping Ratio – (Section III-2) – The ratio of the viscous damping coefficient to the critical damping.

Degrees of Freedom – (Section III-1) – The number of independent coordinates required to completely define the position of the system at any time.

Design Response Spectra – (Section III-12) – Simplified Response Spectra as a kind of average of the actual spectral response of several selected past earthquakes.

Differential Equation of a Beam – (Section 1.12) – The differential equation expressing the lateral displacement function of a loaded beam. For small deformations and considering only the effect of bending, this differential equation is given by

$$\frac{d^2 y(x)}{dx^2} = \frac{M(x)}{EI}$$ (1.38) repeated

in which

$y(x)$ = lateral displacement at coordinate x
$M(x)$ = bending moment at coordinate x
E = modulus of elasticity
I = cross-sectional moment of inertia

For a beam of uniform cross-sectional area, eq. (1.38) may be expressed as

$$\frac{d^4 y(x)}{dx^4} = \frac{w(x)}{EI}$$ (1.2) repeated

where $w(x)$ is the applied external load per unit of length along the beam.

Direct Method – (Section III-9) – A method by which the coefficients in the Element Stiffness Matrices of the structures are transferred and added appropriately to assemble the System Stiffness Matrix.

Duhamel's Integral – (Section III-8) – An integral providing the response of an undamped or a damped linear one degree of freedom system subjected to any given excitation.

Dynamic Magnification Factor – (Section III-4) – The ratio of the maximum displacement of a single degree of freedom excited by a harmonic force to the deflection that would result if a force of that magnitude were applied statically.

Eigenproblem – (Section IV-1) – The problem of solving a homogeneous system of equations: containing a parameter which should be determined to provide nontrivial solutions.

Elastic Supports – (Section 1.15) – Linear elastic supports of a structure may be considered in the analysis by simply adding the value of the spring constant (stiffness coefficient) to the corresponding coefficient in the diagonal of the system stiffness matrix.

Element Displacement Function – (Section 1.12) – The total lateral displace-ment function $y_T(x)$ for a beam element is given by

$$y_T(x) = y(x) + y_L(x) \qquad\qquad (1.37) \text{ repeated}$$

in which $y_L(x)$ is the displacement function of the loaded beam element assumed fixed at both ends and $y(x)$ is the lateral displacement function produced by displacements at the nodal coordinates of the beam element.

The function $y(x)$ is given by

$$y(x) = N_1(x)\delta_1 + N_2(x)\delta_2 + N_3(x)\delta_3 + N_4(x)\delta_4 \qquad (1.6) \text{ repeated}$$

where $N_1(x)$, $N_2(x)$, $N_3(x)$ and $N_4(x)$ [eqs. (1-5)] are the *shape functions*, respective-ly, corresponding to a unit displacement at the nodal coordinates δ_1, δ_2, δ_3 and δ_4 of the beam element.

Element Force Vector – (Section 1.7) – A vector containing the element nodal forces either in reference to the global system of coordinates (*X, Y, Z*) or in reference to the local system of coordinates (*x, y, z*).

Element Stiffness Matrix – (Section 1.4) – A matrix relating the nodal displacements and the nodal forces at the nodal coordinates of an element (i.e., beam element). This relationship may be written in general as

$$\{P\} = [k]\{\delta\} \qquad\qquad (1.12) \text{ repeated}$$

in which

    $\{P\}$ = element nodal forces

    $\{\delta\}$ = element nodal displacements

    $[k]$ = element stiffness matrix

The components of the vector $\{P\}$ and $\{\delta\}$ as well as the coefficients in the matrix $[k]$ will depend on the specific type of structure, such as *Beam, Plane Frame, Space Frame, Plane Truss* or *Space Truss.*

End Releases in Beam Elements – (Section 1.10) – The condition by which one or the two ends of a beam element are connected at a node through hinges.

Equivalent Nodal Forces – (Section 1.7) – Forces at the nodal coordinates of a beam element producing the same displacements at the nodal coordinates as those displacements resulting from the loads applied on the element. The equivalent nodal forces are calculated as follows:

1. **For loads applied on the elements** – (Section 1.7) – The equivalent nodal forces $\{Q\}$ may be calculated either from eq. (1.18) or by determining the element fixed end reactions and reversing their direction.

2. **For imposed displacements** – (Section 1.11) – The equivalent element nodal forces, $\{Q\}_\Delta$ due to imposed nodal displacements, are calculated by

$$\{Q\}_\Delta = -[k]\{\Delta\} \qquad (1.36) \text{ repeated}$$

in which

  $[k]$ = element stiffness matrix
  $\{\Delta\}$ = imposed displacements at the element nodal coordinates.

3. **For temperature effect** – (Section 1.14) – The equivalent vector for nodal forces $\{Q\}_T$ for a linear temperature variation on the cross-section of a beam element is given by

$$\{Q\}_T = \left\{ \begin{array}{c} 0 \\[2mm] \dfrac{\alpha\,EI}{h}(T_2 - T_1) \\[2mm] 0 \\[2mm] \dfrac{-\alpha\,EI}{h}(T_2 - T_1) \end{array} \right\} \qquad (1.50) \text{ repeated}$$

in which

  T_1 = temperature at the bottom face of the beam

  T_2 = temperature at the top face of the beam

  α = coefficient of thermal expansion

  h = height of the cross-sectional area

Equivalent Viscous Damping – (Section III-6) – Damping determined by equating the energy dissipated during one cycle of the structure harmonically excited at a resonant frequency and the theoretically calculated energy that the structure, assumed viscously damped, would dissipate in a cycle of vibration.

Finite Element Method (FEM) – (Section 9.1) – A powerful method for the analysis of structures. The main fixtures of the FEM are:

1. Modeling the structure with elements such as triangular or quadrilateral plates, shells and solid elements interconnected at selected points defining nodal coordinates.

2. Establishing through the Element Stiffness Matrix the relationship between forces and displacements at the nodal coordinates of the elements using approximate interpolating displacement functions.

3. Transferring the coefficients in Element Stiffness Matrices to appropriate locations in the System Stiffness Matrix.

4. Solving the system stiffness equations for the unknown nodal displacements.

5. Determining the element nodal forces and stress distributions.

Fixed End Reactions {FER} – (Section 1.9) – These are the reactions of a loaded beam element assumed to be fixed for translation and rotation at its two ends.

Forced Motion – (Section IV-2) – Motion in which the response is due to external excitation of the structure.

Free Body Diagram – (Section III-1) – A sketch of the system, isolated from all other bodies, in which all the forces external to the body are shown.

Free Vibration – (Section IV-1) – The vibration of a structure in the absence of external excitation.

Frequency Ratio – (Section III-4) – The ratio between the forcing frequency to the natural frequency for a system excited by a harmonic load.

Fundamental Frequency – (Section IV-1) – The lowest natural frequency of a multi degree of freedom vibrating system.

Gauss-Jordan Reduction or Elimination – (Section 1.6) – A computational technique in which elementary row operations are applied systematically to solve a linear system of equations.

Global System of Nodal Coordinates – (Section 4.5) – Element nodal coordinates in reference to the global system of coordinates.

Grid Frame – (Section 5.1) – It is a planar frame with the loads applied normally to the plane of the frame.

Harmonic Force – (Section III-4) – A force expressed by a sine, cosine, or equivalent exponential function.

Initial Conditions – (Section IV-1) – The initial values of specific functions, such as displacement, velocity, or acceleration, evaluated at time $t = 0$.

Interpolating functions – (Section 9.1) – These are functions (generally polynomial functions) used to approximate the displacements at an interior point of a finite element.

Isolation – (Section III-7) – The reduction of severity of the response, usually attained by proper use of a resilient support.

Logarithmic Decrement – (Section III-3) – The natural logarithm of the ratio of any two successive peak amplitudes of the same sign obtained in the decay curve in a free vibration test.

Lumped Mass – (Section 2.2.1) – A method of discretization in which the distributed mass of the elements is lumped at the nodes or joints.

Mathematical Model – (Section III-1) – See Analytical Model.

Matrix Structural Analysis – (Section 1.2) – A method for the analysis of frame type structures (beams, frames, and trusses) using matrices.

The main features of this method are:

1. Modeling the structure into beam (or truss) elements interconnected at selected points defining the nodes.

2. Establishing through the Element Stiffness Matrix the relationship between forces and displacements at the nodal coordinate of the elements.

3. Transferring the coefficients in the Element Stiffness Matrices to appropriate locations in the System Stiffness Matrix.

4. Solving the System Stiffness Equations for the unknown nodal displacements.

5. Determining the element nodal end forces.

Member or Element End Releases – (Section 1.10) – Refers to the introduction of hinges at the ends of a beam element. The modified element stiffness matrices and modified nodal equivalent force vectors for releases (hinges) are given by the following equations:

Case 1: Beam element with a hinged at node 1, eqs. (1.28) and (1.29)

Case 2: Beam element with a hinge at node 2, eqs. (1.30) and (1.31)

Case 3: Beam element with hinges at both ends, eqs. (1.32) and (1.33)

Modal Superposition Method – (Section IV-3) – A method of solution of multi degree of freedom systems in which the response is determined from the solution of independent modal (or normal) equations.

Mode Shapes (also Normal Modes) – (Section IV-1) – The relative amplitude of the displacements at the nodal coordinates of a multi degree of freedom system vibrating at one of the natural frequencies.

Natural Frequency – (Section 2.3) – For a single degree of freedom system is the number of cycles that it will oscillate in one second (*cps*). For a multiple degree of freedom system, the natural frequency is the number of cycles that the system oscillates in one second, vibrating at one of its modes of motion.

Natural Period – (Section 2.3) – The reciprocal value of the natural frequency (sec per cycle).

Newton's Law of Motion – (Section III-1) – States that a force F acting on a particle of mass m results in change of motion (acceleration a) [$F = m\ a$].

Nodal Coordinates – (Section IV-1) – Designate the possible displacements at the nodes of an element or at the nodes of the structure. For example, for a plane frame, there are three nodal coordinates at each node or joint, two components of a linear displacement in the plane of the frame and a rotational displacement around an axis normal to that plane. Nodal coordinates are classified as follows:

1. **Free nodal coordinates** – (Section 1.2) – Nodal coordinates at which displacements are not constrained.

2. **Fixed nodal coordinates** – (Section 1.2) – Nodal coordinates at which displacements are constrained.

Nodes or Joints – (Section 1.2) – Points located at the ends of a beam element (member) or at selected points joining elements in the structure.

Non-Conforming Functions – (Section 9.5) – These are interpolating functions that not fully satisfy the continuity conditions of displacements or slopes between adjacent finite elements modeling the structure.

Orthogonal Matrix – (Section 4.5) – A matrix for which the transpose matrix is equal to its inverse.

Plane Frames – (Section 4.1) – Structural frames in which the members as well as the loads are in the same plane.

Plane Trusses – (Section 7.1) – Plane structures assembled of longitudinal members assumed to be connected at their ends by frictionless pins.

Plate Bending – (Section 9.4) – A structural plate on which the loads are applied normally to the plane of the plate.

Primary Degrees of Freedom (or Independent Degrees of Freedom) – (Section 1.6) – Those degrees of freedom selected for the analysis as independent displacements that are left after the reduction or condensation of the secondary or dependent degrees of freedom.

Principle of Virtual Work – (Sections 1.16) – The Principle of Virtual Work states that for an elastic system in equilibrium (or in dynamic equilibrium), the work done by the external forces is equal to the work of the internal forces during an arbitrary virtual displacement compatible with the constraints of the structure.

Pseudo-Velocity – (Section III-11) – The velocity calculated by analogy with the apparent harmonic motion for a system seismically excited.

Reduced System Stiffness Matrix – (Section 1.8) – A matrix $[K]_R$ establishing the relationship between the nodal displacements $\{u\}$ and nodal forces $\{F\}_R$ at the free nodal coordinates of the structure: $\{F\}_R = [K]_R\{u\}$.

Resonance – (Section III-4) – The condition in which the frequency of the excitation is equal to one of the natural frequencies of the system.

Response – (Section III-5) – The force or motion that results from external excitation on the structure.

Response Spectrum (Shock Spectrum) – (Section III.10) – A plot of maximum response (displacement, velocity, or acceleration) for single degree of freedom systems defined by their natural frequency (or period) subjected to a specific excitation.

SAP2000 – (Section 3.1) – SAP2000 is an interactive, menu driven computer program for the analysis and design of structures. The student version used in this book is limited to the analysis of linear structures modeled with no more than 30 nodes or 30 elements.

Secondary Degrees of Freedom – (Section 1.6) – Those degrees of freedom selected to be condensed or reduced in order to define the system in terms of the selected primary degree of freedom.

Shape Functions $N_i(x)$ – (Section 1.3) – These are functions giving the lateral displacements resulting from a unit of displacement applied to a nodal coordinate of a beam element.

Shear Force and Bending Moment Functions – (Section 1.13) – The shear force $V(x)$ and bending moment $M(x)$ functions along a loaded beam element are given, respectively, by

$$V(x) = P_1 - \int_0^x w(x_1)dx_1 \qquad (1.42) \text{ repeated}$$

and

$$M(x) = -P_2 + P_1 x - \int_0^x (x - x_1) w(x_1) dx_1 \qquad (1.43) \text{ repeated}$$

in which P_1 and P_2 are, respectively, the fixed-end force and the fixed-end moment reactions at the left node of a loaded beam element, and $w(x)$ is the external load per unit of length applied along the beam element.

Shock Spectrum – See Response Spectrum.

Skeletal Structures – (Section 9.1) – These are structures with unidirectional elements, such as beams, frames or trusses.

Space Frame – (Section 6.1) – These are frames in which the members and the forces may be oriented in any direction of the three-dimensional space.

Space Trusses – (Section 8.1) – Three-dimensional structures with longitudinal members assumed to be connected at their ends by frictionless hinges.

Spring Constant – (Section 1.15) – The change in load on a linear elastic structure required to produce a unit increment of deflection.

Square Root Sum of Squares (SRSS) – (Section IV-4) – A method of combining maximum values of modal contributions by taking the square root of the sum of the squared modal contributions.

Static Condensation – (Section 1.6) – The process of reducing the number of free displacements or degrees of freedom.

Stiffness Coefficients, k_{ij} – (Section 1.3) – The force at nodal coordinate "i" resulting from a unit displacement applied at nodal coordinate "j" with all other nodal coordinates fixed with no displacements.

Support Reactions – (Section 1.8) – Forces at the supports of the structure. These forces may conveniently be determined from the end-member forces of those elements that are linked to a particular support.

System Force Vector – (Section 1.7) – A vector containing the forces at the nodal coordinates of the structure. This vector includes the forces applied directly at the nodes and the equivalent nodal forces for loads applied on the elements of the structure. It also includes the equivalent nodal forces for displacements imposed at the nodes of the structure.

System of Coordinates – (Section 1.1) – Two basic systems of coordinates are used in Matrix Structural Analysis:
1. **Global system of coordinates** – (Section 1.1) – A Cartesian system of coordinates with axes X, Y and Z fixed in the space and used to refer any point in the structure.

2. **Local system of coordinates** – (Section 1.3) – A Cartesian system of coordinates with axes x, y and z fixed on a member or element of the structure and used to refer any point on that element.

System Stiffness Matrix – (Section 1.5) – A matrix $[K]$ establishing the relationship between displacements $\{u\}$ and forces $\{F\}$ at the nodal coordinates of the structure:

$$\{F\} = [K]\{u\} \qquad\qquad (1.13)\ \text{repeated}$$

Transformation Matrix – (Section 4.5) – A matrix that transforms the element nodal displacements or the element nodal forces in reference to the global system of coordinates to the local system of coordinates.

$$\{\delta\} = [T]\{\bar{\delta}\} \qquad\qquad (4.15)\ \text{repeated}$$

and

$$\{P\} = [T]\{\bar{P}\} \qquad\qquad (4.13)\ \text{repeated}$$

where

$\{\delta\}$ = element nodal displacement vector in local coordinates

$\{\overline{\delta}\}$ = element nodal displacement vector in global coordinates

$\{P\}$ = element nodal force vector in local coordinates

$\{\overline{P}\}$ = element nodal force vector in global coordinates

$[T]$ = transformation matrix

Transmissibility – (Section III-7) – The non-dimensional ratio, in the steady-state condition, of the peak response motion to the amplitude of the harmonic input motion; or the non-dimensional ratio of the amplitude of the force transmitted to the foundation of the structure to the amplitude of the harmonic excitation force applied to the structure.

Transient Solution – (Section III-4) – The initial portion of the motion which vanishes due to the presence of damping forces in the system.

Uncoupled Differential Equation – (Section IV-3) – A set of independent differential equations obtained through a linear transformation applied to the coupled system of differential equations.

Viscous Damping – (Section III-2) – Dissipation of mechanical energy in which it is assumed that the motion is resisted by a force proportional to the velocity but in the opposite direction.

Selected Bibliography

STRUCTURAL ANALYSIS

Bathe, K. J. (1982), *Finite Element Procedures in Engineering Analysis,* Prentice-hall, Englewood Cliffs, New Jersey.

Guyan, R. J. (1965), *Reduction of Stiffness and Mass Matrices,* AIAA, J., 13,380.

Hibbeler, Russell C. (1995), *Structural Analysis, 3rd Edition,* Prentice-Hall, Englewood Cliffs, New Jersey.

Kassimali, Aslam (1999), *Matrix Analysis of Structures,* Brooks/Cole Publishing Company, Pacific Grove, California.

McCormac, Jack C. and James K. Nelson, Jr. (1997), *Structural Analysis: A Classical and Matrix Approach,* Addison-Wesley, Reading, Massachusetts.

Paz, Mario (2001), *Integrated Structural Analysis: Theory and Computation,* Kluwer Academic Publishers, Norwell, Massachusetts.

Paz, Mario (1983), Practical reduction of structural problems, *J. Struct. Eng., ASCE,* 109(111), 2590-2599.

Sennett, Robert C. (1994), *Matrix Analysis of Structures,* Prentice-Hall, Englewood Cliffs, New Jersey.

Timoshenko, S. P. and J. N. Goodier (1970), *3rd Edition, Theory of Elasticity,* McGraw-Hill, New York.

STRUCTURAL DYNAMICS

Bathe, K. J. (1982), *Finite Element Procedures in Engineering Analysis,* Prentice-Hall, Englewood Cliffs, New Jersey.

Berg, Glen V. (1989), *Elements of Structural Dynamics,* Prentice-Hall, Engle wood Cliffs, New Jersey.

Biggs, J. M. (1964), *Introduction to Structural Dynamics,* McGraw-Hill, New York.

Blevins, R. D. (1979), *Formulas for Natural Frequency and Mode Shape,* Van Nostrand Reinhold, New York.

Cheng, Y. Franklin (2001), *Matrix Analysis of Structural Dynamics,* Marcel Dekker Inc., New York.

Chopra, A. (1981), *Dynamics of Structures: A Primer,* Earthquake Engineering Research Institute, Berkeley, California.

Chopra, A. K. (1995), *Dynamics of Structures: Theory and Applications to Earthquake Engineering,* Prentice Hall, Englewood Cliffs, New Jersey.

Clough, R. W., and J. Penzien (1993), *Dynamics of Structures,* 2nd Edition McGraw-Hill, New York.

Gallagher, R. H. (1975), *Finite Element Analysis,* Prentice-Hall, Englewood Cliffs, New Jersey, p. 115.

Guyan, R. J. (1965), Reduction of stiffness and mass matrices, *AIAA J.,* 13,380.

Hart, C. Gary and Kevin Wong (2000), John Wiley & Sons, Inc., New York.

Harris, Cyril M. (1987), *Shock and Vibration Handbook,* 3rd ed., McGraw Hill, New York.

Housner, George W. and Donald E. Hudson (1959), *Applied Mechanics Dynamics,* D. Van Nostrand Company, Inc., Princeton, New Jersey.

Humar, J. L. (1990), *Dynamic of Structures,* Prentice Hall, Englewood Cliffs, New Jersey.

Newmark, N. M. (1959), A method of computation for structural dynamics, *Trans. ASCE,* 127, 1406-35.

Newmark, N.M. and W.J. Hall (1973), "Procedures and Criteria for Earthquake Resistant Design," *Building Practices for Disaster Mitigation, Building Science Series 46,* National Bureau of Standards. February.

Paz, Mario (1973), Mathematical observations in structural dynamics, *Int. J. Comput. Struct.*, 3, 385-396.

Paz, Mario (1984a), Dynamic condensation, *AIAA J.*, 22(5), 724-727.

Paz, Mario (1984b), in *Structural Mechanics Software Series,* The University Press of Virginia, Charlottesville, Vol. V, pp. 271-286.

Paz, Mario (1985), *Micro-Computer Aided Engineering: Structural Dynamics;* Van Nostrand Reinhold, New York.

Paz, Mario (1989), Modified dynamic condensation method, *J. Struct. Eng., ASCE,* 115(1), 234-238.

Paz, Mario and William Leigh (2004), *Structural Dynamics: Theory and Computation – 5th Edition.* Chapter 20. Kluwer Academic Publishers. Norwell, Massachusetts. 02061 USA. ISBN 1-4020-7667-3

Seed, Bolton and I. M. Idriss (1982), Ground Motions and Soil Liquefaction during Earthquakes (Engineering Monographs on Earthquake Criteria, Structural Design, and Strong Motion Records). Earthquake Engineering Research Institute. Oakland, California., USA. ISBN-10 0943198240.

Smith, J. W. (1988), *Vibration of Structures: Applications in Civil Engineering Design,* Chapman and Hall, New York.

Timoshenko, S. P. and J. N. Goodier (1970) 3rd Edition *Theory of Elasticity,* McGraw-Hill, New York.

Index